Sustainable Aviation

The Sustainable Aviation book series focuses on sustainability in aviation, considering all aspects of the field. The books are developed in partnership with the International Sustainable Aviation Research Society (SARES). They include contributed volumes comprising select contributions to international symposiums and conferences, monographs, and professional books focused on all aspects of sustainable aviation. The series aims at publishing state-of-the-art research and development in areas including, but not limited to:

- Green and renewable energy resources and aviation technologies
- Aircraft engine, control systems, production, storage, efficiency, and planning
- Exploring the potential of integrating renewables within airports
- Sustainable infrastructure development under a changing climate
- Training and awareness facilities with aviation sector and social levels
- Teaching and professional development in renewable energy technologies and sustainability

Luiz Manuel Braga da Costa Campos
Editor

Accelerating Sustainable Aviation Initiatives

Technology, Markets and Social Issues

Springer

Editor
Luiz Manuel Braga da Costa Campos
Instituto Superior Técnico
University of Lisbon
Lisboa, Portugal

ISSN 2730-7778 ISSN 2730-7786 (electronic)
Sustainable Aviation
ISBN 978-3-031-73992-7 ISBN 978-3-031-73993-4 (eBook)
https://doi.org/10.1007/978-3-031-73993-4

This Springer imprint is published by the registered company Springer Nature Switzerland AG
The registered company address is: Gewerbestrasse 11, 6330 Cham, Switzerland

Foreword

Civil air transport represents an important motor of the global economy and has doubled in the past about every 15 years despite any crisis of global scale. The rising concern about global warming and the need for effective actions for reducing the emission of greenhouse gases require intensifying measures also by the growing high-tech sector of aviation and air transport, even if the related share of the aviation contribution to the CO_2 emissions is relatively small.

Beside innovative technologies for novel aircraft and related implementation measures of future air transport, the aviation market situation and the different boundary conditions of the society play an important role for the global aviation development. This book addresses the need for accelerating sustainable aviation initiatives. Its authors aim to illuminate the interdependencies of industrial und technological development of aviation with the global market growth and societal constrains. They also show how the European actors in aviation play their role on a global market and address the emerging issue of greening aviation. One of the ambitious Flightpath 2050 goals of the Advisory Council for Aviation Research and Innovation in Europe (ACARE) is a 75% reduction of the CO_2 emissions of aviation by 2050. Also, the impact of a global crisis such as the COVID-19 pandemic is addressed in this book. It becomes clear that intensive measures for sustainable aviation are required on a global scale, whereby Europe can be a frontrunner.

Several chapters address the duopoly situation of Airbus and Boeing that developed in the last 25 years. Here the focus is on the single aisle aircraft of the A320 and the B737 families, because of their rising production numbers and of the attempt of China to break this duopoly by its new but quite classical C919 airliner. It is illustrated that in the past Europe's industry could gain its leading position of civil aircraft by an enhanced cooperation and the introduction of innovative technologies.

The editor and co-author Luis Manuel Braga da Costa Campos, emeritus professor of the Instituto Superior Técnico in Lisbon, Portugal, is involved in aeronautics research since more than 40 years. In 1989 he performed a study about aeronautics research and development activities in Portugal for the European Commission and became also the representative of the Portuguese Government in the Member States'

regulatory committee for aeronautics research on European level for many years. Since this time, he was involved in numerous activities and research projects on European level.

I would like to congratulate Luis Campos and his co-authors as well as the entire team of the PARE project for their initiative and thorough investigations, which are analyzed and summarized in this book. The research work for this book had been performed from 2018 to 2021 within the coordination and support action PARE (Perspectives for the Aeronautical Research in Europe) with a team of 14 partners from 10 countries within Horizon 2020, the 8th Research & Innovation Framework Programme of the European Union.

The reader will enjoy the facts and figures and their analyses in this book. But they have to keep in mind that this can be only a description of the temporary situation of the complex dynamic development of global aviation. It is appreciated that the authors analyze different scenarios and developed perspectives for the future development in sustainable aviation.

Independent Aeronautics Consultant
Brussels, Belgium
June 2024

Dietrich Knoerzer,

Preface

The contents of this book are based on the final Third Year Report of the Project PARE (Perspectives for Aeronautical Research in Europe) supported in the period 2018–2021 by the 8th Framework Programme Horizon 2020 of the European Union. The book consists of a selection of 10 among the 20 chapters of the original 2000-page PARE Final 3rd Year Report, completely reviewed, not only to summarize the extent from 1000 to 300 pages but also to update for events since 2021 and for the change in outlook and future prospects since then. The PARE report was firmly focused in the future up to 2050, which probably as far as can be reasonably predicted based on current knowledge looking forward three decades ahead. This book should be seen in the broader context of Aeronautics Research Program—lead by the European Commission, with a budget including contribution from industry and other organizations, that has increased over four decades by a factor of about 100, from 36 million euros as part of Brite-Euram in the 2nd Framework Research Program 1983–1991 to 3.4 billion euros in the 9th Framework Program Horizon Europe 2020 covering the period 2021–2027. The priorities for the successive Aeronautics Programs of the European Union have been influenced most by ACARE (Advisory Council for Aviation Research and Innovation in Europe), notably the 24 ACARE Goals to be aimed for and achieved by 2050. The main objective of the PARE project was to review 30 years in advance the progress toward the 24 ACARE 2050 Goals and assess whether current or additional measures were sufficient or needed to reach the targets in time. The 24 ACARE Goals can be seen as a comprehensive set of vertical integration measures for aviation in the 2050 timeframe and are complemented by 34 PARE Objectives supporting them as horizontal cross-cutting activities. The implementation of the 24 ACARE Goals plus the 34 PARE Objectives was the subject of the 58 PARE Recommendations, prepared at the request of the European Commission, during the 8th Framework Program 2014–2020 Horizon 2020, as one of multiple inputs contributing to the preparation of the current 9th Framework Program 2021–2027 Horizon Europe. This brief account of the history and genesis underlying the present book acknowledges the fundamental role and input of European Programs, although the conclusions in fact

have a broad worldwide scope, since they apply to the future of aviation in general as an endeavor of mankind.

The book starts with a focus on the largest sector of aviation in the terms of social impact, namely the airliner market (Chap. 1) with an historic perspective of the four generations since the dawn of the jet age up to the current status and future prospects. The long-term growth of air transport at a rate comparable to the standard living poses a continuing challenge to ATM (Air Traffic Management) to provide (Chap. 2) increasing capacity with undiminished safety, to ensure timely airline services while contributing its share to reducing the environmental impact of aviation. The jet airliner market has remained an Airbus-Boeing duopoly for some time, and the possibility of future competitor from China is discussed (Chap. 3). The two Boeing 737 Max accidents lead (Chap. 4) to a grounding of the aircraft for 20 months and will affect certification of all jetliners much longer into the future, well beyond the COVID-19 crisis that had a partial time overlap. The COVID-19 pandemic has caused the biggest crisis in the history of aviation, which is analyzed from the points of view of (Chap. 5) a specific regional and international airline and (Chap. 6) more generally on all sectors of aviation comparing the pre-COVID status, the pandemic impact, and subsequent recovery. The B737 Max (Chap. 4) and COVID (Chaps. 5 and 6) overlapped with the growing attention to the environmental impact of aviation of which one if not the most visible objective is (Chap. 7) the decarbonization of aviation by 2050. The multiple challenges that aviation faces in the future as a safe and affordable means of transport that respects the environment depend on research breakthroughs and operational advances, that ultimately rely upon (Chap. 8) attracting a pool of young talent including and (Chap. 9) increasing the participation of women in a more diverse workforce with equal opportunities for all. The conclusion (Chap. 10) attempts to summarize the many challenges facing aviation in the 2050 timeframe at two complementary levels: (i) a broad introductory text on key findings and recommendations outlining the current status and issues to be addressed; (ii) a set of more detailed specific recommendations of which challenges to focus on, how and by whom, including a four-level scale of priorities. The editor of this book was the coordinator of the PARE project and has also contributed to four—Chaps. 1, 4, 6, and 10. The contributions to the six remaining chapters—Chaps. 2, 3, 5, 7, 8, and 9—come from the PARE partners that lead the preparation of the corresponding chapters in the PARE Final Report. Ultimately this book is a product of the collective work of all PARE partners, including those not explicitly mentioned as authors or co-authors of specific chapters.

Lisboa, Portugal Luiz Manuel Braga da Costa Campos

Introduction

The first period of pioneering efforts for flight can be divided in five subperiods. The first "dreams" of flying like the birds, as in the Greek myth of Icarus, who built wings of wax, and unwisely "flew too close to the sun," so that the wax melted, and his fatal flaw became a stern warning to the excessive ambitious of men. Those warnings were not heeded during the second phase, as daring individuals jumped from building or cliffs with improvised wings, with records of the tragic fate of such attempts spanning centuries into the past. The third phase of successful flight came with the lighter-than-air in the XVIII-th century, but the balloons, and their powered analogues the dirigibles, never quite overcame their vulnerability to storms and adverse weather. The principles of flight of heavier than air, with forward speed to generate enough lift to balance weight, were only understood in the nineteenth century, and successfully applied to motorless gliders and powered aircraft in the early twentieth century. It took somewhat longer to reach the fifth period of vertical flight, predicted by Leonardo da Vinci in the XV-th century due to the power and vibration problems of rotary wings in helicopters.

The success of the pioneers in achieving safe flight was followed by a second period of increasing performance, again with five subperiods. The first subperiod was to startle and impress the public with the new "flying machines," flying increasing distances, although going around the world required tens of "hops" and stops. The First World War was second subperiod when the military broadened the use of aircraft as fighters, bombers, and reconnaissance and transport aircraft. The third subperiod between the wars saw the start of regular civil air transport on a scale limited by cost and risk, mainly for the wealthy and adventurous. The fourth subperiod of the Second World War saw a massive use of aviation, operating from land bases and aircraft carriers to not only dominate the airs but also play a pivotal role in land and sea battles. The fifth subperiod of superpower rivalry after the Second World War saw the greatest technological advances, including supersonic and orbital flight and much larger aircraft than before.

The third period of democratization of air transport was the result of technology advances allowing building of large aircraft able to transport passengers safely and comfortably over long distances again with five subperiods. The first subperiod was

due to the availability post-Second World War of thousands of surplus military transports, readily adaptable to carry passengers in commercial flights. The second subperiod saw flying boats being discarded as more airports were built for more efficient designs of propeller driven aircraft. The third subperiod saw propeller driven replaced by jet airliners, some related to military in-flight refueling tankers, able to cross oceans in few hours without intermediate stops. The fourth subperiod saw a wide choice of airliners developed, from "jumbos" to regional aircraft, providing safe and affordable air transport everywhere around world. The fifth subperiod saw the technological success of supersonic civil transport discarded in favor of more economical highly efficient subsonic jetliners providing affordable, safe, fast, reliable transport everywhere.

The current fourth period is dominated by environmental concerns, and five points may be mentioned, though they are by no means sub-periods. The preservation of the earth's environment is a task for all activities of mankind, and transportation and aviation are no exception. Although aviation is far from the largest polluters at 2%, the expansion of air travel could increase this to 5% by 2050, and in any case, it must contribute its share to "Net Zero." The safety, efficiency, affordability, and wide availability of aviation, civil and military, must be preserved while meeting new environmental requirements. On the one hand, aviation is one of the most difficult sectors to "clean," requiring advanced technologies be safe and efficient, with stringent limits on weight and volume. On the other hand, aviation is regulated by international standards, and the goal on "Environmentally Friendly Aviation" can be reached globally and worldwide, avoiding different paces of progress dependent on local priorities and politics.

The following ten chapters address some of the technological operational, market, economic, and social issues involved in adding "Environmental Friendliness" to the record of safety, efficiency, and affordability of aviation as a means of transport and defense.

Contents

List of Figures

List of Tables

Chapter 1
Competition over Four Generations of Airliners Past and Future

Luiz Manuel Braga da Costa Campos

Contents

L. M. Braga da Costa Campos (✉)
Instituto Superior Técnico da Universidade de Lisboa, Lisboa, Portugal
e-mail: luis.campos@tecnico.ulisboa.pt

L. M. Braga da Costa Campos (ed.), *Accelerating Sustainable Aviation Initiatives*, Sustainable Aviation, https://doi.org/10.1007/978-3-031-73993-4_1

1.1 Introduction

The current airliners represent the culmination of aeronautical technology [1–7] developed over a century [8, 9], and the airline market [10–13] has become the biggest sector of aviation. The following account covers the airliner market from the dawn of the jet age to the present time and considers also future prospects. The focus is on subsonic airliners since supersonic flight was a major technological success [14–16] without the expected commercial impact, and the current situation is a duopoly between an incumbent Boeing with a long history [17] and a challenger Airbus with proven technical prowess, inherited from its constituent companies.

The long-range airliner market is dominated by the Airbus–Boeing duopoly (Sect. 1.3) that arose at the end of a long competitive period in which Airbus steadily gained ground (Sect. 1.2) starting from a newcomer status. The possibility of other competitors emerging in the long-range air transport market (Russia, China, or cooperation) is also considered (Sect. 1.4). The long-range airliner routes are fed by regional airliners that have a significant market of their own. Europe is strongly competitive not only on long-range and regional airliners but also in other categories like business aircraft and helicopters.

1.2 The Growing Airbus Challenge to Boeing

Airbus as a newcomer to the airline market (Sect. 1.2.1) started with the A300 by filling an empty niche: the short-haul wide-body aircraft (Sect. 1.2.2). A further advance in competitiveness was achieved with the increased use to electronics in the forward-facing crew cockpit (FFCC) of the A310 that reduced flight crew from three to two (Sect. 1.2.3). An even more significant innovation in the A320 as the first fly-by-wire airliner in the world that was countered by Boeing with an improvised second-generation B737 that managed to compete only through the expedient of "grand-father rights" (Sect. 1.2.4). The fourth stage of the Airbus challenge with the twin-engine A330 and four-engine A340 pair was met head-on by the Boeing 777 (Sect. 1.2.5). Airbus decided it had to challenge the remaining Boeing monopoly, the jumbo market of the B747, but the A380 as the world's largest airliner, has had a troubled development and poor market history (Sect. 1.2.6). Boeing took advantage of Airbus troubles with the A380 to launch a new challenge in another sector with the 787 to which Airbus responded late but successfully with the A350 and A330neo (Sect. 1.2.7). Since the A350 was designed to outperform both the Boeing 787 and 777, Boeing reacted with the stretched 777X out of reach of further A350 developments, effectively killing both the Boeing 747 and Airbus A380. The Bombardier temporary challenge with the C-series to companies 10 times larger, together with development delays, led to the massive success of the second-generation Airbus A320neo and the late response of the third-generation Boeing 737max (Sect. 1.2.8). Also, the greater stretch potential of the A320 allowed Airbus to

develop the A321/LR/XLR (Sect. 1.2.9) for which B737 stretches offer weak competition; together with Boeing's safety, production and certification troubles, this leaves Airbus to forge ahead in an ever stronger position in the single-aisle jetliner market, although Boeing retains some lead in the twin-aisle sector.

1.2.1 The European Pioneers and American Successors

The first generation of jetliners (Table 1.1) came mostly from Europe as pioneers, but their American competitors, despite coming later, achieved greater commercial success:

- The Sud-Aviation Caravelle and British Aircraft Corporation One-Eleven were the pioneer twin jets, although the market came to be dominated by the later entries from the Douglas DC-9 and Boeing 737.
- The Hawker Siddeley Trident was the first tri-jet, but the market was subsequently dominated by the later entry into the market of the Boeing 727.
- The Vickers VC10 was a pioneer four-engine aircraft with pioneer Rolls-Royce Conway turbofans, but like the Convair CV-880/990, it lost market position to the Douglas DC-8 and Boeing 707/720.

As for "government" versus "commercial" airliners:

- The private venture De Havilland Comet was the first jet airliner and a commercial failure due to structural fatigue problems unknown at the time.
- Before Boeing sold the first commercial Boeing 707, it had several years of development funded by the US Air Force that had ordered hundreds of similar KC-135 in-flight refueling tankers.

The case of Britain illustrates why the Europeans were the pioneers in commercial jets, but later American designs came to dominate the market.

Even not considering the ill fate of the pioneering jetliner, the De Havilland Comet, there were British jetliners in each category of the first generation: (i) the short-range twin engine BAC One-Elven, (ii) the medium-range three-engine HS Trident, and (iii) the long-range four-engine Vickers VC10.

Table 1.1 First-generation jet airliners (all single aisle)

Type	Short range twin engine	Medium range tri-jets	Long range four-jet
British	DH Comet BAC One-Eleven	HS Trident	Vickers VC 10
French	SA Caravelle	–	–
Boeing	B737	B727	B707
Douglas	DC-9	–	DC-8
Convair	–	–	CV-880/990
Tupolev	Tu – 104	Tu – 154	Tu – 114
Ilyushin	–	–	Il – 62

However, these aircraft were optimized for very specific requirements, for example:

- The HS Trident for the London–Athens route of British European Airways (BEA).
- The Vickers VC10 for the London–Nairobi route of British Overseas Aircraft Corporation (BOAC).

These narrow focus requirements made the aircraft less adaptable to other routes compared with the American designs catering to broader airline requirements.

For example, of two South African airlines one chose the Vickers VC10 and the other Boeing 707. Being designed for the London–Nairobi route, the VC10 had a large wing to land at hot and high Nairobi and did not need a too long range to fly from London.

When because of apartheid nearly all African States banned South African aircraft, the Vickers VC10 could not fly nonstop between Europe and South Africa, and the airline that bought it became bankrupt. The Europeans were pioneers in other areas of the market ultimately dominated by the Americans. The Rolls-Royce Conway of the Vickers VC10 was the first turbofan in service with better fuel economy than the Pratt & Whitney JT3 turbojet on the rival Boeing 707.

Although the RR Conway could be fitted to the Boeing 707, most were sold with the Pratt & Whitney JT3D, a turbofan development of the JT3C. Next Pratt & Whitney developed the smaller JT8D turbofan to power the Boeing tri-jet 727 and the bi-jet 737. The Boeing 737 was the last of the first-generation twin jets but ultimately displaced the first twinjet the SA Caravelle and the BAC One-Eleven. They sold well in the United States before the Douglas DC-9 and Boeing 737 became available and were practically ousted from the US market by their American competitors.

1.2.2 Airbus A300 and the High-Capacity Short-Range Niche

Just before the Airbus entered the airliner market, there were two disjointed sectors:

- Small single-aisle short-range narrow bodies up to 190 seats and 6000 km range.
- Large double-aisle long-range wide bodies up to 450 seats and 14,000 km range.

Airbus identified a niche for short high-density routes: the A300 twin-aisle wide-body carrying up to 300 passengers up to 8000 km. Boeing arrogantly dismissed (i) this Airbus niche market as nonexistent and (ii) the A300 as another government airplane doomed to failure. Although ultimately proved utterly wrong, both these charges could not be easily dismissed at the time and were the statement of two major challenges.

The Franco-German initiative to develop the Airbus A300 was met by the British government with the demand to use a new Rolls-Royce high bypass ratio (HBPR) turbofan, the RB207. This was similar in configuration but distinct from the RB

211 under development at the time for the Lockheed L-1011 TriStar, whose composite fan failures ultimately led to the bankruptcy of both Rolls-Royce and Lockheed, followed by their rescue respectively by the British and US governments. Airbus refused to develop at the same time a new aircraft (A300) and a 'political' engine (RB 207) and opted for an American well-proven engine, the General Electric CF6. The British still participated in the A300, with Hawker building the wing as a 'private' partner 'supported' by the government. The Airbus decision to choose the existing General Electric CF 6 over the new Rolls-Royce RB. 207 was proof that the A300 was not a 'government airplane', and Airbus had the courage to resist political pressures and aim squarely at the market ensuring its success. The Airbus A300 was one of the first of the second-generation jet airliners (Table 1.2).

1.2.3 The Airbus A310 and the Forward-Facing Crew Cockpit (FFCC)

Although the A300 achieved a 10% market share, Boeing refused to consider it in its market projections, allowing Airbus to strike the next coup unchallenged. At the time, the airliner cockpits consisted of large arrays of mostly duplicate analog flight instruments facing the pilot and co-pilot plus large side panel(s) with system instruments facing the flight engineer. Since the flight navigator of the preceding era of propeller-driven airliners had been dispersed earlier, at this time the minimum flight crew was 3. Airbus proposed the introduction of multifunction electronic flight displays capable of showing different information at different stages of flight or in different circumstances, such as system diagrams in the case of a failure or emergency; together with a much-increased level of automation, this would allow the flight engineer and his side panels to be dispensed with, hence the publicity-kind designation forward-facing crew cockpit (FFCC). The elimination of the flight engineer was rather controversial, and it did not escape anyone that the ultimate motivation was to reduce flight crew costs from 3 to 2. Nevertheless, the electronic flight instrument displays were an innovation to stay, providing a much more flexible way of selecting among a wide array of information and the increased efficiency of the A310 added to the market penetration of the A300 reaching 20%. Still, Boeing

Table 1.2 Second-generation jet airliners

		Medium range		Long-range
Type	Short-range single-aisle	Single aisle	Twin aisle	
Boeing	737NG	757	767	747/777
Airbus	A320	–	A300/A310	A330/A340
McDonnell Douglas	MD-80	–	–	MD-10
Lockheed	–	–	–	L-1011
Russian	Tupolev TU-134	–	–	Ilyushin Il-86

continued to refuse the inclusion of Airbus in its market projections, making the third stage of the Airbus challenge, the A320, the rude awakening.

1.2.4 The Airbus A320: The First Fly-by-Wire Airliner

The introduction of fly-by-wire technology from fighters in the first airliner allowed the A320 to be designed with a smaller tail-plane area, less cruise trim drag, and higher efficiency. Boeing initially dismissed the A320 diffidently as the "wonder plane" until realizing that it made the existing B737 uncompetitive. Boeing rushed to upgrade the 737 into a 737 NG ("New Generation") with some limitations. For example, the lack of ground clearance below the wing limited the diameter of the turbofan engine used and required a noncircular flat-bottomed nacelle; the short nacelle pylon also caused transonic flow problems. After all the possible modifications, Boeing realized the second-generation 737 could still not compete with the A320 and resorted to a certification expedient: "grandfather rights." It claimed that the second-generation Boeing 737 only had to meet certification rules held at the time of the first generation and need not comply with current certification rules that applied to new aircraft like the A320. Consequently, the Boeing 737 could seat more passengers in a cabin smaller than that of the A320 by not meeting current emergency evacuation standards. The granting of grandfather rights to the second-generation Boeing 737 by the Federal Aviation Administration (FAA) drew considerable criticism from outside the United States as a poorly disguised protectionist measure, and it was eventually agreed internationally that this would not be repeated in the future. With the extensive redesign and the unfair economic advantage of carrying more passengers in a smaller cabin, the second-generation Boeing 737 managed to compete the all-new A320 and Airbus reached a 30% market share. The fourth stage of the Airbus challenge, the A330/A340, no longer benefited from the disdain or complacency of Boeing that by now realized it had a serious rival and had to compete on merit rather than by expedients like "grandfather rights."

1.2.5 The Airbus A330/A340 and the Family Commonality Concept

The Boeing family of airliners consisting of the short-haul 737 and medium-haul 757 single-aisle and wide-body medium-range 767 and long-range 747 was a mixed bunch of designs with different ages and standards preventing economies of scale in crew training and maintenance tasks across different types. The more recent and up-to-date Airbus range besides consisting of models competitive individually against their Boeing counterparts had the additional advantage of commonalities requiring less maintenance infrastructure and shorter crew retraining across a mixed fleet. Airbus followed conventional belief (A) that long overwater flights required

four engines, an assumption that may have been its only questionable choice and would haunt it for two decades. The fourth Airbus challenge (B) shared the fuselage cross-section of the A300/A310 in the twin-engine medium-range A330 and four-engine long-range A340, using different wings with the same gross weight capability that would later enable the A330neo. The A330/A340 completed the Airbus family started with the A300/A310 and continued with the A320 and enabled a 40% market share. Contrary to the past, Boeing had shaken off its former complacency and was back in its best competitive talent.

The earlier Boeing complacency was supported by the demise or decline in the market of its main American competitors:

- The Convair CV-880/990 transcontinental four jets lacked the transoceanic range potential of the Boeing 707 and were the last Convair airliners.
- The failure of the carbon fiber fan in the Rolls-Royce RB.211 engine without a titanium fan alternative led to the end of the (Lockheed L-1011) Tristar as the last Lockheed airliner.
- McDonnell Douglas was limping along in a survival mode with low production rates of the DC-9/DC-10 and derivatives, surviving on spares revenue from existing fleets, which was doomed to decline.

The success of Boeing over Douglas was a testimony to boldness and innovation:

- The Boeing 707 held the lead over the Douglas DC-8 up to the series 50 until the DC-8-60 series stretch from 190 to 260 passengers gave a 25% seat-mile advantage that could not be matched.
- Boeing countered with the first jumbo, the 747 that needed a new factory and did not fit existing airports and made a success out of gambling on its existence.
- Douglas countered with the DC-10 as a wide-body tri-jet and canceled the DC-8-60 series because it was too much of internal competition.
- The decline of the DC-10/MD-10 versus B747/B767 and DC-9/MD-80 versus the B737/B757 allowed Boeing to take-over its last American competitor.
- The Boeing attempt to salvage the McDonnell Douglas MD-80/90 as a Boeing 717 was an abysmal failure since no airline wanted an aircraft having nothing in common with the Boeing family except the name.
- Airbus had vowed not to make an aircraft smaller than the A319, but did an A318 to counter the B717, and managed relatively modest but still better sales than its rival.

By now the only significant remaining Boeing rival in the world airliner market was coming from an unexpected quarter: 'government inspired' airplanes spun a truly market-oriented competitor: Airbus. NASA complained that a complacent Boeing was no longer interested in new technologies, which were not needed for the sole market leader. Everything changed as Boeing decided to counter head-on the Airbus A330/A340, focusing on its two weak points (A) and (B) above. Boeing made the bold assumption that twin-engine aircraft could fly over water over the whole world, including both the Atlantic and Pacific oceans. The realization that modern turbofan engines seldom failed in cruise led to the prospect of extended

overwater flights more than 90 minutes away from the nearest airport. ETOPS allowance increased steadily from 90 minutes to over 3 hours, allowing unrestricted transatlantic and transpacific flights. With ETOPS, Boeing needed only one twin-engine B777 to counter the A330/A340 pair. With the benefit of the hindsight of coming later, Boeing adds to (A) the further gauntlet (B) choosing a fuselage cross-section wide than that the A330/A340 "inherited" from the A300/A310. The larger capacity of the B777 versus the A340 had this time Airbus scrambling back to the drawing board with the A340–600. Bound to the same cross-section as the A300/A310/A330, all that Airbus could do was to increase the length of the fuselage to the airport limit; the A340–600 had the longest fuselage of any airliner and was the first whose flying qualities were affected by the elastic modes of the fuselage. The A340–600 was almost a totally new aircraft relative to earlier A340 variants, and the large effort and expenditure in its development served only to prolong its life for some years. The end of production of the A340 left the airliner market only with twins except for the old B747 and new A380 inherited from the time it was still believed that four engines were needed for long overwater flights.

1.2.6 The Airbus A380, the World's Largest Airliner

Airbus launched the project of A380 in 2000. The initial rationale for it was quite compelling:

- Airbus had reached a 40% market share and was competing with Boeing in every sector except the jumbo aircraft.
- Boeing was using its "jumbo monopoly" selling a B747 for twice the price of a B767, collecting profits to finance the rest of the range.
- To reach 50% or parity, Airbus had to challenge Boeing in every sector, breaking its last monopoly with the jumbo aircraft.

As the world's largest airliner, the A380 was not short of challenges:

- Its size and weight were well beyond the optimum of minimum weight per passenger of an airliner at about 350 seats.
- As a consequence, the A380 had 30 novel weight-saving features, like 4000 psi hydraulics instead of the 3000 psi standard up to then.
- Given the larger number of in-flight entertainment systems for passenger seats with different arrangements for each airline, the replacement of copper wires by aluminum wires was one of the 30 significant weight savings.

This last apparently innocuous change was the start of a far-reaching chain of events:

- The aluminum wires could not be bent to the same small radius as copper wires, leading to installation problems that were reported late.

- The Airbus method of sending fully equipped sections for final assembly maximizes added value to partners (the opposite of Boeing that maximizes value in final assembly).
- The arrival of unfinished sections at final assembly caused a production bottleneck, with a large number of workers brought to Toulouse from factories in other countries.
- The rate of expenditure would have bankrupted Airbus in 2 years unless a new strategy was adopted.

The detailed redesign of elements of the A380 was not helped by another contrast:

- All Airbus partners, like Boeing, used the CATIA design software developed by Dassault.
- Except for its former Aerospatiale rival that used software from a firm in Seattle, the home of Boeing.

The Airbus A380 is certainly the most spacious and probably the most comfortable airliner ever flown. It did upstage the Boeing 747, which was in any case undermined by the Boeing 777 as an in-house rival. It is not only the largest airliner ever but also one of the most sophisticated. However, some important weak points made for its failure as a sound business case and finally determined the decision in February 2019 to terminate the production at the end of 2021.

1.2.7 The Sonic Cruiser, B787, and A350

Boeing recognized that the market could not support two jumbos, and thus decided not to develop a direct rival to the A380. The Airbus troubles with the A380 were an opportunity not to be missed to strike back in another sector. The Boeing counter, the "sonic cruiser," was a daring and brilliant design: it managed to have a fuselage of constant cross-section while obeying the area rule for low shock wave drag by clever blending with the canard foreplane and main wing. It was capable of cruising efficiently at Mach 0.95 versus Mach 0.85 for a conventional airliner, but sales may depend more on attractive economics than brilliant engineering. Boeing argued in vain that a 10–12% decrease in flight time in long-haul routes would allow more daily flights with the help of shorter turnaround at airports. Airlines were not convinced, and Boeing brought along as a fall back to the 7E7: a conventional design using the advanced technologies of the sonic cruiser, like a composite fuselage, to reduce cost. The 7E7 was the preferred choice for every airline and became the Boeing 787. The B787 gained the largest order book of an airliner in history before its first flight. Boeing predicted that the 787 would be its quickest development and certification program in recent times with entry into service 3 years from start-up. This optimism was not supported by a development period more than twice as long, with perhaps not surprising problems in the production of the single-barrel all-composite fuselage. After entry into service, the Boeing choice of a type of

Table 1.3 Third-generation jet airliners

	Single aisle			Long range	
Type	Regional	Short range	Medium range	Long range	Jumbo
Airbus	A220	A320neo	A321LR/XLR	A350/A330neo	A380[a]
Boeing	–	B737-Max-7	B737-Max-9/10[c]	B787/777X[c]	B747-8
Russia	Sukhoi Su-100	–	–	Ilyushin Il-86	–
China	ARJ21[b]	Comac C919[b]	–	CR-929[c]	–
Brazil	Embraer E-series		–	–	–
Japan	Mitsubishi MRJ[c]	–	–	–	–

[a]Production stopped
[b]Chinese certification only
[c]Project initially with Russia

lithium-ion battery with high-energy density for low weight turned out to cause battery fires that were cured by a redesign adding weight. Notwithstanding the usual development hurdles, the Boeing 787 remained a considerable success, this time helped by the reluctance of Airbus to offer a matching competitor. The Boeing 787 may be considered the first third-generation jet airliner (Table 1.3).

Saddled with the A380 issues, Airbus could hardly cherish another totally new aircraft development. In contrast with the past, when Boeing was slow or inadequate in its responses, this time it was Airbus that delayed a competitive answer. The initial Airbus argument that the A330 was an already available B787 competitor did not convince airlines. The reluctant further developments of the A330 were not too well received either, although the second-generation A330neo later proved to be a moderate market success. When Airbus finally accepted the reality that nothing short of a new clean-sheet design would convince airlines, the A350 was the resulting clever design "splitting" the B787 and B777:

- Larger than the B787 with a comparable cost.
- Comparable in size to the B777 with lower cost.

This clever killing of two rabbits with a single stroke drew the predictable matching reaction from Boeing:

(a) Trying to improve the competitiveness of the 787.
(b) Stretching the 777 to a second-generation 777X beyond the growth potential of the A350.

The move (b) meant that the B777X effectively superseded the B747 and brought stronger competition with the A380 as the world's largest airliner and nearly the sole remaining four-engine type. Concerning move (a), the re-engineering of the Airbus A330 as the second generation, A330neo proved after all that a cheaper-to-buy and cost-effective to operate aircraft could be rapidly developed as an alternative to the B787. The A330neo was inspired by the success story of the A320neo but not nearly as competitive with Boeing rivals.

1.2.8 Bombardier C-Series A220, Airbus A320neo, and Boeing B737max

The third and fourth largest airliner manufacturers in the world, Embraer of Brazil and Bombardier of Canada, are about one-tenth of the size of the two world leaders: Boeing of America and Airbus of Europe. Only the brave or the foolish would try to bridge a gap of one order of magnitude. Embraer has wisely stayed below the Airbus–Boeing market, perhaps scraping it at the lower end, without ever venturing into a direct challenge. Bombardier claimed it was not trying to compete with Airbus or Boeing but in fact tried to do just that with the C-series. The main advantage of the Bombardier C-series was a new generation of geared turbofans from Pratt & Whitney; this advantage would easily disappear if the same engine was adopted by the A320 or B737. The PW1200 geared fan had development problems and delays of its own and comparable performance was achieved by the Snecma/General Electric LEAP engine, which used advanced technology in a more conventional design to come to market earlier. Not only the Bombardier C-series lead in engines was lost, but also the lead in time was compromised by development problems delaying entry into service. The Bombardier C-series was completely upstaged by the rival developments of the Airbus A320neo and Boeing B737max. Following commercial attacks by the US government meant to cut access of the plane on the US market, the acquisition of a majority in the C-series by Airbus marked the end of the adventurous challenge. The C-series was rebranded as A220. In fact, the reactions of Airbus and Boeing to the Bombardier C-series were somewhat different.

Airbus would not let the Bombardier C-series erode the A320 market; bearing in mind the first-generation A320 was relatively modern and efficient, a second-generation re-engineering in the A320neo would completely upstage the challenger and consolidate its market. The A320neo was a runaway success, collecting orders at a rate never seen before and forcing Boeing to enter the fray with the third-generation B737max. Boeing was understandably reluctant to go into a third generation of the B737:

- The first generation was by now over 50 years old.
- The second generation had struggled but managed against the first-generation A320 with the expedient of grandfather rights.
- Without the help of grandfather rights, a third redesign of the B737 would be an even greater struggle against a second generation of the newer Airbus A320.

The Boeing decision to counter the new A320 with a second-generation B737 rather than a new design came back to haunt it again with the third-generation B737 versus the second-generation A320. For example, the low ground clearance of the flat-bottomed nacelle of the second-generation B737 could only become a more difficult problem with the increased diameter of more recent higher bypass ratio engines. Boeing understood the engine under the wing of the third-generation B737 would have a smaller diameter and lower bypass ratio than for the second-generation A320. This shortfall in engine efficiency would have to be compensated in other

areas like aerodynamics, yet Airbus would hardly be left behind in any area and allow Boeing to recover its weaker starting position. The preference of Boeing was to let some years pass to design an all-new single-aisle replacement for the B737. For this reason, Boeing did not feel it necessary to counter immediately the Bombardier C-series, but after the runaway success of the Airbus A320neo, it had no choice but to join the fray. Just as Airbus had been reluctant to launch the A350 to counter the B787, Boeing was reluctant to launch the third-generation B737max against the second-generation A320neo. In both cases, there was no choice, with the A350 recovering only partially the order book lost to the B787, and similarly the B737max trying to narrow the initial gap to the A320neo.

1.2.9 Success of the A321/LX/XLR

In addition, the gap between the third generation of the old B737 and the second generation of the not-so-old A320neo created another market impact:

- The A320neo had enough stretch potential into the A321neo, which is an effective replacement for the Boeing 757, still used on long thin routes.
- The stretched A321LR was an immediate success, amplified by the further stretch to the A321XLR.
- The lack of stretch potential of the B737 leaves Boeing with a weak alternative to the B737Max-9/10 or no alternative other than to develop an all-new middle-of-the-market aircraft (MMA) of a future single aisle (FSA).

The A321/LR/XLR may be seen as a successor to the B757 and B767, or more appropriately a new third class of middle-of-the-market aircraft (MMA). Before, the choices were mostly:

- Narrow-body single aisle smaller airliners for transcontinental routes, excluding crossing oceans.
- Wide-body twin-aisle larger aircraft for intercontinental routes, like crossing the Atlantic or Pacific oceans.

The lower fuel consumption of the SNECMA/General Electric LEAP and Pratt & Whitney PW 1200 engines relative to the SNECMA/General Electric CFM50 and Pratt & Whitney Rolls-Royce V2500 allowed the A321 to cross the Atlantic, which had never been expected of the A320 family and cannot be matched by the B737max due to its lack of stretch potential.

The success of the A321neo/LR/XLR is that it can fly the "long thin" routes of the B757/B767, allowing distances like crossing the Atlantic with narrow-body costs, and avoiding the need for wide bodies. This effectively leads to three classes of aircraft:

- Short-range narrow bodies for the traditional transcontinental routes with moderate capacity.

- Medium-range narrow bodies with a similar capacity capable of transatlantic distances for long thin routes.
- Long-range wide bodies with a large capacity capable of transpacific and global distances for long thick routes.

The A321neo/LR/XLR has thus replaced wide bodies for transatlantic distances when large capacity is not needed. The Airbus A321/LR/XLR is in production, whereas an eventual Boeing MMA could be offered much later when the need for 757/767 replacements would have been satisfied by Airbus products. Although Boeing has toyed with the idea of an MMA/FSA, which is what it would have liked to do instead of the 737max, the prospects are challenging. A new aircraft can be amortized only if it provides at least a 10–20% improvement in fuel consumption to lower direct operating costs. This is unlikely to be achieved bearing in mind that the A320neo and B737max use state-of-the-art engines and a new engine generation might be needed. The development of an MMA would be costly and time-consuming, and it is questionable if the market left unfulfilled would allow a break even. Boeing is, at last, having to live with the consequences of soldiering for too long with the B737 and may after all have not gained much from the expedient of grandfather rights in the long term; it would be doing better now with a less old design. Airbus is benefiting from a more recent A320 design that was an inevitable consequence of being a newcomer to the market. However, with a production lifetime of 10 years longer than the 8 years of the A320, the B737 still has a larger number of units sold, although the gap is closing, and may be reversed by increasing sales of the A321 LR/XLR for which Boeing has no competitor. This sets the background for the status of the Airbus versus Boeing competition in the airliner market (Sect. 1.3).

1.3 The Current Status of the Airbus–Boeing Competition

The situation is different in the single-aisle narrow body (Sect. 1.3.1) and twin-aisle wide body (Sect. 1.3.2) market, leading to an approximate balance with Airbus leading the former and Boeing the latter. Both Airbus and Boeing are in the healthy situation of having the largest order books in history and face challenges in achieving higher production rates (Sect. 1.3.3). The profits may be invested in evolutionary developments of existing aircraft (Sect.1.3.4) or in totally new designs that will require years of maturation to incorporate new technologies (Sect. 1.3.8). In the meantime, the production of the A380 ceased, together with the cargo B747-8, both as the world's largest airliners and last surviving with four engines (Sect. 1.3.5). The prospects for a middle-of-the-market aircraft look slim (Sect. 1.3.6); the absorption of the Bombardier C-series into Airbus (Sect. 1.3.7) left both at an advantage relative to their rivals Boeing and Embraer that failed to reach a similar deal.

1.3.1 Single-Aisle or Narrow-Body Market

The prompt and decisive response of Airbus to the Bombardier C-series challenge led to a large order book for the A320neo as the second-generation A320. The delays and hesitations of Boeing to commit to the third generation of B737 (i.e. the Max) led to a response only when there was a large gap to recover. Boeing proposed the third-generation B737Max to avoid the risk of losing traditional and faithful airline customers. This customer base amounted to a substantial backlog reducing but not closing the gap to the A320neo. The stretched A321neo dominates the long, thin route market, without a clear match in the B737 family, so this part of the gap is not readily closed.

Boeing would need an all-new middle-of the-market aircraft that might come too late to an already partially filled market, with questionable profitability. A bold and risky decision to develop the MMA aircraft by Boeing may be necessary to remain a strong supplier of narrow bodies in competition with Airbus. The Airbus lead in the orders of narrow bodies may be balanced by a Boeing lead in wide bodies, with higher unit values compensating the smaller number to lead to comparable profitability.

1.3.2 Twin-Aisle or Wide-Body Market

The situation here is the reverse of the A320neo lead over the B737max, with Boeing amassing a large backlog of B787 orders before Airbus replaced its evolved A330 plans with an all-new A350. The A350 effectively targeted the B787 offering higher capacity for a comparable cost. And the second-generation re-engined A330neo proved after all that it also had a market as a lower-cost B787 competitor. The double-pronged counterattack with the A350 and A330neo allowed Airbus to narrow the gap to the B787, helped by the development delays of the latter to reduce the effects of a late response. The A350 had the second task of matching the B777 capacity at a lower cost, which it did at the then status. But Boeing could hardly be expected to sit still and let two of its products be challenged by a single Airbus product. The second-generation stretched 777X moved beyond the stretch potential of the A350, and at the risk of killing the already moribund B747. However, as in many other large projects, 777X met some obstacles in the development, which introduced some costly delays. Due to what Boeing euphemistically calls "GE9X anomaly," the maiden flight of the aircraft, initially expected at the beginning of 2019 was postponed, followed also by delays with the airframe, and entry into service still pending. The GE9X, the largest jet engine ever built based on revolutionary technologies, revealed during the endurance tests and fan blade out test the need for some configuration modifications to satisfy the certification requirements. Although the B777X has faced other delays and is yet to enter airline service, it still

ensures Boeing's lead in the wide-body market. The current situation of the Airbus–Boeing duopoly can be summarized as follows:

- Airbus has achieved a 60:40 advantage versus Boeing in the total number of aircraft on order due to the success of the A321/LR/XLR for which Boeing has no rival in its narrow-body airliner range.
- The Boing advantage over Airbus in orders for wide-body aircraft is smaller, but the higher unit value of wide bodies means that in terms of total value of the order book there is closer to a 50:50 balance Airbus–Boeing.

1.3.3 Production Ramp-Up to Meet Large Order Books

One point that Airbus and Boeing share is their biggest order books in history, to the extent some skeptics point to a historic risk of a collapse of the order bubble. This time around with a backlog of several years and a steady growth of air transport it was hard to see how a calamity of such scale could occur. An unlikely big decline in the airliner market would have several years of production as a cushion and large-scale cancellations do not seem to fit airline plans. Rather the reverse, the industry is still trying to recover from an unexpected order boom. Tom Enders, the ex-CEO of Airbus, said that if some years ago someone suggested a production rate of 30 A320 per month he would have thought this was outright foolishness; but later Airbus was aiming at 60 A320 per month, now is aiming at 75, and even at this rate it will take several years to fulfill its current backlog. Although many orders are "nice to have a problem," it does pose many challenges, not only at expanding the final production line but also with suppliers: some may be wary of growing over-size, especially when Boeing and Airbus try to squeeze lower prices. Others may see expansion plans limited by access to credit. In some cases, Airbus has bought suppliers mainly to ensure that they have the resources to deliver the required quantities. Several months after the Max grounding crisis, Boeing reduced 737 production rates from 52 to 42 per month, and then stopped production altogether, leaving suppliers in a difficult situation or need for support.

The COVID-19 pandemic (Chap. 6) was an unprecedented crisis no one could predict and did lead to order cancellations of about 1/3 of the Airbus and Boeing backlogs and reduced production rates that affected the supply chain the most. However, the quick rebound of airline traffic after the COVID pandemic and the pressure toward more environmentally friendly airliners have restored the order books. The production ramp-up to meet increasing demand has been slowed by a supply chain, in a particular for engines but also the components, unable to change rapidly from a strongly depressed to the rapidly booming market. Thus, apart from the painful COVID-19 period of 2020–2021, both Airbus and Boeing are back to fat order books and struggling to raise production rates, for example, from 30 to 52 per month for the B737 family and from 50 to 75 for the A320 family. Airbus

production ramp-up is delayed by supply chain issues, not the production quality issues that repeatedly emerge at Boeing further limiting aircraft deliveries.

The A321neo/XLR and future developments are the current success story. In the past, single-aisle aircraft had smaller capacity (up to 200 seats) and transcontinental range but were unable to cross oceans. Twin-aisle aircraft (300 seats or more) were needed for transoceanic routes. Retrofitting the more efficient engines developed for the Bombardier C-series to the A320 family led to the A320neo version A321LR able to cross the Atlantic. With further stretches, the A321XLR has really become a middle-of-the-market aircraft able to replace wide bodies in medium-range routes with lower costs and higher frequencies. Thus, the A321neo/XLR is not only extending sales of the A320neo family to longer routes but also taking away sales from wide bodies, leaving for them only long-range routes. The A321neo/LR/XLR is currently receiving most of the airliner orders, with Boeing 737/9–10 focusing on larger capacity for the traditional transcontinental routes.

1.3.4 Continued Competition Through Evolutionary Developments

A heavy workload in increasing final production rates and shoring up the supply chain to much higher outputs has left Airbus and Boeing with little spare capacity for major all-new designs. The current aircraft are selling at unprecedented rates, so there is no need or incentive to come up with radical improvements. A radical improvement might require much more efficient engines, but the re-engined second- and third-generation aircraft already rely on the latest and most advanced technology in engines, that still have reliability problems, so there is little scope for a major improvement in a few years. Besides, the engine manufacturers are enjoying an order boom in proportion to that of airframe manufacturers: Snecma and General Electric have challenges similar to Airbus and Boeing in increasing the final production rates and shoring up the supplier base; in spite of problems with the geared fan at Pratt & Whitney and the "self-limitation" of Rolls-Royce to wide-body engines, only the latter seems to be in a less than healthy state. With both airframe and engine manufacturers busy filling order books, incremental improvements are the feasible option, waiting for major progress that could take 5–10 years to mature to enable new designs.

1.3.5 The Fall of the Airbus A380

Apart from the B747 as a freighter, the A380 was the last four-engine airliner, both no longer in production. The management at Airbus announced the termination of the production in 2021, and Boeing followed with the 747 in 2023.

Since the launch of the project in 2000, the demand for very large airplanes faded slowly. No US airline purchased the type. Lufthansa bought only 14. Air France-KLM, Qantas, and Virgin Atlantic canceled orders. Emirates, the larger operator, lost enthusiasm more recently and some calls for ab A380neo did not have much response from Airbus. The economic disadvantage relative to long-range wide-body twins is exacerbated by retaining an older generation of less efficient engines. Two flights of the B787/A350 carry as many passengers as a single A380 flight at a comparable or lower cost, with greater flexibility of schedule but using twice as many airports' slots. The A380 would require the thrust of three existing engines of the size of the B777/A350, but this is not a feasible configuration. The development of a new twin-engine configuration would have required much larger engines than those currently available. Re-engining with four engines of the newer generation, suggested by the largest Middle East operators, would have implied development costs that engine suppliers were not prepared to shoulder due to limited market prospects.

1.3.6 The Boeing New Mid-Market Airplane (MMA)

Generally, middle-of-the-market segment is defined as a mid-size segment, located between the narrow-body and the wide-body market, and which encompasses aircraft carrying 200–270 passengers and a range that can vary from 3000 nm to 5000 nm, as defined by Boeing executives. Due to the poor definition of this market, some aircraft that are found in the limit can be considered or not part of this market such as the A321neo or some variants of the 737 family. However, the main aircraft that have represented the competition within this market until the date are the B757/B767 from Boeing and the A330 from Airbus, taking into account their different variants.

Due to its performance improvements and flexibility, the A330neo is being considered a strong candidate for replacing the aging fleet of Boeing 757/767 s. However, as of July 2019, the backlog for A330neo was of only 207 units, all of them for −900 neo, none for the proposed longer-range -800neo. Meanwhile, Boeing does not have a clear candidate to replace the 757/767 fleet, which may lead in the coming years to a drop in orders in favor of Airbus. It is possible that the new Boeing MMA aircraft might change this scenario if its introduction to the market is not delayed, and it offers performance advantages versus its competitors, but the current difficult financial and engineering resources at Boeing may make such a prospect unlikely before 2030. A good promise is the capability of the A321neo/LR/XLR to cover the MMA segment. It still enjoys a good growth potential (the composite wing in A321neo-plus exercise). Its demonstrated current range is over 4000 nm, and its maximum capacity has still reserved to be increased from the existing 206 seats to perhaps 250. B737max lacks this stretch potential. The A321XLR project, launched in June 2019, declares a maximum range of 4700 nm, covering a majority of the routes MMA is expected to target.

A clean-sheet design of the MMA would be a premiere at Boeing for the last two decades. Their commercial aircraft subsidiary seems to have preferred upgrading older models, a conservative strategy. A state-of-the-art solution for MMA is supposed to contain composite wing and fuselage, a hybrid cross-section, and next-generation engines, everything promising a low $/seat/nm index. If all this can be achieved at a reasonable list price, it can be a winner in its market. However, a development cost somewhere around $15bn might need a higher value for the program break-even volume, much more than the $1–2bn cost of A321 upgrade did imply.

The prospects for the Boeing MMA have declined in the current difficult situation at Boeing and were never promising without a new generation of more advanced engines available not before 2030. The lower end of the market is well covered by A321/XLR. The upper end is covered by the A330neo and would reduce the sales of wide bodies operating in the sector. Boeing has canceled the MMA to focus on the future single-aisle (FSA) replacement for the B737 and that project has been suspended too. The B737max is challenged at the middle of the short-haul market by the A320neo, at the top by the A321, and at the bottom by the A220/Embraer C-series.

1.3.7 The Bombardier C-Series as Airbus A220

The challenges that Boeing faces at the small capacity end of the market with the B737Max are rather serious. For a start, the A321neo stretch beyond the capacity of the B737 to which Boeing must respond to continue to compete in the long thin routes. Boeing may have brought another challenge to itself by opposing the 'subsidized' sales of the Bombardier C-series in the United States. The cost-free acquisition of majority rights in the C-series leaves Airbus with a formidable and comprehensive line-up of state-of-the-art airliners in the 100–300-seat range:

- The A220, ex-Bombardier C-series, is optimized for 100–150 seats and is a good alternative to the slow-selling A319 shrink of the A320.
- The A320neo remains a strong and efficient contender for the 150–200 seats.
- The A321neo dominates the long thin routes with 200–250 seats, with its stretch potential confirmed in 2019 by the launching of 321XLR.
- Beyond 250 seats, the A330neo has a competitive price and good economy.

The link with Airbus benefits Bombardier C-series in several ways:

- Not only in circumventing possible US protectionism by having a final assembly at the Airbus facility in Mobile, Alabama in America.
- Also, Airbus's clout can bring supplier prices to levels that Bombardier would not be able to obtain, making the C-series more competitive in capital cost.

- The Airbus sales and support network and the integration in a product line ranging all the way up from 100 seats give a dimension to the C-series that Bombardier could not hope to achieve on its own.

Against the formidable line-up of the A220/A319/A320neo/A321LR/A321XLR/A330neo, Boeing has only the third generation of B737Max that manages to compete in the 130–200-seat range, leaving an important slice of the market for the replacement of the B757/B767, currently used on long thin routes, almost unchallenged to Airbus. The reaction to the Airbus–Bombardier link with a Boeing–Embraer link was predictable but not equally effective:

- The Embraer E2 series does not go beyond 120 seats compared with the 150 seats of the Bombardier C-series.
- This leaves Boeing with the challenge to cover the 100–300-seat range with the existing B737max below 200 seats and the new MMA above.

The collapse of the Boeing–Embraer deal is a blow to both:

- Boeing is left without a competitor to the A220 in the small end of the short haul sector.
- Embraer is left alone to compete against the former competitor Bombardier C-series now as an A220 with the full backing of Airbus.

While it is widely believed that Boeing must launch an MMA, the challenges it faces are considerable:

- To differentiate from Airbus offerings and compensate for late arrival to the market, it could be a twin-aisle, but this adds drag and reduces efficiency.
- To recover the investment in an MMA, airlines need a 10–20% increase in efficiency that will require a new engine for which all major suppliers (General Electric/Snecma and Pratt & Whitney, Rolls-Royce) will be competing.
- The development time and time to market of a totally new clean sheet Boeing MMA will give Airbus plenty of opportunities to upgrade its current range at a lower cost to try to reduce the market impact of its rival.
- If Boeing succeeds in putting the MMA beyond the nearly exhausted potential of the A321XLR, Airbus can still, with the benefit of hindsight, decide to develop an all-new aircraft possibly superior in at least some aspects.

Overall, Airbus should be able to keep an upper hand in the 100–300-seat airliner market until significant new technologies are implemented in a new generation of single-aisle airliners in the post 2030 time frame. In any case, the healthy and still growing order books of both Airbus and Boeing, if they can rise to the challenge of increased production, bring revenues to do more than evolutionary developments. Both Airbus and Boeing have the financial resources and engineering talent to keep up their healthy competition to the benefit of airlines and their passengers, although Boeing may remain hampered, both technically and financially, for several years, by production quality problems.

1.3.8 The Promise of Radical New Configurations

Recent years have seen a growing interest in configurations distinct from the classical tube-and-wing of current airliners dating back to the vision of Sir George Cayley in the mid-nineteenth century. The sometimes called radically new configurations like flying wings or joined wings were envisioned about 80 years ago in the period between the two world wars, though use has been sporadic until recently. After early post world war II designs with stability challenges, Northrop Grumann B-2 Spirit stealth flying wing bomber and a variety of drone designs have proved the viability of these concepts in several contexts. However, airliner design is subject to a long list of technical, certification, economic, and operational requirements that extend beyond military and drone requirements. The case of the flying wing serves as an illustration of this point. There are clear advantages:

- The higher lift-to-drag ratio can lead to a reduction of 20% in fuel consumption and emissions.
- The large internal volume of the thick wing provides plenty of passenger and cargo space.
- The choice of overwing engine nacelles, or flush or buried engines, or distributed propulsion would reduce noise at airports.

All these benefits do not come without challenges, some of which can erode the result:

- The engine nacelles in the accelerated airflow above the wing are affected by compressibility effects at lower speeds than underwing nacelles and could reduce cruise speed for the same drag.
- The engines at the rear top of the wing would lie in a thick boundary, whose ingestion could cause a stall or operating problems, that could become more serious for flush or buried engines.
- The distributed propulsion raises issues of transmission of power, like high-pressure losses in ducted flows or resistive dissipation of high-electric currents.
- The top-mounted rear engines create a large pitch-down moment opposing rotation at take-off, requiring a large lift and angle of attack and a long undercarriage to avoid tail scrapes.
- The trimming of the large pitch-down moment in cruise with upward deflection of trailing-edge control surfaces would reduce lift and lower lift-to-drag ratio benefits.
- The wide fuselage would place most inboard passengers far from the windows, whereas outboard passengers would have large displacements in roll maneuvers.
- The certification requirement of emergency evacuation in darkness from one random side of the fuselage in 90 seconds is more difficult to meet in a wide and short cabin than in a long and thin one.
- The easy isotropic pressurization of a cylindrical fuselage does not apply to a flying wing that must be either divided into tubes or need extra bracing or both.

This sample of issues shows that it is not a straightforward conclusion to decide on the overall benefits of new aircraft configurations, although they will inevitably come, sooner or later.

Given the current large-order backlog of modernized aircraft and the time to mature further significant advances in technology, the next generation in technology may not come in less than 5–10 years' time, with market competition making the Boeing/Airbus MMA the most likely all-new aircraft no sooner than 2030–2035. It will be a tube-and-wing aircraft, a configuration whose potential is far from exhausted since it is possible to incorporate several improvements. On the other hand, the radical transition to a new configuration could be too risky an experiment in a market-oriented product. A half-scale demonstrator of a new configuration like a flying wing or joined wing might be useful to:

- Convince the industry that all design trade-offs have been mastered and a reliable design database has been created.
- Allow certification authorities, service providers, airlines, airports, and maintenance organizations to prepare for upcoming changes.
- Familiarize and educate the public about upcoming progress out of past precedent.

The prospect of an evolved tube-and-wing next generation and a flying wing generation to follow may depend on a major effort for in-flight validation of the latter in parallel with the development of the former. The precursor half-for full-scale Boeing flying wing aircraft could arise out of a NASA experimental aircraft program or a tanker/transport aircraft for the US Air Force or a combination of both. In this case, Europe should not fall behind and must support a comparable flying wing demonstrator.

1.3.9 Prospects for New Airliners in the 2030s

The wide-body market is well served by the Airbus A350 and A330neo and Boeing B787 and in development B777X, so no new aircraft may be expected soon.

The situation is different in the narrow-body market:

- The B737Max and new variants remain effective competitors for the A320neo for transcontinental flights.
- The Airbus A220, born Bombardier C-series, is a more recent design with some advantages over smaller variants of the B737max for shorter distances or lower capacity.
- The A321neo/LR/XLR has no rival in the B737 family and is biasing the 50:50 split to 60:40 split in favor of Airbus versus Boeing.
- The single-aisle market is now two different aircraft rather than a single model, which cannot possibly cover in an efficient manner the wide gap between the A220 and A321 XLR.

Whereas some observers insist that Boeing must urgently develop an A321 XLR competitor, to avoid further market shift toward Airbus, former Boeing CEO David Calhoun clearly stated that there would be no new Boeing aircraft before the 2030s. There are four good stated or unstated reasons for this:

1. The capital investment in a new generation of airliners requires at least a 10– 20% reduction in fuel consumption to be recovered. The gains in efficiency from improved aerodynamics, lighter structures and better controls do not amount to this, without a significant contribution from engine technology. The latest generation of turbofans for single-aisle airliners, the CFM LEAP and PW 1200, are still in the maturation phase, with reliability and production issues. A new generation of higher efficiency engines will not be available before the 2030s. If Boeing started development of an FSA in 2024 with current technology, it would come into service shortly before the 2030s and would have a short life before new technologies would leave it outdated, with a small market risking not to reach break-even of development cost.
2. Boeing is not short of work on all its three families of airliners B737/B787/B777X and could hardly cope with a major simultaneous all-new development. The B737Max is now out of a 20-month grounding following two fatal accidents, but consequences remain as well as new problems. Boeing has a backlog of stored aircraft to return to service. Some of these aircraft are being sold to different airlines, thus requiring modifications. A notable case is Chinese aircraft since China was the last country to re-certify the B737Max and most airplanes there have not returned to service. New problems are still being found with the B737Max. The last structural issue with the fin attachment to the fuselage affects hundreds of aircraft. The latest incident the in-flight separation of a cabin door. The more recent redesign the anti-icing of the engine nacelle to avoid break away of pats. The certification of new variants B737–8 and B737–10 also faces problems and delays. Boeing got an exemption from the new cockpit emergency alerting standards mandated by the FAA for 2023 and beyond, but there are other certification issues. The B737–10 with its novel complex undercarriage is an important major effort to increase capacity and competitiveness for transcontinental routes. All these issues do not help Boeing to raise the monthly production rate from less than 30 to the pre-COVID level of 52, with the FAA recently imposing a lower production limit of 38, until production quality issues are resolved, and Boeing lately failing to reach even the reduced limit.

   The other two Boeing airliner families also have problems of their own. The B787 has structural problems with its all-composite fuselage, which led to a 1-year halt in production. Although production has resumed, not all structural issues have been resolved, and the FAA will make independent inspections of every new aircraft. The B777X is facing at least four challenges: (1) development problems with the new General Electric engine, not surprising for the most powerful jet engine ever; (2) following the B737 accidents and Boeing's unsatisfactory transparency on its design, the FAA is taking a much closer and assertive scrutiny of the B777X, which was not expected in earlier times of greater confidence: (3) the human factors and cockpit issues related to the B737 accident have

led to a much closer scrutiny of these aspects on the B777X; (4) the kind of structural problems that have plagued the B737/B787 have also surfaced in the B777X. In conclusion, Boeing has plenty of work to do to bring the B737/B787/B777X out of problems and does not need a parallel all-new development.

3. Boeing is in the worst financial position in history, with a debt of 40 billion dollars. Also, the development problems and delays with the B737/B787/B777X add to costs. Military programs like the B767 tanker and T-7 advanced trainer also face losses before eventual profits. The Boeing Starliner module for crew access to the international Space station is well behind schedule and above budget. The strong order book of Boeing commercial aircraft leaves no doubt that Boeing should eventually emerge from its current difficult financial situation, certainly when it reaches normal production and deliveries of all three families B737/B787/B777X. Thus, it may take a few years before Boeing can turn debt to surplus to the extent of financing a 15–30 billion minimum new aircraft program.
4. In the current situation, there is no incentive for Airbus to develop a new aircraft before the 2030s, so Boeing will just have to accept its weakness relative to the A321neo/LR/XLR. The latter is the biggest success of Airbus if it can raise production rates, despite supply chain limitations. Also, the stretch or development potential is not exhausted beyond the A321XLR, leaving Airbus in a strong position. Airbus has an additional option in a long-range A220LR, which would be a very efficient aircraft, although too much of an internal competition for the standard A320neo, or perhaps a replacement if the sales of the latter fall. Airbus has stated it will not wait for a Boeing rival to develop an A320 family replacement when the necessary technology is available, for service entry around 2035, and Boeing may not be able to field a major improvement over existing aircraft any time earlier.

Since neither Boeing nor Airbus will develop a new airliner before the 2030s, what will be the new fourth generation?

1.3.10 The New Fourth Generation in the 2030s

The apparently sole option of Boeing for the next-generation B737Max successor, or future single aisle, is quite clear: the transonic braced wing for which Boeing recently received a contract to build and flight test a large-scale model as a risk reduction effort. The large-span high aspect ratio improves aerodynamical efficiency and the resulting aeroelastic effects are countered by braces, which add to lift, and inevitably to drag. Airbus also has a conventional aircraft design, with a wing of large span and high aspect ratio with comparable aerodynamic efficiency, using a different approach to counter aeroelastic effects and flutter: flapping other wing panels to reduce aerodynamic loads, that can be folded on the ground to reduce span and parking space. The flight demonstrator is modified business jet. The flapping wing should have less drag than a braced wing, which may be a back-up option for Airbus instead of the only choice for Boeing.

Where Boeing and Airbus differ markedly is on the timescale for airliners fueled by liquid hydrogen (LH). Boeing believes that neither the technology, nor the airlines nor the airport infrastructure will be ready for the next generation, and maybe only for the generation after: it could be argued that (i) Boeing is busy with sorting out issues with its three current airliner families B737/B787/B777x, that (ii) its large debt and high current costs could not allow expensive major research and development efforts, and that (iii) there is not too much pressure from the US government or commercial operators to go the LH route.

The situation is quite different with Airbus. There are not too many issues with current aircraft families A220/A320neo/A321neo/A330neo/A350 apart from the already resolved question about the centerline fuel tank for A321LR/XLR. Although Airbus and Boeing face similar challenges with the supply chain and availability of engines to ramp up production, Airbus is in a much healthier financial position to invest in future research and development. Perhaps the biggest differentiator is a third factor, namely the high priority and intense pressure from the European Union to move away from fossil fuels to the hydrogen economy, and a number of incentives that Airbus could hardly ignore and justify a major effort toward LH-fueled aircraft. Airbus has three zero-e aircraft designs of which the flying wing or blended wing body is the most radical. The transition from fossil to LH-fueled airliners faces major challenges on (a) aircraft design and certification, (b) airport supply infrastructure, and (c) general availability and cost.

Concerning aircraft design and certification:

- Compressed hydrogen at 700 bar is not viable due to the weight of tanks and long refueling/turnaround time, leaving as only alternative liquid hydrogen, requiring cryogenic temperatures below 20 K.
- Although the energy density per unit weight of hydrogen is higher than for fossil fuels, it is the lightest chemical element and gas, and the energy density per unit volume of liquid hydrogen is less than one-third of fossil fuels, requiring a larger tank volume to be accommodated within the airframe.
- Being the lightest of gases, hydrogen has issues of leakage, evaporation, and corrosion, which must be addressed to avoid the risk of explosive mixtures with air, requiring safety and certification standards.

Concerning hydrogen supply infrastructure at airports:

- On-site production will require new facilities.
- Supply of large quantities by land vehicles may not be feasible due to congestion of roads near airports, with alternative of pipelines and large storage tanks.
- If aircraft refueling on the tarmac near passengers is deemed unsafe, there are the options of (i) refueling at remote location or (ii) replacing empty with full LH tanks, with supporting logistics.

Concerning supply quantities and prices aviation will be competing with other users in the transport and other sectors:

- Hydrogen production is energy-intensive, and the clean path is to use wind farms and solar panels, whose energy is otherwise wasted at off-peak time.
- The required energy supply far exceeds current capacity, and to the investment in production infrastructure must be added the cryogenic distribution network.
- The current price of green hydrogen is several times that of fossil fuel and will have to be reduced as production volume increases.

If despite all these challenges Airbus succeeds by the mid-2030s to have a certified LH-fueled A220 replacement widely adopted by airlines and supported by airports, it will have gained a lead that will be hard for Boeing to recover in less than one generation. An LH-fueled jetliner available not earlier than 2035 may have the payload-range to replace the A220 at the lower end of the Airbus single-aisle airliner family, but not the A321 at the higher end, and Airbus states that no single aircraft design can cover the whole family. A possibility is (1) a fossil fueled flapping wing tube-and-wing design for larger payload-range and a (2) LH-fueled flying wing design for smaller payload-range, with the former (1) a stronger prospect if both are not affordable or the second (2) is insufficiently mature.

Where Airbus and Boeing totally agree is on the use of synthetic aviation fuels (SAFs) as drop-in replacements for current fuels, raising the current 50% certification limit to 100% by 2030 for their entire current range of models. The term "drop-in" means that SAFs replace fossil fuels without change to engines, fuel systems, or airport infrastructure. SAF standards are still being formulated, and long-term effects as lubricants and seals and joins have to be assessed. The main challenge with SAFs, as for LH, is the production of the vast quantities needed at a moderate cost:

- Biofuels should not compete with food production for agricultural land and could only use other land.
- Available feedstocks like cooking oil and urban or forest waste are not sufficient, and others (Algae blooms) are less efficient.
- Energy consumption is high for e-fuels combining green hydrogen with carbon dioxide from the atmosphere.
- Investment in production infrastructure is costly and only viable with long-term assured demand.
- Costs exceed those of fossil fuels by a factor of 3–5 and must be reduced as production volumes increase.

SAFs are carbon neutral, that is, return to the atmosphere the carbon dioxide consumed during their production. They are an intermediate stage, not a replacement for hydrogen, that is, carbon-free since its combustion produces no carbon dioxide, and instead "mostly" water. However, there are other emissions from both SAFs and LH, like NOx, sulfur, particles, aerosols, and contrails, that must also be addressed. Hydrogen fuel cells (HFCs), which are emissions-free, may reach in the next generation sufficient power for regional aircraft, but not jetliners above 100 seats.

The current Airbus–Boeing duopoly also raises the issue of whether another competitor can emerge in the next generation(s).

1.4 Potential Competitors in the Airline Market

The Soviet Union had alternatives to offer in every airline market sector (Sect. 1.4.1), although the closed and sheltered nature of its clients made a difficult transition to the post-Soviet partial collapse of the Russian aircraft industry. Japan's experience in contributing substantially to the design and production of Boeing airliners has led to the design of complete regional aircraft, with no signs of higher ambition (Sect. 1.4.2). That ambition clearly exists in China but may need foreign collaboration, which appeared more likely from Russia than from Ukraine (Sect. 1.4.3). The sorry state of the Russian aviation sector following the invasion of Ukraine and Western sanctions may force China to press ahead alone with its civil aviation programs.

1.4.1 The Transition from Soviet to Russian Airliners

The closed monolithic posture of the Soviet Union required that there must be an airliner in every sector, to be also used by satellite countries and supported by mostly politically motivated exports of:

- The Tupolev Tu-114 long-range four turboprop.
- The Tupolev Tu-104/124/134 twin jets.
- The Tupolev Tu-154 tri-jet.
- The Ilyushin IL-62 four jet.
- The Ilyushin Il-86 wide body.

The Tupolev Tu-114 was an airliner adaptation of the four turboprop Tu-20 bomber that could fly nonstop from Moscow to Havana during the Cold War. The Tupolev Tu-144 Koncordski supersonic airliner never entered airline service. It needed afterburning for supersonic cruise, limiting its range: it never carried passengers, flying mail from Moscow to Alma Ata instead. All others in the list above were produced in sometimes large numbers and operated over the years without having to face the competition of Western airliners. The collapse of the Soviet Union led to an abrupt change, with Russian airlines buying Western airliners that were more efficient and reliable, besides having much better after-sales support. The poor record of Soviet-era aircraft for reliability and inadequate spares support was only made worse by the near-collapse of the former Soviet civil aircraft industry.

The former military aircraft industry survived better the collapse of the Soviet Union and transition to the Russian Federation, mainly through the Sukhoi family Flanker fighter variants, which gathered enough export orders to overcome the

decline of the home military market. Its former fighter house rivals, Mikoyan-Gurevich fared worse with limited sales of the Mig-29 Fulcrum and Mig-25/31 Foxbat/Foxhound relying mainly on the home market. The outcome of the reorganization of the former Soviet into the "new" Russian aircraft industry was that

- Mainly civil design bureaus like Tupolev, Ilyushin, and Yakovlev tried to soldier on the basis of updated Soviet-era designs that were not too competitive to start with.
- Of the predominantly military design bureaus, only Sukhoi had the market and resources to lead the consolidation and rationalization.
- Consequently, Sukhoi became the leader in new civil aircraft designs in the Russian Federation, an area in which it had never participated in Soviet times.

There is no hint of doubt of the ability of Sukhoi to design most types of competitive military or civil aircraft if the resources, engines, and avionics are available. The Sukhoi Superjet 100 (SSJ100) regional aircraft had limited success with slightly more than 150 units delivered during a decade, but hardly better could be expected in the circumstances.

Another civil challenge, a targeted competitor for the Airbus A320/Boeing B737 was developed by Yakovlev Design Bureau and three prototypes were built at Irkutsk. The aircraft, currently known as Irkutsk MC-21 but expected to be renamed Yak-242 when deliveries to customers have started, is at least on paper, an impressive design, matching the Western aircraft in most respects, and perhaps having some edge in some areas by virtue of a clean sheet design versus second–/third-generation redesigns. With a very high percentage of composites in its structures of the fuselage, wings, and fins, sophisticated aerodynamics, and an efficient P&W-geared turbofan, MC-21 is promising a state-of-the-art level of performance. In 2019, the certification program with the local regulator was reported to be quite advanced, while EASA certification is expected in 2020. However, a long series of factors dim commercial prospects for this impressive initial effort:

- There was no evidence of a production line let alone a complete chain.
- The maturity of Russian-made systems was an open question.
- The Russian occupation of Crimea and interference in Ukraine cast doubts on the supply of Western systems and materials.
- This included Western engines and avionics for which Russia has no equivalents.
- The "Russification" of engine and avionics technologies to Western standards would be costly and slow.
- The Western sanctions and low oil prices reduced the Russian budget for all sectors, including aviation.
- The militaristic and aggressive Russian policy gives priority to the military over civil aviation.
- Given the cuts in military aircraft development and production, it would be surprising indeed if civil aviation would fare any better.
- Even if a fully Russianized aircraft could be produced, the necessary spares support for exports might not be credible.

It appears that Russia planned to make great efforts to sell types with limited export potential or failing that rely on Western airliners or slowly develop more modern types in collaboration with China.

If the prospects for the Russian civil aviation industry were not too good before the invasion of Ukraine, they only turned far worse after:

- Although Russia "nationalized" hundreds of Western airliners previously on lease, without any compensation for legal owners, the lack of spares limited their operation.
- Cannibalizing some aircraft for spares and avoiding sanctions by sending aircraft for maintenance in other countries like Iran will only delay the decline.
- Placing massive orders for Russian aircraft depends on the ability to produce them, replacing all Western components.
- The lack of Russian companies able to produce and maintain jigs and other major production infrastructure will limit factory output.
- The Russian economy weakened by sanctions, and massive expenditure supporting a prolonged war, which takes absolute priority, leaves little scope for civil aircraft programs.

Although the Russian aircraft industry has the technology and expertise to produce airliners, it may lack the resources to satisfy adequately the home market, and even less to achieve the export success that has eluded it for decades. Recently Russia has been seeking foreign airlines to operate some of its domestic routes.

1.4.2 Japan's Market Captive to Boeing

The large trade surplus of Japan relative to the United States has led to a longstanding policy of the Japanese government to keep the airline industry aware of the patriotic duty to buy Boeing. Not content with having a sizeable captive Japanese market, Boeing has subcontracted to Japanese industry the design, development, and production of major parts of its aircraft. It has been mentioned that up to 40% of the B787 development was financed by Japan. In these circumstances, Japan is most unlikely to launch a Boeing competitor and will instead continue to blend in, share, and finance Boeing designs to give Airbus a lesser role in the Japanese market. The Japanese ambitions for the design of complete civil aircraft are thus limited to regional airliners. The Nanc YS-10 twin turboprop was an example in a distant past stretching to the current Mitsubishi MRJ-110, which has been recently canceled, after a long development period, and despite a respectable order book. The Japanese have maintained a significant research effort in supersonic airliner design in the hope of becoming a full partner if such a project ever sees the light of the day after the memory of Concorde. The Airbus penetration of the Japanese airliner market has improved gradually.

1.4.3 China's Turboprop and Turbojet Certification Hurdles

China is aware of the value of the aircraft industry. When the EU considered taxes or sanctions against Chinese 'dumping' practices, there was a reminder that "a single jet airliner is worth 200 million t-shirts." There is an Airbus A320 final assembly line in China using entirely imported components. When Airbus proposed sourcing locally some components, the Chinese airlines reacted negetivelly to the idea. Embraer planned business jet production in China; however, the Chinese government taxed components imported for the Phenom business aircraft, making the whole operation unprofitable; Embraer had to scrap the production plan at a loss. These two opposite events show the Chinese government's emphasis on local production and airlines' mistrust of the same.

The Chinese are currently in the process of development and certification for a turboprop (Xian MA700), a regional jet airliner (ARJ21) plus a single-aisle competitor to the A320/B737 (designated C919). MA700 project started the prototype building in 2017 and plans to certify with the local authorities in 2021 and subsequently with FAA and EASA. Considering the experience with ARJ21, this ambitious planning seems rather unrealistic. ARJ21 program started in 2002, and the certification process with local regulator and FAA was initiated in 2011. FAA certification is not yet completed at present. CAAC certification obtained in December 2014 allowed the operation of the type in China and seven units were delivered to local airlines. When asked about these long-delayed development programs, Chinese officials replied that their efforts are much younger than the decades of Airbus and Boeing experience. The Chinese lag most in engine technology. The engineer who developed the first Chinese military engine for the J-10 fighter received the highest decoration of the country, although the power plant appears to have been initially unreliable, eventually doing better than imported Russian engines.

The Chinese difficulties in certifying its three civil aircraft, if and when eventually overcome, could lead to a significant change in the local and global markets:

- The Chinese domestic market is large enough to support airliner development on its own.
- The availability of Chinese aircraft, say a single-aisle airliner, could partially close the local market for the A320/B737.
- A Chinese aircraft in this class would use Western engines and avionics, so the advantage of low labor rates on final cost could be small.
- It cannot be excluded that the Chinese would promote exports by the use of dumping or subsidies as they have been accused in the past on lower value products.
- The focus of Chinese trade in Asia, Africa, and South America and its lack of concern with local ethics could increase penetration of those markets.
- Even if unable to compete on fair terms in Europe and North America, the Chinese could significantly erode the Airbus and Boeing markets.

- Current Chinese production capacity for airliners is much smaller than Airbus or Boeing, at about one-tenth, and could take time to ramp-up.

China has given at least partial or limited local certification to the C919, even without being close to EASA or FAA certification, showing both inabilities to meet and disregard for international standards. Sure enough, the C919 has a large order book from domestic airlines and the challenge is to ramp up the low production rate to meet demand. In the meantime, China has to keep ordering Western airliners, recently more from Airbus than Boeing, including A320 from the Chinese production line.

The hardest sector for the Chinese to crack is wide-body airliners, initiiated in cooperation with Russia.

1.4.4 China's Path Toward Wide-Body Airliners

The Chinese are clearly quite far from being able to produce a wide-body airliner, both from the airframe and propulsion points of view. The signature of a Sino-Russian agreement at the highest head-of-state level to jointly develop a wide-body airliner demonstrates the high priority in both countries not to depend on Airbus and Boeing and to compete with them. This agreement could be seen as a partnership of Russian aviation technology and Chinese finance and market. The available Russian wide-body technology is the updated Ilyushin IL-86 converted to IL-96, with engines less efficient and reliable than Western engines; it is not competitive now and would be less in the future. Besides, the former Soviet experience with large aircraft was with Antonov military transports powered by Ivchenko high bypass ratio engines both located in Ukraine, outside Russia now. The logical step of a Chinese approach to Ukraine would be most welcome for all sorts of reasons ranging from political, to economic, financial, and technical. The pair of Antonov Aircraft/Ivchenko Progress engines has a long record of producing some of the largest aircraft and engines at the time:

- The Antonov An-12 Cub was the standard Soviet tactical transport, as the Lockheed C-130 Hercules in the United States and Transall C-160 in Europe, and is currently produced in China.
- The Antonov An-70 with its contra-rotating propfan pioneered 20 years ago the technology now used in the Airbus A400M Atlas.
- The Antonov An-22 Antei was a four-turboprop large transport that preceded the An-124 Ruslan as a rival to the Lockheed C-5 Galaxy and Boeing C-17 Globemaster in the United States.
- By growing the An-124 from 4 to 6 underwing engines, the payload was increased from 150 to 250 tons in the single prototype of the An-225 Mryia, which with a gross weight of 600 tons is the world's heaviest aircraft.

The tale of Antonov and Ivchenko after the collapse of the Soviet Union was a complex one:

- The close links between the Russian and Ukrainian aircraft industries meant that Russia remained an important but reluctant client, trying to do without Ukrainian products whenever possible.
- The 20-year lead of the An-70 was lost in protracted development of a Ukrainian aircraft and engine that Russia did not want to fund or depend on.
- Antonov/Ivchenko explored their smaller regional aircraft, but this was a limited source of revenue.
- The An-124 Ruslan operated by Dnieper airliners become a leader in worldwide air transport of outsize loads, for example, in United Nations humanitarian missions.
- The annexation of Crimea and Russian military intervention in the Donbass area was the final strain in soured relations and a loss to both sides.
- Russia lost a part of its aviation supply chain that it would be costly and long to replace.
- Ukraine had to replace facilities lost in the Donbass region and look for new markets.
- The Ukrainian recovery strategy has included a plan to produce regional aircraft in Saudi Arabia.
- License production of the An-225 in China has been reported in the press in September 2016, with details concerning plans to start building two prototypes in Chengdu and Shaanxi.

The latter would be heaven's send that could allow China to overcome all its current limitations:

- Its military transports are based on the Antonov An-12 equivalent to the C-130 Hercules and Ilyushin Il-76 equivalent to the C-141 Starlifter; it lacks a strategic transport like the Lockheed C-5 Galaxy and Boeing C-17 Globemaster and is behind the Airbus A400M Atlas.
- The An-225 would give China the world's largest military transport aircraft and a power projection capability worldwide to exceed Russia and Europe and rival the United States.
- Some analysts believe that the eventual use of the aircraft to transport and launch space vehicles is not excluded.
- Ivchenko can provide not only the engines for the An-225, but also for Chinese fighters, helicopters, and other aircraft.
- The availability of a full range of modern engines would give the Chinese time to develop their own with less dependence on Russia or the West.

The briefly reported news in December 2017 of license production of the AN-225 in China has not been confirmed and would hardly have pleased Western powers like the United States.

Instead, the focus has returned to cooperation with Russia on a CR-929R wide body with three variants having ranges of 6500 to 7570 nm with 250–320

passengers. It is estimated to have similar capability and size to A330–900 and 787–9. The designation CR-929R indicates the Chinese intent:

- The Chinese "C" to follow the Airbus "A" and the Boeing "B."
- The 929 wide body to follow the 919 narrow-body family.
- The "R" in CR-929R for collaboration with Russia.

The project is based on the split of work between partners: China will produce the fuselage and will perform the assembly at COMAC plant in Shanghai, while the composite wings, empennage, and tail section are to be produced in Russia. The engine would have GE and Rolls-Royce as the main contenders. Subsequently, Chinese and/or Russian alternatives are expected to be developed. The CR-929 project schedule was agreed to target the completion of the certification in 2027.

China had tried to buy the troubled Bombardier C-series program that could have paved the way for the long-delayed C919 certification. This was prevented by the cost-free acquisition of a majority stake by Airbus in the C-series that was a masterstroke in several aspects:

- It prevents Chinese access to Western airliner design and certification expertise.
- It extends the Airbus range down to 100–150 seats with an alternative to the slow-selling A319.
- It compounds the challenge that Boeing faces with the B737Max and future MMA.

Thus, China may have to rely on cooperation with Russia on the CR-929R wide body, which may not help with the certification of the C919 and ARJ21. The lack of some advanced technologies and wish to satisfy the home market with local production have put the Chinese at odds with the Russians in the CR-929:

- The Chinese want transfer technology on high bypass ratio engines and wide-body composite structures that Russia wants to retain.
- The Russians want an export market and do not agree on Chinese local production for the home market.

The disagreements have delayed the project over the years. Russia might have a fallback solution in the Ilyushin Il-96 wide-body replacing four turbofans by two larger ones; China has no immediate solution.

The sorry state of the Russian civil aircraft industry following the invasion of Ukraine and Western sanctions means that Russia may no longer be a viable partner for the CR-929, even if the long-running Sino-Russian disputes over workshare, production, and market could be resolved. China is not shy to go alone as proved by the domestic only near certification of the C919. It may have to follow the same path on the C929 with potentially long development time.

Following the "success" of the military turbofans powering the most recent large J-20 stealth fighter, China may be developing civil engines either for the C919 or a large turbofan for the C929, or both. China is developing a flying wing stealth bomber to rival the American B-2 Spirit and B-21 Raider, and similar technology could be applied to a long-range airliner. Flight test of a small-scale blended wing-body (BWB) airliner has been carried out in the United States, Europe, and China.

Could China try to bypass the traditional wide-body twin-aisle design for which it has no experience and attempt to leapfrog Airbus and Boeing with a long-range blended wing-body (BWB) airliner?= China has the market, the resources, human and material, and the ambition, but has still a long way to go to become an Airbus–Boeing rival, although it may eventually overcome the long list of technology, design, airframe, engine, systems, production and certification challenges to reach that stage. The C929 may take long to develop, the production rate of the C919 of a few dozen per year contrasts with the 1000-plus internal orderbook and China may remain dependent on Western airliners for several years to come.

References

The development histories, operational use and technical specifications of airliners can be found in the following 17 references:

1. *All-the-world's aircraft.* Jane's since 1913, last edition 2022/2023.
2. Chant, C. and Taylor, M.J.H. *The world's greatest aircraft.* Aerospace Publishing 1997.
3. Donald, D. *Encyclopedia of world's aircraft.* Aerospace Publishing 1997.
4. Jackson, P. *Encyclopedia of world's aircraft.* Amber Books 2004.
5. King, H.F and Taylor, J.W.R. *One hundred significant aircraft*, Jane's 1969.
6. Marshall, C. *World's greatest aircraft.* Aerospace publishing 1991, 3 vols.
7. Taylor, M.J.H. and Mandey, D. *The Guiness book or aircraft.* Guiness Publishing 1970, 6th ed. 1992.
8. Taylor, J.W.R. *The aerospace chronology.* Jane's Publishing 1983.
9. Chemel, E. *Chronique de l'aviation.* Éditions Cronique 1991.
10. Taylor, J.W.R. *World's commercial airlines.* Regency Publishing 1997.
11. Green, W and Swanborough, G. *Commercial aircraft.* Salamander 1978.
12. Donald, D. *Civil aircraft guide.* Aerospace Publishing 1999, 3 vols.
13. Klee, U. *Airline fleets international.* Bucher & Co. 1966, 30th ed. 1996.
14. Hygate, B. *British experimental jet aircraft.* Argus Books 1990.
15. Barfield, N. *Aeropatiale/BAC Concorde. Profile Publications* 1970–1990, 257 vols.
16. Marsden, J. *Flight cut-away drawing.* Flight Publications 1907–2006.
17. Bowers, P.M. *Boeing aircraft since 1916.* Putnam 1980.

Chapter 2
Seamless Air Traffic Compatibility and Operations Worldwide

Pablo Lázaro Galán and Raquel Martínez Arnáiz

Contents

2.1 Introduction

For decades, as the world has become more global, air traffic has experienced growth as it is a global industry. In spite of that, aviation has suffered the effects of every recession that occurred in the past. In particular, the latest recession was caused by COVID-19 in 2020. Once the pandemic was over, the growth was expected to continue during the following years and, as air transport plays a key role within transport networks, both current and future requirements of all agents involved must be accomplished.

Looking back, aviation has experienced an intense growth since the Airline Deregulation Act came into force in 1978 in the United States, and subsequently since similar measures were taken in Europe during the 1990s. Indeed, the number of instrumental flight rules (IFR) movements in Europe doubled from five million in 1988 to ten million in 2008. Since then, traffic figures have been hit by two recessions: the real estate financial system and the COVID-19 pandemic. Considering that 9.3 million IFR flights were recorded in 2022, aviation is expected to reach pre-COVID traffic levels (11.1 million IFR flights in 2019) in 2025. Then, traffic is

P. L. Galán · R. M. Arnáiz (✉)
Skyway Air Navigation Services, Madrid, Spain
e-mail: plazaro@ferrovial.com; rmmartinez@skyway-ans.com

L. M. Braga da Costa Campos (ed.), *Accelerating Sustainable Aviation Initiatives*, Sustainable Aviation, https://doi.org/10.1007/978-3-031-73993-4_2

expected to increase steadily until reaching around 16 million IFR flights in 2050, 44% more than that in 2019. Therefore, congested airports, terminal area airspace (TMA), and a route control center will have to accommodate many more daily movements than they do nowadays.

This expected growth will lead the industry to face two big challenges: the accommodation of demand and reduction in the overall time spent by the passengers. Regarding the first, it will be necessary to increase the capacity (including both airports and airspace capacity). Airports' capacity is mainly set by their runways' throughput that is directly related to the time needed to manage each flight safely. In this manner, the challenge to achieve a maximum throughput is to optimize final approach spacing in line with wake vortex, prevailing atmospheric conditions and radar separation requirements so that the spacing is close to minimum runway occupancy time. On the other hand, terminal area airspace (TMA) is the most restrictive airspace in terms of capacity. TMA is the managed airspace where several large and complex airports operate in close proximity to smaller, local airports, and the operations within them are dynamic and heavily influenced by demand, regularly resulting in the need to delay aircraft in established vertical holding stacks and causing other delays in the air and on the ground. Finally, regarding the other big challenge, minimizing the overall time implies that all elements of the chain should perform seamlessly: no take-off queue, no holding pattern at landing, no major weather or ATM disruptions; efficient check-in, passport, and security checks; fast luggage handling; efficient airport ground movements and operations; and uncongested local transport to and from the airport.

Therefore, it is important to identify how the aeronautical industry is facing these challenges nowadays and how it will face them in the future to set the guidelines needed to accomplish all the goals.

2.2 Accommodation of Demand

2.2.1 Air Traffic Capacity

This section addresses the main issue of air traffic capacity and flexibility of operation, including runway and airways terminal capacity, airport ground infrastructure, and air traffic management.

The goal of being able to manage 25 million flights will challenge three main elements in the transport system: (a) the capacity of the runway system, (b) the capacity of the terminal area airspace (TMA), and (c) the en-route capacity. The accommodation of such growth in flights will be determined by the most restrictive of these three capacity limits.

The European air traffic network contains about 170,000 links between airports. Over a network of more than 2100 airports, just 25 out of all airports generate 44% of flights as well as two-thirds of flights come from the top 75 and 90% of all traffic come from the largest 250 airports.

In particular, there is a geographical concentration of airports in the region of London–Amsterdam–Munich–Milan. This creates dense air traffic, with large a number of climbing and descending aircraft, which is a significant challenge for terminal area and en-route capacity.

Regarding airports, its operations depend on the interactions between many factors that affect the capacity of the runways to some degree. Hence, the main limit on airport capacity is the availability of the runways. The runway throughput is directly related to the time needed to accommodate each flight safely. A standard separation of 90 seconds between flights would allow 40 movements (take-off or landings) per hour from a single runway. Careful planning can increase this figure up to 60 movements per hour per runway, depending on the safe separation between aircraft, which is the critical safety factor. In the case of two or more runways, simultaneous operation of runways is permitted if they are parallel (no crossing paths) and spaced more than 400 meters (the vortex wakes of aircraft operating from one runway do not affect operations from other runways).

The separation requirements in segregated mode depend on the most constraining of any one of the three parameters: (1) wake vortex separation, (2) radar separation, or (3) runway occupancy time. From the technological and operational perspectives of the runway operation, the challenge to achieve a maximum throughput is to optimize final approach spacing in line with wake vortex, prevailing atmospheric conditions, and radar separation requirements so that the spacing is close to minimum runway occupancy time. The maximization of runway capacity would be achieved by "dynamic separation" (which sets the separation distance or time appropriate to the characteristics of each pair of aircraft and the prevailing atmospheric conditions), and state of the art in wake vortex, radar separation, and runway occupancy time technology and procedures.

Once the maximum runway throughput has been achieved, the only way to increase capacity at congested airports will be airport expansion through additional new runways and infrastructures. This affects basically the social/human dimension of the target as the growth of airports is severally constrained by social restrictions. Besides the plans for airport expansions, by 2030 no fewer than 19 European airports will be operating at full capacity eight hours a day, every day of the year. This will mean 50% of all flights affected by delays; a system more vulnerable to disruption due to airport congestion and less able to recover from crisis situations; and delays that will persist in the system for longer and will propagate more rapidly and widely.

In addition to physical constraints such as airport layout, there are "strategic" factors such as airport scheduling and "tactical" factors that include the sequencing of aircraft and the sustainability of throughput during specific weather conditions.

Besides runway capacity, the other important factor to accommodate the expected traffic demand is to manage take-offs and landings with the minimum safe separation without (a) having aircraft circling above in holding patterns and (b) queuing on the ground to reach a runway position. The maximum use of available runway capacity requires four-dimensional space-time navigation, so that successive aircraft land and take-off at precise times with the minimum safe separation. This

requires not only efficient management of ground movements but mainly efficient air traffic management in the terminal area around airports that is the most congested. The issues to be resolved include (i) the organization of incoming flights into a landing sequence with optimal separations; (ii) the management of the take-off sequence without waiting or idle times on the ground; (iii) the merging of the take-off (ii) and landing (i) sequences without holding patterns in the air; and (iv) the compatibility of terminal area traffic (take-offs and landings) with another airways traffic. Items (i)–(iv) are among the most important aspects of air traffic management (ATM) often with the greatest impact on capacity.

Accordingly, the terminal area airspace (TMA) is the managed airspace environment created to assist in achieving safety and efficiency where several larger, more complex airports and smaller, local airports operate in proximity. It is characterized by high numbers of aircraft conducting climbing and descending maneuvers in a relatively small volume of airspace. Operations within the TMA airspace are dynamic and heavily influenced by demand, regularly resulting in the need to delay aircraft in established vertical holding stacks and causing other delays in the air and on the ground. The biggest TMAs in Europe are today complex and saturated scenarios where the traffic of the busiest airports in Europe is integrated with the traffic of other airports in their neighborhood. Examples of high-density TMA in Europe are Paris, London, or Frankfurt.

The Paris TMA includes two major airports: Paris Charles De Gaulle (LFPG) and Paris Orly (LFPO), and some secondary airports: Le Bourget (LFPB), Pontoise (LFPT), Beauvais (LFOB), and many other general aviation aerodromes like Toussus-Le-Noble (LFPN) and Lognes (LFPL). Within the TMA, major and segregated arrival and departure flows converge and leave from the two main Paris airports (Paris CDG and Paris Orly). These two airports are close to each other (about 20 NM). Further, the vicinity of Le Bourget, which is dedicated to business jet operations, induces additional traffic complexity within the Paris CDG approach. Considering the average daily movements at both airports (roughly 1300 at LFPG and 650 at LFPO), this TMA has to attend more than 2000 movements daily on average. The London TMA includes two major airports: Heathrow (EGLL), with 1300 movements per day and Gatwick (EGKK), with 800 movements per day, which are close to each other (about 20 NM as well). There are also some secondary airports: Stansted, with 600 movements per day, Luton, with 400 movements per day, or London City, which are in expansion as low-cost airlines operate from these secondary airports. In all, the TMA has to manage more than 3000 movements per day.

Regarding the en-route airspace capacity, the highest concentration of en-route traffic takes place in Europe in the "core area" comprising of Belgium, Northeast France, Germany, and Switzerland, which is the densest and most complex airspace. At this zone, the density of fights is higher than 5 aircraft per hour and square kilometer.

Benchmarks to be achieved in en-route and terminal area will require technological, operational, and also social/human improvements currently under design for the future ATM system. Key to the future ATM concept is the possibility to fly the

optimal flight path seamlessly by all the users of the airspace. Taking full advantage of both existing and newly developed technologies—such as Galileo—Future ATM target concept relies on a number of new key features at three different dimensions:

Technological and operational dimension:

- Trajectory management, reducing the constraints of airspace organization to a minimum.
- New aircraft separation modes, allowing increased safety, capacity, and efficiency.
- System-wide information management, securely connecting all the ATM stakeholders that will share the same data.

Social/human dimension:

- Humans as the central decision-makers: controllers and pilots will be assisted by new automated functions to ease their workload and handle complex decision-making processes.

Network operation dimension:

- The network operation plan, a dynamic rolling plan for continuous operations that ensures a common view of the network situation.
- Full integration of airport operations as part of ATM and the planning process.

Hence, future demand needs to be accommodated by each of the air transport systems components: airport runway system, terminal management area airspace, and en-route airspace.

Regarding the first subject, in the past, a growing gap between capacity and demand at several busy EU hubs has been identified. Congestion at these airports will remain a concern. Traffic will continue to grow in the future, as it has done over the past 50 years despite periods of economic downturn and other disruptions. Although air traffic in Europe will grow slower than in emerging economies, it will reach around 16 million flights in 2050 as stated above.

Some measures have been discussed to reduce the levels of unaccommodated demand, aiming at increasing capacity at airports and improving operations in the airspace.

- Alternative airports

- Shifting to alternative airports is considered a real option for some airlines and airport operators provided that potential issues of environmental acceptance and terminal airspace congestion can both be overcome. The measure would be efficient if it reduces unaccommodated demand by around 30%, whenever passengers and airlines are willing to relocate to such airports, which is directly linked to ground transportation links.
- SESAR improvements
- The SESAR program to increase system capacity by developing and implementing new technologies, approaches, and procedures is perceived as the strongest enabler for sustaining future long-term demand. SESAR plus investments to

bring airports to the performance level of the best-in-class have the potential to increase airport capacity by a significant margin, reducing unaccommodated demand by around 40%.

- Reducing runway occupancy time
- A key indicator for runway capacity is runway occupancy time (ROT).
- During the arrival of an aircraft, ROT is defined as the time interval between the aircraft crossing the threshold of the runway and the tail of the aircraft leaving the runway. Runway capacity is often limited by ROT because only one aircraft can use the runway at any given time. The leading aircraft must first vacate the runway before the trailing aircraft is allowed to cross the threshold.
- ROT can be reduced using high-speed exits. These exits are not perpendicular to the runway, but instead use a smaller angle allowing aircraft to vacate sooner and at higher speeds, reducing ROT. These are already implemented in most of the major airports all around the world.
- In addition, the SESAR project will investigate the use of satellite navigation and augmentation capabilities, such as GBAS and satellite-based augmentation systems (SBAS), to enhance landing performance and facilitate advanced arrival procedures (e.g., curved approaches, glide slope increase, or displaced runway threshold). By doing so, noise is reduced while runway occupancy time (ROT) is optimized. The aim is to also reduce the need for separation for wake vortex avoidance.
- Time-based separation
- When there are strong headwinds, aircraft ground speed is reduced on the final approach. This results in a reduced landing rate causing delays.
- The concept of time spacing is based on the performance of an aircraft in strong headwinds conditions, where wake vortex is quickly dispersed, permitting then to reduce the distance between aircraft, while maintaining safety levels. Consequently, airports can operate with the same landing and capacity rates as in light wind conditions.
- Time-based separation aims at reducing the gap in landing rates in light and strong headwind conditions. It will help maintain airport capacity at the same level in all wind conditions.
- Recategorization of wake turbulence categories
- Runway capacity is directly linked to the minimum separation between aircraft. This minimum separation is constrained by ATS surveillance capabilities and wake turbulence.
- During recent years, knowledge about wake vortex behavior in the operational environment has increased thanks to recorded data and improved understanding of physical processes. It is mainly for this reason that it was possible to revise wake turbulence categorization and corresponding separation minima to enable optimization of airport capacity and efficiency while maintaining acceptable levels of safety. A safe minimum separation implies considering wake vortex gener-

ated by an aircraft but also the wake encounter impact and resistance of the following aircraft on departure or final approach. Existing ICAO wake vortex separation minima were implemented over 40 years ago and might become outdated. These separations are based on certificated maximum take-off mass (MTOM) and include four categories (super, heavy, medium, or light) allocating all aircraft into one of them. This leads to over-separation in many instances because the separations are defined based on the worst case in each category (Table 2.1).

Each category can cover a wide range of different-sized aircraft that leads to much conservative separations in many cases, and so a loss of efficiency and runway capacity.

As a result, EUROCONTROL developed a recategorization of ICAO wake turbulence scheme and associated longitudinal separation minima on approach and departure, called "RECAT-EU," to the benefits of airports and ATM network performance enhancement.

European Wake Vortex Recategorization (RECAT-EU) is a new, much more precise categorization of aircraft than the traditional ICAO one. It aims at safely increasing airport capacity by redefining wake turbulence categories and their associated separation minima. It divides the current heavy and medium categories into two subcategories and creates a new super heavy one for the Airbus A380 (Table 2.2).

Thanks to this new categorization, several benefits are expected:

- The runway throughput could be improved by 5% or more during peak periods.
- Reduction of the overall flight time for any arrival or departure. This could offer a better flexibility for ATCOs to sequence the air traffic.
- Allow a faster recovery from adverse weather, helping reduce the overall delay and could also help comply with ATFM slot due to departure separations that could be reduced.

Table 2.1 ICAO wake turbulence categories and separation minima

Leader/follower	A380–800	Heavy	Medium	Light
"A380-800		6 NM	7 NM	8 NM
Heavy MTOM ≥136 tons		4 NM	5 NM	6 NM
Medium 7 tons ≤ MTOM <136 tons				5 NM
Light MTOM <7 tons				
"Lower medium" "E"		6 NM	7 NM	8 NM
"Light" "F"				

Table 2.2 RECAT-EU WT distance-based separation minima

Leader/follower		"Super heavy" "A"	"Upper heavy" "B"	"Lower heavy" "C"	"Upper medium" "D"	"Lower medium" "E"	"Light" "F"
"Super heavy"	"A"	3 NM	4 NM	5 NM	5 NM	6 NM	
"Upper heavy"	"B"		3 NM	4 NM	4 NM	5 NM	
"Lower heavy"	"C"			3 NM	3 NM	4 NM	
"Upper medium"	"D"						
"Lower medium"	"E"						
"Light"	"F"						

In this context, it is widely recognized that to increase performance ATM modernization should look at the flights within a flow and network context rather than segmented portions of its trajectory as is the case today. Upcoming research and developments must be studied earlier to make a wide research framework in which each project has both enough funds and duration to achieve its goals. The fact of trying to ensure that this wide framework is defined under previous studies will allow us to close the gaps remaining between the goals set for 2050 and the actual improvements reached in 30 years.

2.2.2 *Ground Infrastructure*

A goal for ground transport is reducing access time to get to an airport, so new means of transportation are under investigation. In this case, vertiports and heliports could be a solution as they could be located much closer to city centers, providing an alternative with faster access than airports, if noise issues can be resolved. If we focus on vertiports, a network of traditional or electric aircraft that take off and land vertically (called VTOL aircraft for vertical take-off and landing) will allow fast transportation between suburbs and cities and, ultimately, within cities, improving urban mobility.

The development of infrastructure to implement a VTOL network will likely be more economic than other usual infrastructure such as roads, rail, bridges, or tunnels. This is due to the tops of parking lots and already existing helipads could be the basis of a network of "vertiports" (VTOL hubs with multiple pads) or "vertistops" (a single VTOL pad with minimal infrastructure). Although this seems a very good approach, it would not be possible to deploy a complete VTOL network with just existing pads at this moment.

2.3 Reducing Overall Travel Time

2.3.1 Multimodal Transport

A constraint may come from the physical limits of transportation infrastructure and the underlying issue of planning: (i) in the expansion of existing airports or addition of more runways; (ii) in the construction of new airports, vertiports, and heliports; (iii) in the road/rail infrastructure that provides fast access; and (iv) in the efficient organization of ground movements within the confines of the airport.

The choice of air travel compared with other means of transport depends not only on flight time but also on the ground movements to and from the airport, which is an issue addressed in this section.

A coherent ground infrastructure implies the design and implementation of an integrated, intermodal transport system as part of which airports evolve into integrated, efficient, and sustainable air transport interface nodes. Airport access has been improved accordingly through an innovative approach toward safe, efficient, frequent, comfortable transport systems and services and connections with other modes of transport must facilitate an easy and quick access to the airplane.

According to EUROCONTROL, the average flight length of 80% of the flights within Europe is 504 NM while the average flight length of the flight outside the regions (20%) is 878 NM. This means that the average flight time of 80% of the flights in Europe does not exceed an hour. This will leave a maximum of 3 hours for the passenger to arrive from its departing point to the plane and to get from the plane to its final destination, including the processing times at the airport and all the connections with other means of transport.

Technological and operational dimensions will be of high relevance to achieve the average 3-hour target time of connection and processing time. However, the social dimension of such integration will become very relevant as the main impact of an airport on the surrounding community comes not only from aircraft operations but also from the ground infrastructure required to access to the airport and to connect the airport with other modes of transport.

In this context, the airport of the future is conceived as the central link of intermodal transport. Inter-modality is understood as the transport of goods and passengers using several modes of transport in one trip and involves the inter-coordination of those different means of transport. This coordination is made thanks to adequate intermodal infrastructure and to intermodal agreements concluded by transport operators. Agreements allow for common reservation for the whole trip, coordinated timetables, a common checking, and the certainty to travel to the final destination despite delays faced by one or several transport modes during the trip, etc. These agreements aim at satisfying the customer's needs and assuring a positive and seamless travel experience is central to the success of intermodal passenger transport.

The interconnectivity at European airports is often still limited to urban transport, with very few (high-speed) train stations located at airports. Some of the existing intermodal links do not fully meet the passengers' expectations, leading to low

usage. As an example, in the United Kingdom, train stations at regional airports have been closed due to the small number of passengers who made use of the facility.

Air/rail intramodality seems to offer promising opportunities for the future of the transport system by limiting the isolated use of road or air traffic (both responsible for congestion and air pollution) and providing combined trips, generally with rail. However, so far there are not many intermodal agreements in Europe. Funding and the possibility of signing exclusive agreements between airlines and other means of transport are essential enablers to foster intramodality. This multimodal transport may include high-speed trains for the national or international network, trains, subways, tramways, or suburban trains at regional airports as well as electric ground vehicles.

One of the most important challenges will be achieving public confidence in automation, although this will demand significant advances in technology. Automation will mean that users are informed about the status of their journey and alternative options, periodically or on demand. Information points will be distributed around the terminals and interactive devices embedded in transport systems so that passengers can access travel information at any time using smart phones or interactive panels/screens situated along the intermodal transport network.

Furthermore, a main goal for the future intermodal transport system is to reduce dependence on the automobile as the major mode of ground transportation and increase use of public transport, especially in the case of the future air transport system.

2.3.2 Overall Ground Plus Air Travel Time

Air travel times can vary significantly in Europe, from 1 hour in Central Europe (Paris–Frankfurt) to 4 hours between extremities of the continent (Lisbon–Bucharest). If most flights do not exceed 2 hours, leave within the four-hour total time frame, 1 hour to travel to and from the airport and go through airport services. This objective is achievable if all elements of the chain perform nominally: (i) no take-off queue, no holding pattern at landing, no major weather or ATM disruptions; (ii) efficient check-in, passport, and security checks; (iii) fast luggage handling; (iv) efficient airport ground movements and operations; and (v) uncongested local transport to and from home or work.

To assess the efficiency of the air transport system, and more particularly once considering the door-to-door time, it is also important to know how far these airports are located from the European city centers. According to data from the European Personal Air Transportation System STUDY (EPATS), for almost 80% of the European cities the nearest airport is located at 20 km. Such a short distance reflects that the general accessibility of European airports is high. In the case of the 30 largest airports in Europe, passengers have a choice between different public transport service providers for access between the centers of the respective cities

and the airports. Currently, 23 out of the 30 largest airports in the European Economic Area (including Switzerland) have a direct rail access at or in the vicinity of the passenger terminal. Besides, several rail access projects are currently being planned or under construction.

In conclusion, what can be deduced is that regardless of the progress achieved up to now in any area described above, it may be necessary that stakeholders, including authorities and companies, redesign the strategies to be followed to be in a better position to accommodate the future demand and furthermore reduce the overall travel time spent by the passengers.

On the one hand, future demand should be accommodated within a framework in which performance-based operations are completely implemented and therefore it allows aircraft to fly the most efficient route and profile, assuring improvements in capacity. Besides, automation and advanced navigation technologies should be introduced to allow us to improve accuracy and quality of the service and the system capacity. Another important matter is the necessity to address the challenges of en-route capacity at the same level than the challenges in terminal area as well as developing airspace integration requirements for enabling safe, efficient low-altitude operations through a new airspace design, dynamic geofencing, congestion management, and terrain avoidance for a UAS traffic management (UTM) system.

On the other hand, airports are usually located far from the city center, resulting in long airport access times for passengers combined with buffer times for uncertainties of durations for airport processes like security checks or even unpredictability of airport access times. Therefore, key enablers to reduce overall travel times are a reduction in airport access times, a higher predictability of times accessing the airport and process times inside the terminal. Related to that, the interconnectivity at European airports is often still limited to urban transport, with very few (high-speed) train stations located at airports. Some of the existing intermodal links do not fully meet the passengers' expectations, leading to low usage. However, air/rail intramodality seems to offer promising opportunities for the future of the transport system by limiting the isolated use of road or air traffic (both responsible for congestion and air pollution) and providing combined trips, generally with rail. In short, the main challenges to overcome to achieve the desired framework for intramodality relate to standardization and funding, but also to remote check-in and luggage handling and better scheduling and decreasing delays.

Chapter 3
Evolution of the Chinese Aircraft Industry

Rosa María Arnaldo Valdés and Victor Fernando Gómez Comendador

Contents

3.1 Introduction

In the last 40 years, the aeronautical industry has evolved from a specialized sector to a leading industry worldwide. Companies, governments, and associations all over the world acknowledge the importance of the aviation industry in supporting global economic development. Aviation improves local access to global markets, creates opportunities for social interchange and development, and supports humanitarian response to emergencies. Prospects for the growth of this sector are optimistic. The main aircraft manufacturers, such as Boeing, Airbus, and Bombardier, foresee an increase in air travel demand per year for the next 20 years.

R. M. A. Valdés (✉) · V. F. G. Comendador
Universidad Politécnica de Madrid, Madrid, Spain
e-mail: rosamaria.arnaldo@upm.es; fernando.gcomendador@upm.es

L. M. Braga da Costa Campos (ed.), *Accelerating Sustainable Aviation Initiatives*, Sustainable Aviation, https://doi.org/10.1007/978-3-031-73993-4_3

Despite the positive perspective of the market, many challenges have populated the road toward this future. To succeed, the aeronautical industry must maintain innovation as one of its primary assets. It must master a wide range of technologies and collaborate to integrate them into an aircraft design and development program. A collaborative approach to innovation is key to achieving these goals. Breakthrough and emerging technologies will continue to be the main development differentiators, and sustained R&D efforts are essential to ensure sustainable growth. Governments and international institutions are preparing strategic responses.

Within the international efforts to cope with these challenges is where the Perspectives for Aeronautical Research in Europe (PARE) comes along. The overall objective of PARE is to trigger collaboration between European stakeholders to support the achievement of the 23 Flightpath 2050 goals. As part of this process, the project has the task of identifying the actions required in the coming future for the proper development of the aerospace research sector that can benefit from a detailed and rigorous analysis of possible political, social, and industrial scenarios by carrying out, among other works, some "what if" analysis.

The "what if" case studied here is based on the perspective of the growth forecasts of the air transport sector in the medium and long term that characterizes the Asian emerging economies, and in particular China, as the ones with the highest growth in air transport. In this regional growth context, China is developing a powerful aerospace industry whose ability to consolidate will condition the worldwide aerospace industry scenario.

In recent years, we have witnessed several attempts by the Chinese industry to develop and certify large transport aircraft, such as the regional jet ARJ21 certified by the CAAC after years of delays; the C919 in the single-aisle segment whose certification has been postponed for years; or the future development of a wide-body model C929. How the success of all these attempts will effectively affect the Airbus/Boeing leadership in the industry is going to be one of the big issues in the industry in the future.

China's ability to certify and produce commercial aircraft efficiently, economically, and on time to take advantage of the country's anticipated development could greatly influence the international scenario and have a major impact on the current balance of the aerospace industry.

Keys in that future will be the capacity of the Chinese industry to certify its current aircraft developments, produce on a large scale, gain the confidence of the airlines, and acquire a significant share of the market. All these aspects are raised and developed in a first case study that aims to shed light on how the possible scenarios can influence the development of the European air transport industry and how Europe can react in each case with the best political, regulatory, research, educational, and industrial strategy.

3.2 Objectives and Scope of the Analysis

The objective of this study is to analyze the potential of the emerging commercial aviation industry and its possibility of affecting, in the long and medium term, the actual duopoly run by the European Consortium Airbus Group and the American company Boeing Corporation. This project provides a set of conclusions and recommendations within different fields to maintain appropriate competitiveness in the sector.

China's first airliner, COMAC C919, is expected to compete with the narrow-body, single-aisle aircraft Boeing 737 and Airbus A320. C919 first flew in May 2017 and, after completing its testing program, was certified by the Civil Aviation Administration of China (CAAC) in September 2022. The first delivery to an airline (China Eastern) was in December 2022. Airlines plan to use C919 on domestic routes, including Shanghai, Beijing, Kunming, Chengdu, Shenzhen, Xi'an, and Guangzhou.

However, in February 2023, China Eastern, the only owner of C919, delayed the inaugural flight without further explanation. Since then, the operation has been discontinued.

This study will analyze the impact of this aircraft's entry into the market and will estimate quantitatively the market share that will be taken from the current duopoly. To do so, the study will analyze several questions in both technical and business fields, such as

1. A study on the single-aisle sector and its relevance, together with future perspectives of the sector. The objective is to identify why this sector has become relevant to airlines and how it is expected to evolve over the next 30 years.
2. An insight into the new Chinese COMAC C919 aircraft, including the evaluation of its features and time frame, to assess its potential competitiveness with the incumbent models. To do so, the study will provide technical data regarding the design of the aircraft as well as the production capacities and business intentions of Chinese companies.
3. An analysis of the possible market share that a new aircraft may obtain. This will be supported by considering different scenarios within certification possibilities and will answer two main questions: If the aircraft achieves FAA/EASA certification, how this will affect Airbus and Boeing orders and deliveries? On the other hand, if the aircraft is only certified in China and in some of its commercial partner countries, will this cause a long-term domestic substitution of aircraft orders in these territories?

This study will develop a SWOT analysis of the new aircraft and will be supported by different scenarios in both the short and long term. An analysis of these scenarios will provide a global vision of the evolution of the Chinese airline market in different external situations.

To achieve the objectives mentioned in the previous section, this study considers the following aspects:

- Certification problems: The Chinese regional jet ARJ21 obtained certification from the CAAC after years of delay. However, without international certification by the FAA or EASA, it is operated by Chinese airlines on domestic routes. C919, an A320/B737 competitor, was finally certified by Chinese authorities in 2022 (after being postponed several years), although the aircraft is not yet fully operational. Chinese CAA has failed to agree to certification with the FAA and EASA. Further, in the future, the project of a wide-body airliner, the C929, will be developed, with Russia using Russian engines. Meanwhile, Chinese airlines have no choice but to buy Western aircraft. Will this change?
- Domestic market: The domestic market is sufficient to justify the development of new airliners. Will the certification by the C919 by the early 2020s and C929 certification later in the 2020s influence the domestic market? How would Airbus and Boeing sales in China be affected? What would be the effect on global Airbus/Boeing sales and prospects?
- Production cost: Chinese aircraft use Western engines and avionics; therefore, the production cost cannot be much lower. How much lower is it realistically? China has been accused of "dumping" a variety of products, mostly of much lower value than that of airliners. When faced with anti-dumping measures by the EU, the Chinese replied that an airliner was worth 200 million T-shirts. If Chinese airliners obtain EASA (or FAA) certification, there is a risk of dumping. What effect does it have on international markets and Airbus's/Boeing's global prospects?
- Production capacities: The Chinese regional jet ARJ21 is already in service and production after being certified by the CAAC. The Chinese airlines' intention seems to be to use this aircraft for domestic routes in substitution of the options of the two main Western regional jets manufacturers, Embraer and Bombardier. The aircraft received 302 orders by August 2018, most of which were from Chinese airlines, as well as 31 from other countries, such as the Republic of Congo or Laos. COMAC's industrial plans were to achieve a production rate of 20 aircraft a year by 2018, but they only managed to deliver 6 units by date, which puts in doubt its capacities to satisfy a high-growing demand. Finally, the company announced that the 100th delivery means COMAC delivered 32 ARJ21 in the second half of 2022, which is not only the best number in any half-year, but also beats the annual delivery record of 23, which was achieved in 2020. Even if the Chinese company achieves international certification for its airlines, will they be able to produce at the same rate as its Western competitors? Will this also affect the C919 sales and orders?
- Customer countries: China has developed close relations with several developing countries, including those with questionable human rights records (such as China) and dubious business practices. In fact, some countries that are not eligible for EU or US assistance due to domestic issues find China as an alternative partner that does not raise such concerns. Could China use these countries as customers, preferring its airliners, even if not the most efficient (such as the satellite countries of the Soviet Union in the past)? How much market share can be obtained in this way, possibly combined with dumping at unbeatable prices?

- Domestic substitution: China is still unable to supply major aircraft elements such as engines or avionic systems that function together. The engineer who developed the first modern Chinese jet engine received the highest decoration in the country. Military variants of the engine are already in use, with reports of poor reliability. China is investing heavily in aerospace technology, with 200,000 students taking university courses in this area. What are the prospects for the domestic Chinese development of major aircraft systems? Which and when? Are most or all Chinese airliners a prospect? In which time frame? Would the international market rely on Russian/Chinese engines and systems? Given the poor record of some countries (such as Russia) in spare and after-sales support, would China be better (including the C929R)?

3.2.1 Methodology

The methodology used in this study is twofold. On the one hand, a study of the market has been developed using qualitative and analytical methods that will be sufficient to provide information about how the market could react to Chinese irruption. On the other hand, game theory analysis was performed to evaluate the results of the entry of a new competitor and its implications for a market currently dominated by the duopoly composed of Airbus and Boeing. The objective of this analysis is to evaluate the possible strategies to face the Chinese company COMAC in a single-aisle market, considering the medium- and long term.

This section discusses the methodology of the "what if" analysis. The proposed methodology was twofold. On the one hand, a study of the market will be developed using qualitative and analytical methods that will be sufficient to provide answers to the questions and cases raised in the objective and scope of this study. Additionally, the "what if" analysis is complemented by the application of game theory to evaluate the results of the entry of a new competitor and its implications for the market that is currently dominated by the duopoly composed by Airbus and Boeing. The outcome of the games will evaluate the possible strategies to be applied by Airbus and Boeing to face the Chinese company COMAC in the SA market, considering a medium- and long-term horizon. In addition, game theory will allow for a sensitivity analysis to consider the impact of various uncertainties that influence the case.

Information and analysis coming in the first part of the analysis will be used to define the structure of the competitive games in the second part of the analysis and the possible scenarios, as well as to synthesize the possible strategies of the three players and construct the aircraft valuation model for the estimation of the payoffs. Figure 3.1 illustrates the main steps of the entire process.

As can be seen from the figure, the initial market analysis consists of five main steps. The first step comprises the definition and analysis of the single-aisle market, which is currently dominated by the Boeing/Airbus duopoly. The competition between these companies is analyzed in this step, and it also includes a characterization of the SA aircraft models that both companies are currently available in terms

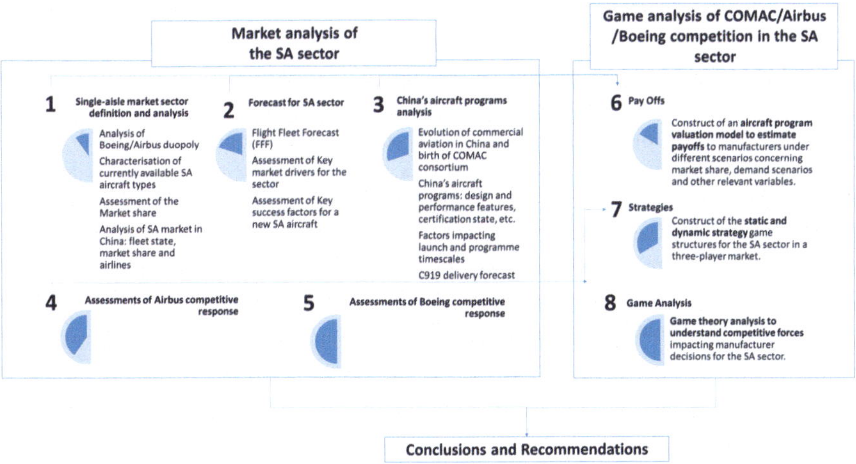

Fig. 3.1 Methodology for the "what if" study on the China case. (Source: PARE Reports; https://www.pareprject.eu/_files/ugd/e00f26_4a3e8c26adf3465a97414598df52986.pdf)

of design and performance features. Finally, an analysis of the SA market in China was performed through an evaluation of the current fleet state, its main operators, and airlines. This analysis allows the quantification of the Airbus and Boeing market share in the region, which is an essential aspect to understand how the presence of a new competitor could have a great influence on both companies in the country.

The second step comprises the following main topics: (i) flight fleet forecast (FFF), (ii) identification of key market drivers for the sector, (iii) identification of key success factors for a new SA aircraft, and (iv) performance of a SWOT analysis for the single-aisle market. Traffic and fleet forecasts are critical elements in evaluating the market share that COMAC can absorb from Airbus and Boeing in the country.

The third step involves an in-depth analysis of China's aircraft programs, starting with a brief description of China's attempts to manufacture commercial aircraft, resulting in the birth of the COMAC consortium as a commercial aircraft manufacturing company. Then, the models that this company has currently available or is already committed to developing are analyzed in detail in terms of design, cost, certification, and performance features. It also includes a valuation of the different factors that could affect the launch and programmed timescales, which serve as inputs for the scenario definition. Finally, market share is estimated by considering different situations that could stimulate or damage the sales landscape.

Steps 4 and 5 are dedicated to outlining and evaluating the Airbus's and Boeing's reaction as well as their possible strategies to compete with COMAC in the single-aisle market, with the goal of not losing market share in China and internationally.

Additionally, as shown in Fig. 3.1, the game competitive research approach comprises three stages (steps 7–9). In step 7, static and dynamic game structures for a three-player SA market are constructed. In step 8, using the outcomes of the

previous sections, an aircraft program valuation model is implemented to estimate the payoffs of manufacturers under different market share and demand scenarios. The purpose of this model is not to precisely determine aircraft manufacturers' profitability but to estimate the rank of payoffs to determine how changes in the market structure may change the equilibrium game outcome. It should be noted that this analysis is hindered by the proprietary nature of the aircraft program economic data. The consortium used reusable assumptions based on publicly available data sources. These assumptions will also be subjected to a sensitivity analysis to determine the impact of the assumptions and proxies on the study's funding. Finally, in step 9, game theory analysis is used to model the competitive forces impacting manufacturers' decisions.

The results of the whole process will allow us to understand how competition can affect manufacturers' decisions, as well as how the entry of a new rival in the market could imply changes in the duopoly composed of Boeing and Airbus. This understanding and the derived conclusions and recommendations may assist policymakers in developing regulatory and incentive mechanisms to improve aviation and inform expectations of the introduction of a new competitor into the global aviation market. The whole approach will also allow us to test policy options to determine their outcomes in a competitive market based on the assumptions of the valuation model.

3.2.2 Single-Aisle Sector

Single-aisle aircraft is a widely accepted term for narrow-body airliners used for short- and medium-haul flights. This section focuses on market characterization. It will review the current single-aisle models as well as the market structure and lessons learned from the competition between Airbus and Boeing in the past.

Boeing B737 is in its third generation since the 1960s, and Airbus 320 in its second generation since the 1980s. Third-generation B737 has a lower stretch potential than second-generation A320. The new generation of more efficient engines has turned the A321 into more than its designers could originally hope, able to cross continents, and some oceans. Additionally, Boeing's 737 requires more engineering work to re-engine than Airbus's A320, which could provide a re-engine advantage to Airbus, as a smaller investment and less technical risk are required, if needed.

With more than 10,000 deliveries of B737 and 8500 of A320, they are the numeric best sellers in the Boeing and Airbus fleets. Current backlog numbers might exceed 12,000 or even reach 15,000 sales. At this point, both manufacturers harvest the profits of their more successful products and plan to do so in the short and medium term. C919 might be a challenge to this long-term duopoly situation or at least an extra impulse for new competition in the market.

In the following sections, we analyze in detail the current models produced by Airbus and Boeing in this segment. Table 3.1 summarizes the models and their main characteristics.

Table 3.1 Current Airbus and Boeing single-aisle models for short–medium haul

Aircraft model	First flight	Length (m)	Wingspan (m)	Range (nm)	Capacity (two-class configuration)	Maximum capacity	Total orders	Total deliveries	List price (USD millions)
A319neo	2017	33.84	35.80	3700	120–150	160	35	0	101.5
A320neo	2014	37.57	35.80	3400	150–180	194	4143	641	110.6
A321neo	2016	44.51	35.80	3600	199	230	2327	184	129.5
A321neo LR	2018	44.51	35.80	4000	206	240			
A321XLR	2023	44.51	35.80	4500	206	240			
B737–800	1997	39.47	35.79	2935	160	189	4991	4979	102.2
B737-900ER	2000	42.11	35.79	2950	177	220	505	504	112.6
B737 MAX 8	2016	39.52	35.90	3550	178–193	210	2590		
B737 MAX 10	2019	43.80	35.90	3300	188–204	230	579		

The single-aisle market is integrated by a narrow-body aircraft with more than 100 seats, which operates on short- and medium-haul routes. In the past, this market was split among various manufacturers: Boeing, McDonnell Douglas, Lockheed, and Airbus. Boeing's 727–200, with 150–185 seats, dominated the short–medium-range market until 1980 when MD-80 entered service with a 37% fuel burn improvement. Boeing and Airbus responses were Boeing's 737–400 and Airbus's A320 in 1998, which offered significant performance improvements over MD-80, and since then, both manufacturers have been updating these product lines. Finally, with Boeing's acquisition of McDonnell Douglas in 1997, the market became a global duopoly, where Airbus and Boeing divided the market. Figure 3.2 illustrates this situation with the evolution of seat market shares and fuel burn performance of the three manufacturers during the period 1980–2009.

In this sector, incumbent manufacturers are protected from competition through the following entry barriers:

- The cost of developing a new aircraft is extremely high, and experts estimate it to be between $3 and $14 billion, depending on aircraft size and technology level.
- New aircraft development requires a high level of expertise over a long period.
- Production learning effects result in a 20-cost drop-in unit costs every time the quantity produced doubles. This provides a competitive unit cost advantage compared with models with long production runs [3].
- Airlines are increasingly demanding lower operating costs and prices.

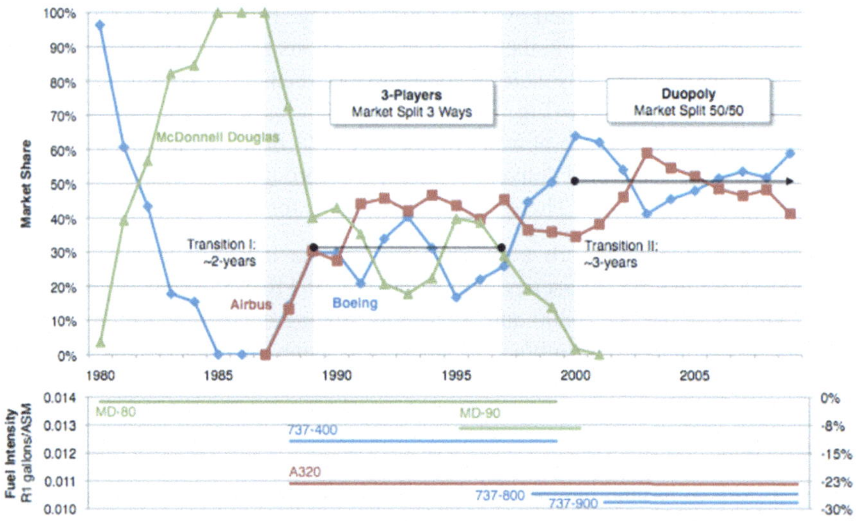

Fig. 3.2 Single-aisle, 150–185-seat market shares and fuel burn performance, 1980–2009. (Source: Game Theory Analysis of the Impact of Single Aisle Aircraft Competition on Fleet Emissions" James K.D. Morrison, R. John Hansman, and Sgouris Sgouridis. 2011 Massachusetts Institute of Technology (adapted))

- Airline switching costs are significant and tend to keep airlines locked into one product family. Fleet commonality also reduces operating and maintenance costs, increasing the reluctance of airlines to switch manufacturers. To gain market share, the new manufacturer needs to offer a product with significant performance improvements to overcome the switching cost or any other advantage that compensates for the switching cost.

To maintain market share, manufacturers must develop aircraft with performance equivalent to those already established. If a manufacturer wants to gain market share, it needs to produce an aircraft with significant performance improvements to overcome airline switching costs. If a manufacturer develops a superior aircraft, its competitors risk losing their market share. This leads manufacturers to an action–reaction game, where each responds to others' moves to prevent an inferior aircraft in a market segment from losing market share and profit potential. This reaction might imply a re-engineering of existing airframes, or even the development of a new, clean sheet design.

3.2.3 Lessons Learned from Past Airbus–Boeing Competition in the Single-Aisle Market

The competition between Airbus and Boeing has been characterized as a duopoly in the large-jet airliner market since the 1990s. In fact, the global aircraft industry today is increasingly controlled by Airbus and Boeing. The prize for this competition is the dominant position in the market in continuous growth. A total backlog of 14,816 aircraft [4] is currently distributed mainly between the "Two Big," representing 89% of the total industry (Fig. 3.3).

Large commercial jets are now approximately 60% of the total industry output by value, not only at the final delivery level but also through most of the component and structure supply chains. There are several reasons for this duopoly:

- Airbus and Boeing absorb a greater share of the industry. In 2018, Airbus acquired Bombardier's C-series with a new line of 110/130-seat jets, provisionally known as the A220-200 and A220-300. Boeing is creating a joint venture with Embraer, covering Embraer's E-Jet series, spanning 75–120 seats.
- Extremely high entry barriers.
- Extreme concentration at the top of the market in terms of major revenue production.

The history of jetliner competition is characterized by certain significant examples of competition between these two companies. Since the early 1980s, every move of one competitor has been mirrored by the other. As can be observed in Fig. 3.4, nearly each of the airplane types in the portfolio of Boeing had, and probably will have, a counterpart in Airbus, and vice versa.

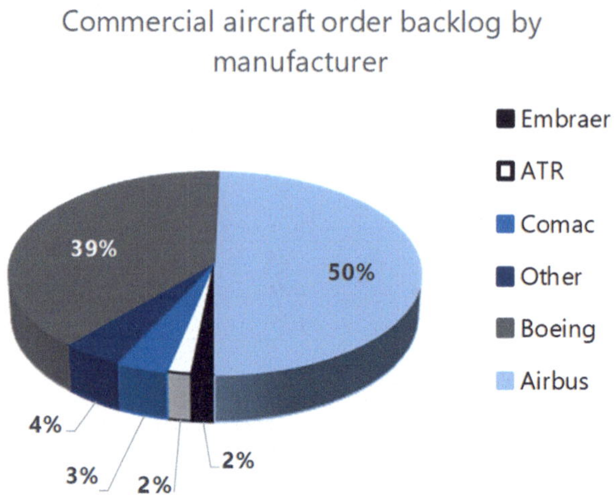

Fig. 3.3 Commercial aircraft order backlog by manufacturer. (Source: https://b4b15208-f1b4-47fa-8e27-54a3371ab3c0.filesusr.com/ugd/e00f26_7a835de893834b80ba94b9c11dcbdf98.pdf)

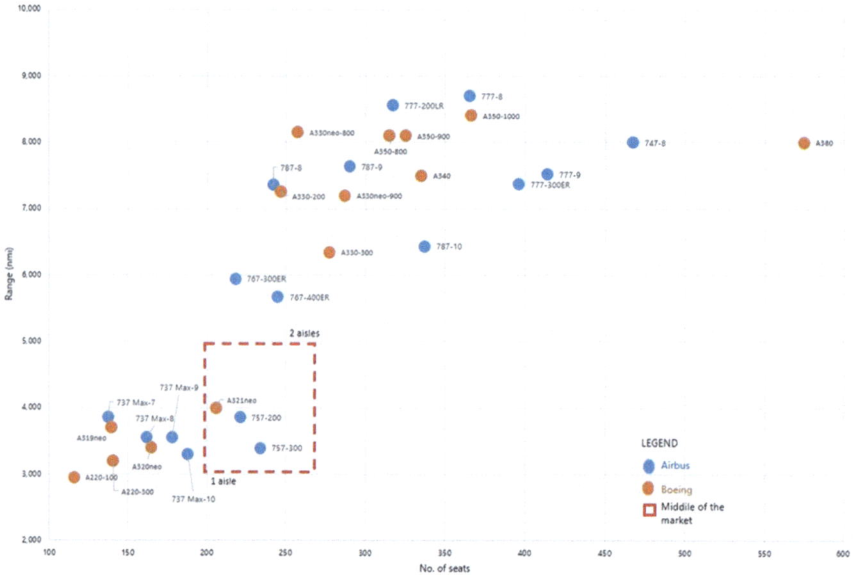

Fig. 3.4 Boeing and Airbus payload range diagram. (Source: PARE Reports)

A review of the Boeing and Airbus duopoly can bring hindsight to what was relevant for the dominance of the single-aisle market in the past and might also be in the future.

Traditionally, the market has followed a cyclic pattern: a growing period of roughly 7 years followed by a dropping period of approximately 3 years, with

Table 3.2 Airbus and Boeing 2018 orders [6–8]

	Deliveries		Net orders		Backlog 2018
Model	2018	2018/2017	2018	2018/2017	
A220	20	n/a	135	n/a	480
A320ceo	240	63.7%	10	7.8%	165
A320neo	386	213.3%	531	57.3%	5981
Total Airbus	646		676		6536
737NG	324	n/a	−24	n/a	88
737 Max	256	56.3%	699	53.3%	4675
Total Boeing	580		675		4763

deliveries falling by 30–40%, or more in the bad period. However, the industry has experienced continuous growth since 2004 and slowed down during the 2016–2017 parenthesis (single-aisle deliveries pause before A320neo and 737 MAX deliveries ramp up). These expectations call for continued growth by 2020. Long-term demand drivers include a strong passenger traffic growth trend projected over the next two decades but a slowdown in market demand for new aircraft by carriers, industry original equipment manufacturers (OEMs), and industry value chain ramping up production to deliver on the huge, accumulated order backlog. The 12,000 jetliners on backlogs at Airbus and Boeing alone are estimated to be worth over 7–8 years of production. The A320 family was on course for 63 planes per month, with 737 headed for 57 per month. Considering these figures, for the first time, the jetliner market will have a 16-year growth cycle, and possibly longer, more than twice as long as the usual 7-year boom[1].

Some consultants estimate that the segment generates a vast majority of the profits as they represent the bulk of the historic volumes delivered (around 10,500 for Boeing and 8500 for Airbus) and the existing orders (4763 for Boeing and 6536 for Airbus), according to the data provided by manufacturers [6, 7]. However, as shown by the data in Table 3.2, the order rush appeared to have decelerated in 2018. The strong cyclicality of the industry is still to come, and the growing cycle is becoming longer than ever. It is highly recommended to closely track the evolution of recent single-aisle order decelerations in such a growing market as the single-aisle segment to anticipate changes in any key driver affecting the sector.

Both companies still maintain a BB ratio (book-to-bill ratio, the ratio of orders received to the amount billed for a specific year) at a value higher than 1, which is characteristic of a boom period. In addition, the B737 and A320 families, with their latest versions, are models with higher backlogs, illustrating the success of the single-aisle segment.

[1]Assuming the new rates were achieved in 2023, the single-aisle segment would have seen 450% growth (by value of deliveries) over 19 years, in constant year dollars. More single-aisle jets will have been delivered between 2010 and 2024 than were delivered in the first 51 years of the jet age, 1958–2009.

Historically, Boeing 737 first entered the service in 1968. A variety of derivative aircraft based on the initial design, with different ranges and seating capacities, have been produced over the years. Airbus, after being successful with the A300 series, planned its narrow-body family starting in 1978 as Jets 1 and 2, a project targeted to compete with B737 and DC-9, the uncontested leaders of the market at the time [5]. The engineering capacity freed by the completion of the first project (A300) was put to work for the second project.

The consortium introduced its A320 family into service in 1988. It was the first airliner conceived with fly-by-wire controls. The aircraft's fuselage has been stretched and shrunken to fill different market niches with the introduction of the A321, A319, and A318. A variety of engines have been used on the Airbus airplanes allowing for incremental improvements in fuel efficiency. As an answer, Boeing launched the members of the Next Generation 737 family in the late 1990s and early 2000s with updated engines, cabin interiors, and flight deck avionics as well as winglets and changes to the airframe.

Categorically, within the vintage generation, the Boeing 737-600 competes directly with the A318; the Boeing 737-700 competes directly with the A319; the Boeing 737-800 competes directly with the A320, and the Boeing 737-900 competes directly with the A321.

The battle between both companies continued with the introduction of the Airbus Neo generation and the Boeing MAX generation. On the one hand, the 737 MAX series is offered in four lengths, typically configured for 138–230 seats and 5954–7084 km. The 737 MAX 7, MAX 8, and MAX 9 replaced, respectively, the 737–700, −800, and −900. Additional length is offered with the further stretched 737 MAX 10 (scheduled to be delivered from 2020 but still not certified). On the other hand, the Airbus Neo series is offered in three variants, which are based on the previous A319, A320, and A321. The passenger capacity varies between 140 and 244 seats and a range of up to 7400 km (Table 3.3).

The price list for two competing models belonging to rival families is comparable and the discount policies of both companies are similar (sometimes even 50–60% discount on the list price) [11].

Table 3.3 Narrow-body current market [9, 10]

Model	No. of seats	MTOW (t)	Range (nm)	List price (MUSD)
A220-100	116	60.80	2950	79.5
A220-300	141	67.60	3200	89.5
737 MAX 7	138	80.30	3850	96.0
A319neo	140	75.50	3700	101.5
737 MAX 8	162	82.19	3550	117.1
A320neo	165	79.00	3400	110.6
737 MAX 9	178	88.31	3550	120.2
737 MAX 10	188	92.00	3300	129.9
A321neo	206	97.00	4000	129.5

In today's aerospace market, the MAX and the Neo are the best-selling products for both manufacturers. These models are more fuel efficient, longer ranged, and have enhanced passenger interior and enhanced passenger comfort than previous B737/A320 families, both of which have sold very well since their introductions. A comparison of the evolution of the order backlogs for each of the competing families as of September 2018 is shown in Fig. 3.5.

From this representation, it can be easily understood that the attraction of the new generations (third for Boeing and second for Airbus) was causing the slow extinction of the previous one and that Airbus managed to absorb a larger proportion of the market demand growth. Comparing both families in terms of total orders, Airbus rules the market of re-engined single aisle, with 60% of the market. It is certainly possible that Airbus will maintain the 60–40 advantage in the following years, but this scenario may change in the future.

The continuous Boeing–Airbus rivalry has been in place for decades, since Boeing began, in the late 1980s and early 1990s, to take seriously the challenge of the small outsider established in 1970. The only notable competitor at that moment for Boeing, in the commercial airliner market, was McDonnell Douglas, manufacturer of both narrow bodies and wide bodies. But McDonnel was strongly affected by recession in the early 1990s and was to be merged into Boeing in 1997. Some analysts even consider that the Airbus presence on the market accelerated the decline of McDonnel and its absorption by Boeing [2]. For a better understanding of the development circumstances of the duopoly, a short review of the conditions in the industry is useful.

The significant achievements in the aerospace industry (and here the airliners' production is relevant) are based on three ingredients:

1. *Strong financials*: It is well beyond the possibilities of a normal-size company to spend the multibillion dollars necessary to develop a new type of airliner. Producing such machines is an act of large-scale economics, so it needs to be supported, more or less explicitly, by governments. This happens mainly because the private capital is reluctant to approach very large investment with a rather long recovery horizon (they prefer early repayment profiles) [2]. The capital markets are also less inclined to take the risk of failed projects and assume its painful consequences.

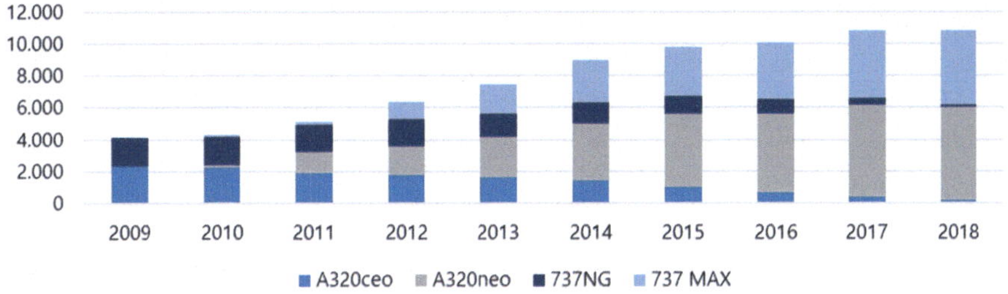

Fig. 3.5 Evolution of order backlogs [12]

2. *Powerful science and engineering resources*: Resources that need to be based on an existing wide base of STEM (science, technology, engineering, mathematics) education output, on a systematic experience accumulated in any of the contributing fields, as well as on a good capability of invention and innovation.
3. *Efficient industrial organization*: Developing a product means also a proper industrialization. Reaching appropriate production volumes at competitive costs and quality levels to satisfy the market demand is probably the most difficult task. It requires a rather rich experience, a strong discipline, a quality approach well implemented, and a science of managing a large supply chain. Every such component of the industrial system is to be built and maintained using a careful design and proof process.

The absence of any single one of these three ingredients spoils any chance of contemplating the entrance on this market. This means that a high entry barrier prevents outsiders from threatening the incumbents' positions. Any tentative approach to steal a fraction of the market from the "Big Two" by an outsider not strong enough in all three ingredients is doomed (as in the case of Bombardier, despite the active support of the Canadian government).

Finally, it is worth mentioning that in the recent history of the industry there are several relevant factors and lessons learned that can be extracted and that should be considered in the future in order to maintain competitiveness in the market. Here are some examples:

- Competition between two producers: A duopoly is a market situation in which only two producers exist. The decisions and actions of one producer affect and are affected by the decisions and actions of the other producer. An example of this situation is reflected in the Airbus/Boeing duopoly, in which both companies have maintained an aggressive competition over time. As a result, manufacturers must pay close attention to the actions and reactions of its competitor as well as to respond to each other's moves to prevent an inferior aircraft in a market segment from losing market share and profit potential. This scenario generally fosters relatively high innovation, high production, and low prices in order to maintain advantage in the market. However, there is also a risk that both producers may collude explicitly or tacitly or reach an agreement in order to reduce their risks for investment and new product development.
- Price competition: Due to the intense competition within the sector, it is quite usual that companies such as Airbus and Boeing apply price discounts on their products to gain more market share. This is especially applied to the commercial launch of a new airplane in order to get more orders from airlines. In fact, both companies have accused each other several times of carrying out this type of practice. However, price war hurts the profits of all the companies in a well-established industry and, for that reason, companies should avoid it.
- Innovation: Innovation is a key factor that has been used by Boeing and Airbus in their strategies to achieve success, which has enabled both companies to develop products that attract very high demand in the market. Innovation nor-

mally implies the use of advanced technologies to develop new products with the objective of seeking performance advantages in their products. Developing modern technologies is a very positive factor that helps advance the industry, allowing to obtain more modern and efficient aircraft. As an example, the Boeing 787 Dreamliner was the first large airliner to use composites for most of its construction.

- Commitment in deliveries: In the industry, there have been cases in which manufacturers have been unable to comply with promises made regarding aircraft deliveries. An example is the Airbus A380 case, a superjumbo jet that has received to date 331 firm orders. This aircraft had several delivery delays that caused dismay from its buyers as well as a drop in the earnings expected. In addition, Airbus had to negotiate a compensation with its customers for postponing deliveries. Therefore, to be able to fulfill delivery commitments is an important factor as, in case of not achieving it, it would lead to disappointment and distrust of the company.
- Customer-oriented strategies to reach success: Both Boeing and Airbus carefully research customer needs and strive to satisfy them since they represent a competitive and successful factor for a company. In addition, the airplane purchasing decision criteria of airlines include not only load and range factors and operating costs but also passenger comfort. Airbus has been quite competitive and successful in recent years as a result of developing a clear empathy with its customers, encouraging a two-way flow of views, ideas, and technical feedback on its aircraft in service around the world.

At present, Boeing and Airbus appear equally competitive. Both companies must understand their customer's needs and buying behavior, anticipate how customers' needs will evolve over time, keep a close eye on competition, be innovative in creating customer value, and strive to deliver total customer satisfaction. The company that can consistently and efficiently do all of these will be winner in the end.

3.2.4 SA Market in China

The SA market in China is scrutinized. Although there are more than 40 airlines in China, 30 are part of the "Big Four" airlines and competition has been the same between 1994 and 2012. The three biggest companies are state-owned and are included among the 20 biggest airliners in the world. By May 2019, the A320 family in service in China totaled some 1658 aircraft (55% of the SA market), against the 1353 of the B737 family that stand for 45% of the Chinese SA market. Additionally, in April 2019 Airbus signed a deal for 300 aircraft with the state agency China Aviation Supplies Holding Company worth an estimated $US35 billion at list prices.

Airlines and Operators

Development of the airline industry in China is commonly divided into five periods [13, 14]:

- The pre-reform tight regulation (1949–1978).
- The transitional stage (1979–1987).
- State-led consolidation and privatization (1988–2004).
- Market-driven consolidation and deregulated competition (2005–2012).
- A new wave of deregulation (2013–today).

The actual picture of the air transport airline industry in China was confirmed during the third period because of the consolidation impulse by the CAAC, which significantly reduced the number of airlines: 21 in 1994 vs 10 in 2004. This triggered the emergence of the "Big Four," namely Air China, China Eastern, China Southern, and HNA Group. The first three are state-owned while the fourth one, HNA Group (which owns Hainan Airlines), is a mix of private and municipal, and is currently struggling with economic problems.

In 2013, it started a further liberalization, with the relaxation of minimum price regulation on 31 routes. This was a strong incentive for the low-cost carrier (LCC) and a way to compete with China's new high-speed rail network. In 2015 and 2018, the CAAC relaxed again airfare, enabling airlines to increase ticket prices. The deregulation concerned 101 domestic routes in 2015 [15] and 306 highly competitive domestic routes in 2018 [16]. Additionally, in 2018, they relaxed the "one route, one airline" policy from 2009, favoring more internal competition between Chinese carriers, enabling new private companies to enter the competition. The CAA also approved a second wave of private airlines (Fig. 3.6).

Although there are more than 40 airliners in China, 30 are part of the "Big Four" airlines and competition, according to Wang, Bonilla, & Banister, 2016, has been the same between 1994 and 2012. Privately or locally owned airlines tend to fly mostly in routes with less demand in central and western provinces, where local governments are eager to support regional carriers. Tough monopolistic air routes decline in number, most of them today are operated by such airlines. Many experts claim that the LCC model has a strong potential in China due to the extreme price sensitivity of the air transport market and the quasi nonexistence of airline loyalty [17]. Figure 3.7 presents the size of the low-cost airliners in China.

Single-Aisle Fleet in China

The SA market in China is scrutinized (Fig. 3.8). Although there are more than 40 airliners in China, 30 are part of the "Big Four" airlines and competition has been the same between 1994 and 2012. The three biggest companies are state-owned and are included among the 20 biggest airliners in the world.

We analyze the fleet of the biggest Chinese airlines [18]. According to *Business Insider*, the big 3 were among the 20 biggest airlines in the world [19]:

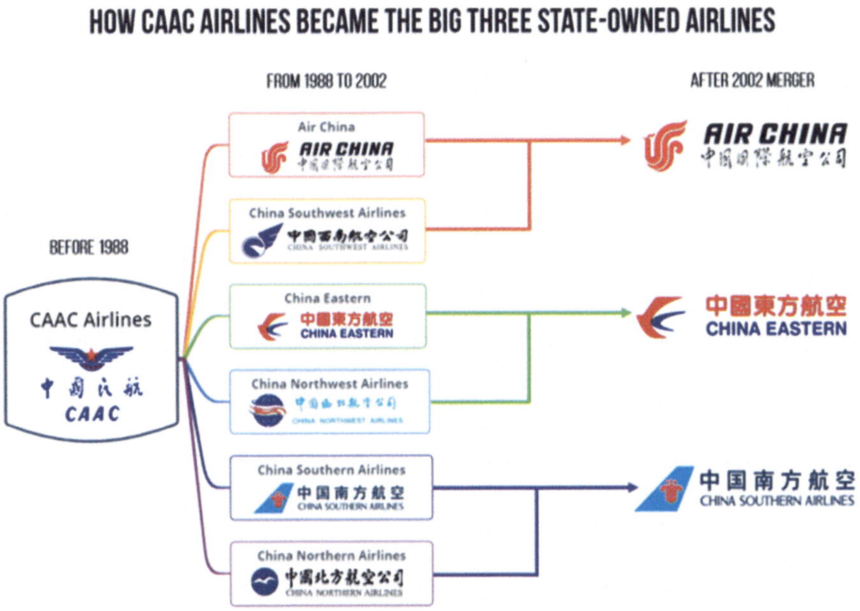

Fig. 3.6 CAAC split-up and later mergers. (Source: China Aerospace)

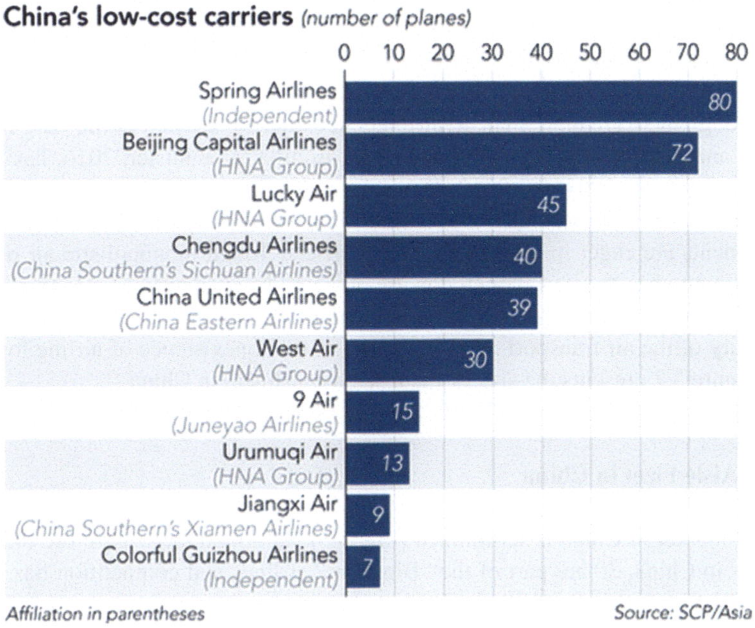

Fig. 3.7 Chinese LCCs by fleet size in 2018. (Graph from [14])

Fig. 3.8 Single-aisle sector in China. (Source: PARE Project)

- Air China is number 10 with a capacity of 90,531,776 seats and a fleet size of 418 aircraft.
- China Eastern Airlines is number 7 with a capacity of 122,917,175 seats and a fleet size of 525 aircraft.
- China Southern Airlines is number 6 with a capacity of 131,972,745 seats and a fleet size of 597 aircraft.

By May 2019, there were about 1658 A320 family aircraft in service (55% of the SA market), against the 1353 of the B737 family that stand for 45% of the Chinese SA market. Additionally, in April 2019, Airbus signed a deal for 300 aircraft with the state agency China Aviation Supplies Holding Company worth an estimated $US35 billion at list prices.

3.2.5 Traffic Forecast

China is on its way to becoming one of the world's largest aviation markets, accounting for nearly 20% of the global traffic by 2037. Traffic is expected to increase in China by 6.2% annually. To cope with that increase, the fleet will grow by 4.5% annually, doubling the fleet by 2038 and accounting for a market value of around 1.190 billion dollars.

China itself will receive 7690 deliveries, 31% of which will correspond to the replacement of existing airplanes. Out of the 7690 deliveries, 5730 will be single aisle, leading to a total fleet of 6100 airplanes at the end of that period. Single-aisle airplanes will represent 71% of the total fleet in China, and it is expected that the flexibility in size and range of this fleet will enable fast growth in point-to-point markets within China and bordering regions.

3.2.6 Evolution of Commercial Aviation in China

The great development, especially during the last 10 years, of domestic flight market in China is supporting the Chinese will to set up an indigenous civil industry. This is reflected in multiple high-level policy papers, such as the "13th Five-Year Plan" and "Made in China 2025."

More than 200 airports operate in China, and the number of aircraft has tripled in the last 15 years. Only in 2014 the country took 20% of all aircraft deliveries. The demand among China's rising middle class for air travel means that the country is likely to displace the United States as the world's largest aviation market (measured by traffic to, from, and within the country) by the mid-2020s.

The prospects for the industry are extraordinarily good. According to IATA, China will account for 1.6 billion passengers annually by 2037 for a total of 1.5 billion in the United States. According to Airbus and Boeing, China will account for 17% of the 40,000 global airplane deliveries expected during the next 20 years. Airbus and Boeing expect China to buy somewhere in the range of 7500 new aircraft over the next 20 years. China is considered the largest market, valued at 1 trillion US$ cumulatively (Fig. 3.9). Figure 3.10 illustrates the foreseen aircraft deliveries in China for the next 20 years.

This playground will be the scenario for incumbent airframes competition to increase market share and China's emerging aviation home-grown alternative to reduce dependency on imported aircraft.

To take advantage of this future growth, the Chinese government decided to structure a domestic commercial aircraft manufacturing industry capable of reducing its dependency on foreign suppliers and, in the long term, of competing with established OEMs on a global scale.

The history of Chinese-made passenger aircraft is summarized in the timeline in Fig. 3.11. Between the 1950s and the early 1980s, the Chinese aeronautical industry

Fig. 3.9 China aviation market evolution. (Source: PARE Project)

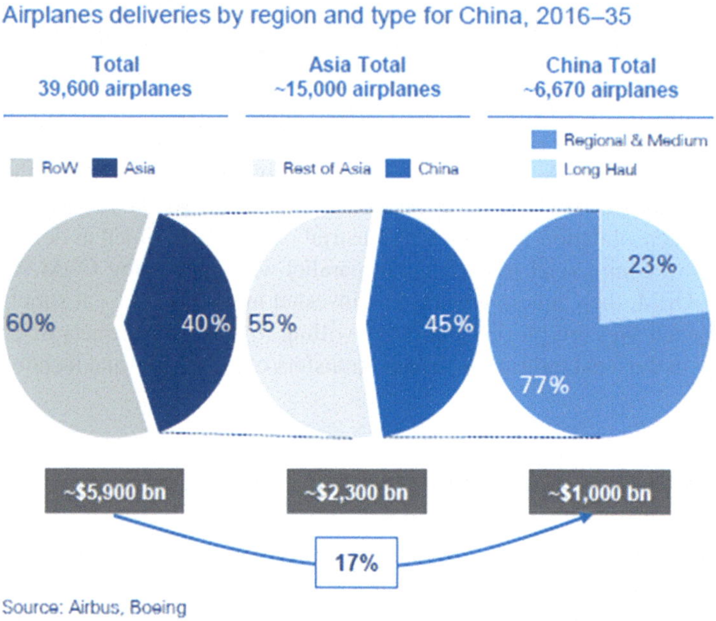

Fig. 3.10 Foreseen aircraft deliveries by region and type for China, 2016–2035. (Source: Airbus, Boeing)

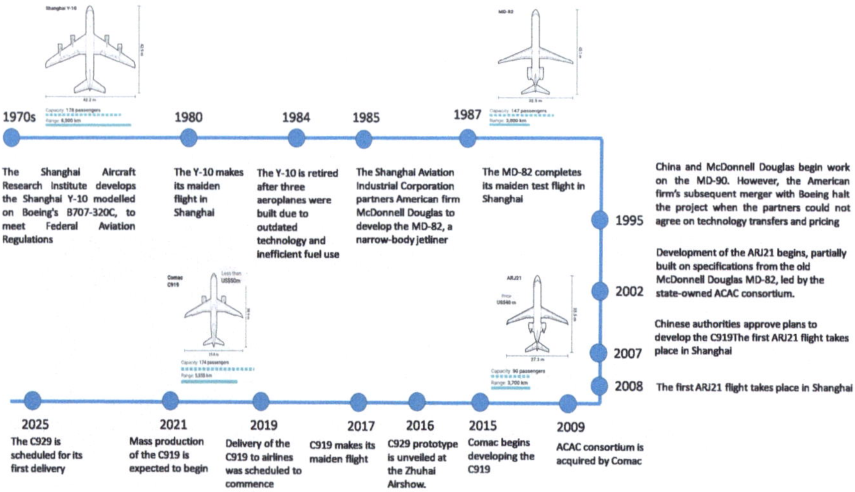

Fig. 3.11 A timeline history of Chinese-made passenger aircraft [23]

mainly focused on the development of military aircraft and a few unsuccessful attempts at manufacturing commercial jet aircraft (e.g., Shanghai Y-10).

In 1985, the partnership established with McDonnell Douglas represented a major turning point, accelerating the Chinese industry's learning process through the joint manufacturing of the MD-82.

The creation of the Aviation Industry Corporation of China (AVIC) in the 1990s and the Commercial Aircraft Corporation of China (COMAC) in the 2000s initiated a phase of consolidation for Chinese industrial capabilities, aimed at developing a full range of commercial jet aircraft. In parallel with developing COMAC as the domestic OEM, the Chinese government invested in building a local supply chain and requested Western players that were willing to capture contracts in China to partner with the local industry, operating transfers of know-how and technologies.

3.2.7 Structure of the Industry and Research Networks

The evolution of commercial aviation in China has also been analyzed, and particular attention has been devoted to the structure of the industry and to the aviation research network structures. The size of Chinese industrial players is large, with 152 enterprises, spread over 22 provinces. However, the benchmark on the number of employees might point out potential inefficiencies and redundancies that the Chinese industry still has to polish.

Total employment in commercial aviation manufacturing has increased from 234,390 in 2005 to 254,844 by 2010, a 9% overall increase and numbers that rival employment in this industry in the United States and other major countries with a large commercial aviation manufacturing industry (see Fig. 3.12). The absolute

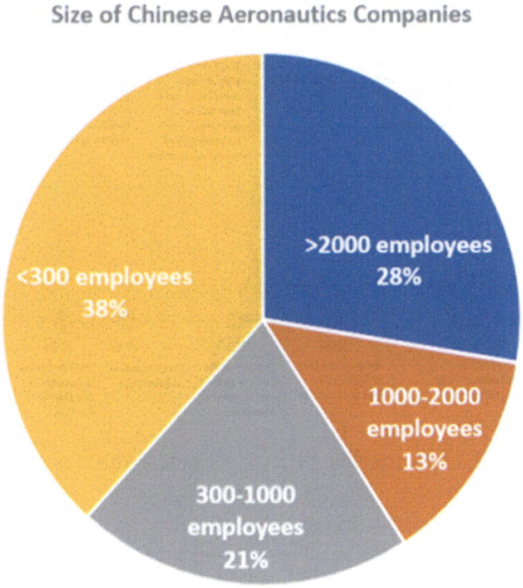

Fig. 3.12 Distribution of Chinese aeronautical companies per size. (Source: China Aerospace)

numbers and shares of employees who are engineers/technicians or are recorded as working in research and development activities have increased in recent years. With 152 companies, the Chinese aviation industry employs directly 325,000 people, on average 2138 employees per company. Out of these, 250,000 workers are employed in the civil aviation sector. In comparison, France had in 2017, 350,000 employees working in over 3000 aeronautical companies: roughly an average of 117 people per company.

Most of the Chinese aviation industry is turned toward the defense industry: only a third (roughly) of the industry's revenue is generated by civil aeronautics. In 2016, from a total of 247.83 billion RMB of total revenues generated by both civil and military aviation industry, the civil activity accounted for 32.4% (80.29 billion RMB) only. Moreover, a very significant amount of the civil revenue comes from Airbus China (FAL in Tianjin). The rest is shared among the following (in decreasing order): aircraft parts and aircraft engine manufacturing, MRO, UAV/UAS industry, and indigenous aircraft manufacturing, as indicated in Fig. 3.13.

The aeronautical landscape is spearheaded by three large state-owned conglomerates: AVIC, COMAC, and AECC, representing respectively 31% (47), 4% (6), and 9% (13) of all Chinese aeronautical companies. The industry is immersed in an important reorganization. Only in 2015, AVIC was still composed of 63 subsidiary companies, just before the creation of AECC. The same can also be said about COMAC, which was split off from AVIC in 2008. Revenues are less significant than expected due to the dwarfing effect of Airbus China: 15.1%, 4.5%, and 4% of the total industry respectively for AVIC, COMAC, and AECC.

Output and employment in China's commercial aviation manufacturing industry have been increasing, but domestic sales, not exports, have been the primary driver. Compared to aviation manufacturing industries in other countries, sales remain concentrated in the domestic market. Cumulative exports ran 17.3% of cumulative

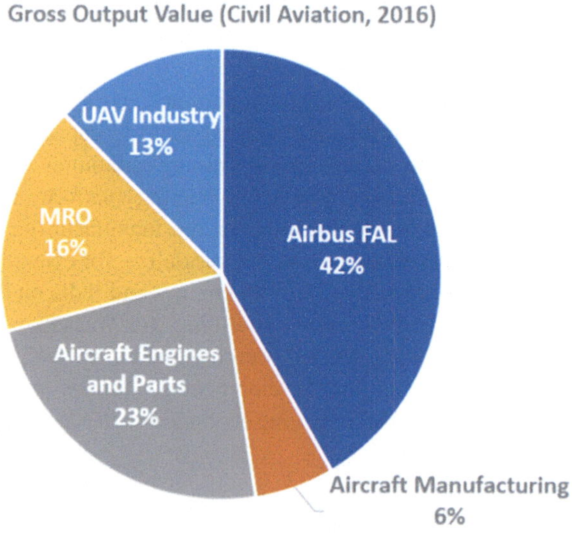

Fig. 3.13 Distribution of Chinese aeronautical companies per size. (Source: China Aerospace)

output from 2005 to 2010; exports as a share of output have fluctuated between 13 and 21%. Significant parts of regional jets produced in China, including avionics and engine components, are Western in origin. In 2014, China exported $3 billion in aircraft and parts [3] vs the $57 billion exported by the United States in civilian aircraft and nearly as much in aircraft engines and parts.

The leading subsector in the Chinese aviation industry is aircraft parts, both manufacture and repair. China's import market for aircraft parts and components exceeded $2.19 billion in 2016 (30% from the United States). China's demand for aircraft parts can be attributed to a number of factors, including increasing capacity utilization rate, the aging and expansion of China's aircraft fleet, and the domestic production and assembly of aircraft. The best immediate opportunity for foreign companies will be in supplying parts for China's commercial aircraft fleet as this is the largest and best-established segment of China's aviation market and is currently dominated by Western aircraft with Western suppliers. China's domestic aircraft parts and assembly manufacturing sector is also growing. Around 200 small aircraft parts manufacturers and a number of regionally based major manufacturers concentrated in Shanghai, Chengdu, Xi'an, Nanchang, Harbin, Shenyang, and Shijiazhuang. Large aircraft and engine manufacturers have committed to expand procurement in China over the long term, although highly technical and sophisticated parts will continue to be imported.

The consolidation of the aviation industry in China is contributing to a more global, more fragmented, more competitive industry that will give Western companies major short-term cost reduction opportunities that they must capture.

Lower labor costs in emerging countries, which are, on average, three to five times lower than that in the developed countries, can provide major economic savings and advantages, even considering transportation, the coordination complexity and supply chain management, supply disruption risks, etc. The cost of manufacturing typical aircraft structures (such as body panels or fuselage sections) can still be roughly 20–25% lower in these emerging markets than in more developed ones. Lower labor costs make these economies also interesting for labor-intensive maintenance and repair services.

China, and other emergent economies, has the potential for increasing amounts of low-cost manufacturing and engineering capacity for the aerospace industry. China leaps forward in learning, gaining economies of scale, and taking its place as low-cost manufacturing and engineering platform for the world. China could become the preferred location for the manufacturing of simple airframes, while other emergent economies might specialize, for example, in other areas (Russia on low-pressure modules of aircraft engines and India on detailed engineering). These changes represent a major opportunity for Western players to improve their cost performance through global sourcing, manufacturing, and engineering.

A dynamic view of the emergent economies and their role in the aerospace industry suggests that emerging economies will accelerate changes in the value chain.

3.2.8 The COMAC Consortium and Aircraft Programs

The COMAC consortium and its aircraft programs are also analyzed. Although COMAC stands out as a company with a heavy support from the Chinese government, both political and financial, this support might not be unlimited. One major improvement made by the Chinese industry is the acquisition and production of technology and know-how, which will allow them to improve in the production of their aircraft models. Despite the improvements they have made so far, they still have a long way to go to even meet the production and quality levels of the incumbent players on the market (Fig. 3.14).

C919 is benchmarked with respect to their Western competitors A320 and B737. The supply chain of the three models is analyzed as well as the technological innovation level of the C919 and the number order received so far. A total of 16 foreign suppliers are involved in the C919 program, including GE, Honeywell, Safran and CFM, Rockwell Collins, GE Aviation, Eaton, United Technologies Corporation, Hamilton Sundstrand, Leonardo, Parker Aerospace, and Liebher; with joint ventures covering avionics, flight control, propulsion system, electric power generation and distribution, composite materials, and fuel and landing gear systems. To date, COMAC boasts of just over 1000 commitments for the C919 from Chinese airlines and lessors. However, according to the *Air Finance Journal* Fleet Tracker database, firm order numbers might be fewer than 400. The promise of C919 selling price, half the price of its competitor, is also analyzed.

On the other hand, the difficulty of getting certified by the FAA or the EASA is still a major holdback on COMAC and the Chinese aspirations to gain the MoM. Using their great and growing domestic market and their political influences with countries attracted by their dumping strategy can, still, develop their new models. However, not being able to go to the United States or the EU is a major issue in order to compete with Airbus or Boeing. Another related issue is that, whenever

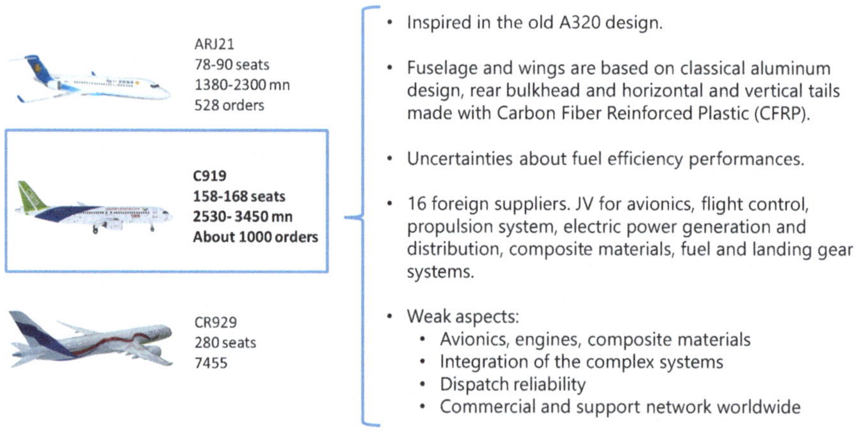

Fig. 3.14 The COMAC family. (Source: PARE Project)

they are able to certify their models, these will not be a state-of-the-art product, which can be a great problem to cope with the competition of the market in that time period.

Special attention has also been devoted to some key factors that will impact C919 program and delivery timescales, including

1. Development learning curve.
2. Certification hurdles.
3. System integration skills.
4. Engine industry.
5. Avionics.
6. Composite materials.
7. Possible customer countries.
8. Production capacity constraints including estimation of production curves.
9. Shortage of skills.

In addition, the incumbent players already have built up over the years a commercial image comes short of assurance to the market that their products and services are the first-quality aircraft available. COMAC does have to prove this. This gives Airbus and Boeing a network of subsidiaries and the ability to get financial support from private entities more easily than COMAC.

3.2.9 Lessons Learned from Other Industries Development in China

Several authors have studied the effectiveness of the Chinese government's industrial development policies in key industries during the last decades. The most relevant cases might be those discussed here.

High-speed trains: China opened its first high-speed train line in 2007. Today, it has the longest high-speed network in the world, and the plans are to expand even more the network, up to 25,000 km by 2020. The first bid was launched in 2004 and awarded to three of the four contenders, Kawasaki Heavy Industries, Alstom, and Bombardier. Companies were required to have a local Chinese partner to manufacture trains. In a few years, state-owned Chinese partners were able to manufacture their own trains and did not purchase the 200 trains foreseen in the contract. In a very short period of time, local companies absorbed the technology and were no longer dependent on foreign companies.

Wind power generation: China became the world's largest manufacturer of wind turbines in 2009, passing in less than 4 years of being just a player to become the major player in the industry. By 2012, almost all the units installed in China were manufactured locally. This powerful industry development responds to a series of measures that include domestic subsidies, licensing agreements, acquisition of foreign companies, and joint ventures with foreign manufacturers.

Automobile manufacturing: China became the world's largest market for new car sales in 2009. To exploit this market, manufacturers have to set up assembly operations in China and a joint venture with a local partner. Other measures to favor local industry include the restriction to purchase foreign vehicles by government agencies or subsidies for electric vehicles only for models produced by Chinese companies. However, in this case, those measures have not yet been able to generate a strong automobile industry, and models manufactured by joint ventures still dominate the market against national pure products. Despite local partners being able to acquire the technology, Chinese consumers still prefer foreign brands because of better reputation and prestige. All these have translated into an increase in the production of joint ventures but very low development of pure local brands. This has since changed with Chinese electric cars dominating the home market and increasing exports.

The main lesson that can be derived from these three sectors is the relevance of the industry structure in the process. Those industries where the customers are state-owned companies (wind power generation) are very much sensitive to Chinese government policies to drive purchases, as well as those industries where state-owned companies have a monopoly (railway sector). In these cases, the Chinese government has been able to induce firms to buy products manufactured by Chinese companies, even when the products are available from joint ventures with foreign manufacturers. The state-owned purchasers have not been concerned about the disputes about ownership of the technologies underlying these products.

Where the decision to buy depends on the final consumers, the situation is different. In the automotive industry, Chinese customers are free to choose the vehicle they prefer, and foreign brands manufactured by joint ventures dominate the market. Automotive brands have been able to maintain and control their intellectual property better than other industries and have made optimum use of their reputation for safety and reliability. They have also created a dealership network and have invested in marketing in China to back up their position in the market. They have also been able to compete in cost by spreading R&D cost over their global operations and reducing the cost per vehicle of developing a new model. The more recent expansion of the Chinese electric car industry is a case of state subsidies/incentives to state and private companies.

The commercial aviation manufacturing industry falls somewhere between the previous cases. On the one side, the Chinese government influences the choice of aircraft purchased by China's state-owned airlines. The CEOs of these airlines are selected by the government, which can pressure them to buy national products. However, Chinese airlines are subject to competition among them and with other companies because they sell airplane tickets directly to consumers. There will be a conflict of interest between government desires to purchase aircraft manufactured by COMAC and the need to ensure that their airlines operate safely and profitably.

Experts claim that because of its outdated design the C919 will be more expensive to operate than next-generation Boeing and Airbus narrow-body aircraft. These differences in operating costs will directly affect the airlines' profitability. The extent to which the CEOs of the three main state-owned Airlines will purchase aircraft that ensure the continued success of their operations, regardless of the pressure to purchase Chinese products, will define different possible scenarios.

3.2.10 Chinese Policies SWOT Analysis

Additionally, the policies deployed by the Chinese government to advance its commercial aviation industry are analyzed in detail and recommendations are derived for those companies willing to develop aviation activities in China. This analysis is backed up by a detailed SWOT analysis covering the aspects of technology, labor, finance, and marketing.

Strengths

- Chinese suppliers are increasingly proficient in process technologies.
- Joint ventures have improved Chinese technological capabilities.
- Acquisition of foreign firms: new product, process, technologies, and markets.
- Chinese machinists/workers are proficient in composite materials.
- Design and engineering talent rates are very high.
- Substantial numbers of well-trained technicians and engineers.
- Quality of institutions of higher education.
- COMAC has not experienced financing constraints.
- China providing resources to create national champions.
- Diplomatic leverage and state financing induce third countries' orders.
- Government compels state-owned airlines purchasing.

Weaknesses

- JV does not guarantee that the Chinese partner improves its capabilities.
- Not yet mastered some key advanced technologies.
- Aircraft technologically is obsolete.
- Local focus on technological over profits.
- Chinese project management skills.
- Hierarchical management style.
- Project management imposes substantial costs.
- Financial support from the Chinese state is not unlimited.
- Sales network and the credibility.
- COMAC global logistics network for its new aircraft.
- COMAC faces competition from used aircraft.
- Reliability is an essential feature of an aircraft.

Opportunities

- Certification is not a permanent barrier.
- Concern about intellectual property.
- Wages for production workers in the aviation manufacturing sector are still substantially higher in foreign countries than for similar Chinese workers.
- Incumbent aircraft manufacturers face financial pressures that COMAC and AVIC do not.
- Incumbent worldwide service and distribution networks.

Threats

- Certification is a market entry barrier.
- Importance of proprietary technologies to commercial success.
- Continuous development of cutting-edge new technologies protects incumbent manufacturer from Chinese competitors.
- Clusters all over the world to design and assemble aircraft.
- Incumbent manufacturers still had a cost, quality, and technological competitive advantage in managing technological development and their production lines.
- Boeing and Airbus Group and all the Tier One commercial aviation component suppliers are large, financially sound companies.
- Restricting subsidies available for trade financing among the United States, the EU member states, and other countries.

3.2.11 Analysis and Conclusions

Finally, the results from this research are directed to assess the possible problems that the future market may hold and, furthermore, help Europe fulfill its main objectives regarding Flightpath 2050.

All the previous considerations have helped us define and analyze five relevant scenarios: three short–medium-term and three long-term scenarios (Fig. 3.15).

Scenario 1A: C919 certified only by Chinese CAA with entry into service by 2021 with a conservative production learning curve. C919 performance lags behind B737 and A320 and final selling price is half of the price of the B737 and A320 aircraft. In this case, C919 will be sold primarily in China by governmental influence on the four big Chinese airlines. Demand for the C919 is estimated as a percentage of the overall deliveries of the four big Chinese airlines, between a conservative 10% and a more optimistic 30%. A conservative realistic approach of roughly 400 firm

Short – medium term scenarios			Long term scenarios	
Scenario 1A	Scenario 1B	Scenario 2	Scenario 3A	Scenario 3B
• C919 certified only by Chinese CAA • No Western certification is granted • EIS by 2021 • Conservative production curve • Performances lag behind B737 and A320 • Prices half the price of B737 or A320		• FAA - EASA certification • EIS by 2025 • Performances lag behind B737 and A320 • Prices half of B737 or A320 • Conservative production curve	• New upgraded version of C919 equivalent to its contemporary Boeing and Airbus models. • Free market competition rules for market share allocation • Fuel efficiency main driver guiding airlines buying decisions. • Incumbent manufacturers will either maintain; reengineer or bet for a new clean sheet design .	
• C919 sold in CHINA	• C919 additionally sold in 3th Chinese influenced countries	• C919 could be sold worldwide in a market niche.	• Performance equivalent to A/B reengineered models	• Performance equivalent to A/B new models
Expected C919 demand				
CHINA' MARKET 2018-2037 • A320 60% • B737 25% • C919 14%	• 1-3% of SA in 3th countries: less than 100 additional C919	• 9% of the global worldwide SA deliveries over the total period 2018-2037: 2657 aircraft out of 31175.	• COMAC will only receive positive payoff if neither incumbent develops a new aircraft.	• New entrant's aircraft would take significant market share unless the incumbents decide to move.

Fig. 3.15 Summary of the scenarios and their evaluation. (Source: PARE Reports)

orders is also considered. The resulting Chinese market share will be 60% for the A320, 25% for the B737, and 14% for the C919. This approach is equivalent considering that 20% of the four big SA deliveries will be covered with the C919. A maximum production rate of eight units per assembly line has been estimated, like those achieved by Western manufacturers, when the assembly line is fully operational. However, attending to the Tianjin experience with both A320 and A330, the Chinese industry still will take time to get to this optimum production rate. This production rate will deliver 1260 C919 units in 20 years with only one production line. COMAC could satisfy the foreseen demand with one single line. If COMAC were able to put in place two production lines, its delivery capability would be higher than Boeing's expected demand, but it will require high efforts and investments.

Scenario 1B: Similar to scenario 1A but C919 is additionally sold in countries where the Chinese government has political or economic influence. This market will be small and somehow anecdotic. The number of A320 and B737 that are currently operated by airlines registered at those third countries' potential buyers for the C919 ascends to 863, 18% on the single-aisle fleet in these regions. The possible political and economic influence of Chinese companies in these countries will hardly justify big orders from their airlines. Considering orders and exports of the Chinese aviation industry, this hypothetical demand is estimated to be between 1 and 3% of the single-aisle operating in those countries, which means an additional demand between 22 and 65 additional aircraft. As expected, the figures are minimal and do not change the results of the analysis in scenario 1A significantly.

Scenario 2: C919 obtains FAA and /or EASA certification with entry into service by 2025 with a conservative production learning curve. C919 performance lags behind B737 and A320 and the final price will be half of the price of B737 and A320. C919 will be sold primarily in China by governmental influence on the four big Chinese airlines, but it could be sold worldwide. Its lower price will favor its position in a market niche despite its lower performance. Internal Chinese demand will not be significantly different from the one in scenario 1A. However, C919 demand from external countries might vary between 5% of the market for a high-fuel prices scenario and 10% for a low-fuel prices price scenario. The reaming single-aisle demand to be shared between Boeing and Airbus is not significantly affected. By 2037, C919 would have obtained a demand of less than 9% of the total SA deliveries over the total period 2018–2037; this is 2657 out of 31,175. However, in this case, the C919 selling might be enough to justify economically the investment in the C919 program. To take full advantage of this window of opportunity, the Chinese industry should focus its efforts on improving its production curve and expanding its production chain as quickly as possible without affecting reliability.

Scenarios 3A and 3B are classic examples of the new entrant problem with three players.

Scenario 3: COMAC produces a new upgraded version of C919 equivalent in terms of technologies and performance to its contemporary Boeing and Airbus models. By that time, Boeing and Airbus will have probably also evolved their current 737 and 320 models. Competition will take place under free market rules with fuel efficiency guiding airlines' buying decisions. Incumbent manufacturers will

either maintain their product line with only a minor modification and minimal performance improvements, re-engineer them to improve their performances significantly, or bet for a new clean sheet design to improve even more their performances. Two alternatives are contemplated in scenario 3.

Scenario 3A: COMAC produced a new aircraft with performance equivalent to those of Airbus or Boeing re-engineered models. In this case, the new entrant would only be able to capture 50% of the market from two stagnant incumbents. Based on the work by [1], this scenario presents an of-symmetric equilibrium, in the case Boeing develops a new clean sheet improved design to maintain 50% of the single-aisle market with Airbus and COMAC splitting the remaining 50% of the market with models slightly inferior in performance to the Boeing one. In this scenario, COMAC will only receive a positive payoff if neither incumbent develops a new aircraft.

This result indicates that if COMAC would seek to maximize profits it could decide not to produce a new model if it is only as good as the re-engined A320 and B737 families. In this situation, incumbent manufacturers will not be concerned with new COMAC single-aisle designs unless there is a high probability that the new models could match or exceed the performance of the incumbent new aircraft.

Scenario 3B: COMAC produced an aircraft with performance equivalent to those of the Airbus's or Boeing's new clean sheet design. The results illustrate that an equilibrium might be possible in which one incumbent chooses to maintain while the other decides to re-engineer. The greater investment required by Boeing to re-engineer results in an of-symmetric equilibrium. The superior performance of the new COMAC's aircraft will be able to capture a significant market share while Airbus attempts to maintain market position by re-engining the A320 latest models. In this situation, Boeing's optimal strategy is to avoid the investment and maintain its current aircraft. Once the competitors' new and re-engineered aircraft enter service in stage 2, Boeing suffers from a greatly reduced market share, but it will continue to make small profits due to its unit production cost advantage while harvesting its existing product line. The new entrant has a positive expected net present value in each possible outcome, except if both incumbents develop a new aircraft.

These results suggest that there may be rents available in the single-aisle market, providing an incentive for increased competition if new entrants are able to overcome the significant entrance barriers to develop an aircraft that can compete with the incumbents' new aircraft option.

Finally, an additional assessment can be made to determine the appropriate time for such a decision. The result shows that if the new COMAC aircraft is as good as the new Airbus or Boeing ones, the payoffs for move first or delay are very close, with payoffs that are sensitive to the assumptions of the aircraft program valuation model. However, if the new COMAC aircraft will only match the performances of the Airbus or Boeing re-engined aircraft, both manufacturers will have an incentive to develop a new clean sheet design aircraft as soon as possible. An early decision by an incumbent manufacturer would reduce COMAC's market share up to a point that might not have a positive NPV. This suggests that an early movement of the incumbent manufacturers could prevent the COMAC impulse to evolve C919. The

current situation suggests that, at current low production rates of tens per year, comparable to Airbus and Boeing per month, COMAC will take a long time to fulfil the 1000-plus orders from Chinese airlines, that will remain dependent on Western airliners. Even if i the C919 gains FAA or EASA certification, that remains uncertain, there is little available capacity for exports. The prospects for the twin aisle C929 developed solely by China, without Russian collaboration, may be even more long-term. The possibility of China leapfrogging the West with civil adaptations of military designs, like a stealth bomber or a long range tanker, cannot be totally excluded, but is somewhat remote. In the meantime the Chinese airline market may continue to expand, without much alternative to western airliners, and possibly continuing political interference in the choice between Airbus and Boeing aircraft. The dependence of China on foreign airliners would reduce: (i) for international flights, if Chinese aircraft obtained western certification, needed to fly outside China and increase limited exports; (ii) for domestic flights if COMAC manages to significantly increase production rates to meet its large backlog of orders in less time more than a decade at present level.

References

1. J. K. D. Morrison, S. Sgouridis, and R. J. Hansman, "Game Theory analysis of aircraft manufacturer innovation strategies in the face of increasing airline fuel costs," no. June, pp. 1–113, 2011.
2. D. Neven and P. Seabright, "European industrial policy: the Airbus case," Econ. Policy, 1995.
3. C. L. Benkard, "Learning and Forgetting: The Dynamics of Aircraft Prodcution.," Am. Econ. Rev., vol. 90, no. 4, pp. 1034–1054, 2000.
4. A. Dunn, G. Fafard, "Did backlogs catch an autumn chill?," Flight Int., 2018.
5. T. J.W.R., Ed., Jane's all the world's aircraft. London: Macdonald and Jane's Publishers Limited, 1978.
6. Boeing, "Orders and deliveries." [Online]. Available: https://www.boeing.com/commercial/#/orders-deliveries.
7. Airbus, "Orders and deliveries." [Online]. Available: https://www.airbus.com/aircraft/market/orders-deliveries.html.
8. M. Kingsley-Jones, "Data shows Boeing was 2018 winner," Flight Int., 2019.
9. Airbus, "No Title." [Online]. Available: https://www.airbus.com/.
10. Boeing, "No Title." [Online]. Available: http://www.boeing.com/.
11. T. Hepher, "How plane giants descended into global 'price war.'"
12. World Airlines Census," Flight Int., 2018.
13. J. Wang, D. Bonilla, and D. Banister, "Air deregulation in China and its impact on airline competition 1994–2012," J. Transp. Geogr., vol. 50, pp. 12–23, Jan. 2016.
14. Y. Zhang and D. K. Round, "China's airline deregulation since 1997 and the driving forces behind the 2002 airline consolidations," J. Air Transp. Manag., vol. 14, no. 3, pp. 130–142, May 2008.
15. T. Madge, "Getting The Deal Through—GTDT," Gettind the Deal Through, 2018.
16. Reuters, "Chinese airlines soar on fare price liberalization - Reuters," Reuters, 2018.
17. M. Brekelmans, "China's big 3 airlines must face up to the LCC threat - Nikkei Asian Review," Bussines Insight, 2018.
18. "China Airlines Fleet Details and History," Planespotters, 2019.

19. B. Zhang, "The 20 biggest airlines in the world, ranked I Business Insider España," Busines insider, 2019.
20. S. Creedy, "Airbus signs mammoth Chinese deal for 300 aircraft - Airline Ratings," Airline ratings, 2019.
21. Airbus Commercial Aircraft, "Global Market Forecast 2018-2037," 2018.
22. Boeing, "Current Market Outlook 2017-2036," 2017.
23. D. Wong, "Can 'Made in China 2025' help turn the nation's domestic aerospace industry into a world leader?," South China morning post, 2019 .
24. "CAAC Specifies Five Main Tasks in the 13th Five-year Plan for Civil Aviation Development," Civil Aviation Adminsitration of China, 2017.
25. L. Industrie, "Situation De L ' Emploi En 2014 Situation De L ' Emploi," 2015.
26. "La filière Aéronautique I Conseil national de l'industrie," Conseil natioal del l'indsutrie.
27. "China - Aviation I export.gov," export.gov, 2017.
28. C. Bédier, M. Vancauwenberghe, and W. van Sintern, "The growing role of emerging markets in aerospace I McKinsey," McKinsey& Company, 2008.
29. M. J. Kroes and M. S. Nolan, Aircraft Basic Science, Eigth Edition, Eighth Edi. New York: McGraw Hill Education.
30. J. Alstott, G. Triulzi, and B. Yan Jianxi, "Mapping Technology Space by Normalizing Patent Networks," Scientrometrics, pp. 443–479, 2017.
31. R. Watts, A. Porter, S. Cunningham, and D. Zhu, "Vantage point intelligence mining: analysis of natural language processing and computational linguistics," in Principles of Data Miing and knowledge Discovery, J. Komorowski and J. Zytkow, Eds. 1997.
32. L. Xiuxiu, Mei; Ping, Qin; Xiaotao, "Visualization Analysis on Collaborative Network and Research Hot-Spot for Aerospace Engineering Subject," Adv. Eng. Res. Vol. 118. 2nd Int. Conf. Autom. Mech. Control Control Comput. Eng. (AMCCE 2017), 2017.
33. L. Y. Chen, "China Sets the Pace in Race to Build the Factory of the Future," Bloomberg. .
34. J. Dang and K. Motohashi, "Patent statistics: A good indicator for innovation an China? Patent subsidy program impacts on patent wuality," China Econ. Rev., pp. 137–155, 2015.
35. C. Fisch, P. Sandner, and L. Regner, "No Title," Chian Econ. Rev., pp. 22–34, 2017.
36. L. Aaboen, J. Laage-Hellman, and L. Frida, "University Spin-offs and their roles in business networks," Ind. Mark. Manag., 2014.
37. B. Soffer, "Chairman of Israel Techn Transfer Netwrok adn Head of the T3 tec.".
38. K. Muller, "Discussion Paper No 08-034 University Spin-offs's transfer speed- Analyzing the tiemm from leaving university to venture."
39. [Bradley Perrett, "Chinese Advances I [primary-term] content from Aviation Week," 2009.
40. Global Security org, "Shanghai Aviation Industry (Group).".
41. Ari Magnusson, "COMAC Rising: China's State-Owned Aerospace Manufacturer Aims High," Apex, 2019.
42. "COMAC C919, first analysis - Leeham News and Analysis," Leeham news and analysis.
43. Tu Lei, "C919 powered by majority Chinese technologies - Global Times," Globaltimes.c, 2017.
44. "Airbus A320, Boeing 737, COMAC C919: where do the suppliers come from?—The China Aerospace Blog," China aerospace blog, 2019.
45. R. Cliff, C. J. R. Ohlandt, and D. Yang, Ready for Takeoff: China's Advancing Aerospace Industry. 2011.
46. F. Hashim, "ANALYSIS: Comac marches forward with ARJ21 and C919," Fligth Global, 2018.
47. J. Jiang, "ICBC Financial Leasing Co., Ltd. signs the agreement of 45 C919 orders with COMAC_Commercial Aircraft Corporation of China, Ltd.," ICBC Financial Leasing Co, 2011.
48. K. Cantle, "COMAC wins more orders for C919, ARJ21 I Airframes content from ATWOnline," ATW air transport world, 2014.
49. M. Toh, "PARIS: Comac wins commitments for 64 jets," FligthGLobal , 2015.

50. M. Toh, "AIRSHOW China: Comac wins lessors' orders for 23 C919s," FLigth Global, 2016.
51. B. Goh, "China's ICBC Financial Leasing orders 55 C919 planes - Reuters," Reuters, 2017.
52. B. Goh, "China's HNA, COMAC sign deal for 200 C919, 100 ARJ21 jets - Reuters," Reuters, 2018.
53. E. Lin, "Aviation analyst with the investment bank UBS in Hong Kong." 2018.
54. S. Norihiko, "Real Orders for Chinese Commercial Jet? - China Real Time Report - WSJ," The Wall Street Journal. .
55. Airline watch, "Ethiopian Airlines considers COMAC C919 instead of Boeing 737 MAX 8," Airline watch, 2019.
56. C. Loh, "Air Italy Dumps Boeing 787 Plans In Favour Of Airbus A330 - Simple Flying," Simple Flying, 2019.
57. Cl. Charpentreau, "The A330-800: it doesn't sell, but it flies," Aerotime News, 2018.
58. J. Kasper Oestergaard, "Airbus and Boeing Report December and Full-Year 2018 Commercial Aircraft Orders and Deliveries—Defense Security Monitor," Defense and security monitor. .
59. A. K. Gupte, "Comac: China's Challenge to Airbus and Boeing - Bloomberg," Bloomberg, 2010.
60. "Domestically Manufactured Large Aircraft—The Making of China's Aviation Manufacturing Industrial Cluster," High Tech Sci. Technol. Rev., 2010.
61. M. Mecham and J. C. Anselmo, "A Big Bet on China for Suppliers," Aviat. Week Sp. Technol., 2010.
62. "Overview of Chinese Investments in the Western Aviation Industry—The China Aerospace Blog," The china aerospace blog, 2018.
63. A. (Rhodium) Hanemann, Thilo (Mercator); Huotari, Mikko; Kratz, "Chinese Fdi in Europe : 2018 Trends and Impact," no. March, 2019.
64. C. Ohlandt, L. Morris, J. Thompson, A. Chan, and A. Scobell, Chinese Investment in U.S. Aviation. 2017.
65. Baker McKenzie, "An update on Chinese investment into Europe," 2015.
66. "AVIC International concludes the acquisition of Aritex to become market leader worldwide for automatic assembly systems," Airframer, 2016.
67. S. Scheuer, "Investment fears: Chinese take away another German firm, eventually," Handelsablatt, 2018.
68. Assocaited Press, "Continental Motors Embarks on $75 Million Alabama Project | Alabama News | US News," U.S. News, 2018.
69. S. Wentzler, "Thielert: AVIC International Announces the Founding of the Continental Motors Group and Expands Operations Into China | MarketScreener," MarketScreener, 2014.
70. "A Retrospective on China's Aircraft Cabin Industry—The China Aerospace Blog," The China aerospace blog, 2018.
71. "PARIS: Chinese-owned Gardner looks at aerostructures expansion," FLigth Global, 2017.
72. "Gardner Aerospace announces acquisition of Northern Aerospace—Aerospace Manufacturing Magazine," Aerospace Manufacturing, 2017.
73. W. Spitz, R. Golaszewski, F. Berardino, and J. Johnson, "Development Cycle Time Simulation for Civil Aircraft," Natl. Aeronaut. Sp. Adm., no. January, 2001.
74. V. Der Bruggen, "Dynamics of Implementation of Mitigating Measures to Reduce CO2 Emissions from Commercial Aviation," no. July 2010, pp. 1–115, 2010.
75. L. Dong, "China's Comac Expects Mass Production of C919 Plane After 2021 - Bloomberg," Live Now, 2018.
76. M. C. Kushan, F. H. Cakir, S. Gurgen, and M. A. Sofuoglu, "Technics of Composite Materials in A400M Fuselage," in International Conference on Natural Science and Engineering (ICNASE'16), 2016.
77. S. Gurgen, H. Cakir, M. Alper, and M. Sofuoglu, "Composite Structures of C919 .," in Proceedings of Conferencefora 8th International Conference, 2018.
78. "New models after the C919: 51% composite material use or not?," 2018.

79. "Why China C919 Wings Do Not Use Composite Materials Processing Capacity Needs to be Improved | FRP roofing sheet,flat panel making machine manufacturer," TITAN FRP.
80. M. D. Moses and E. Bruce, "China Closes in on Plane Order From Africa Amid Overseas Push - Bloomberg," Bloomberg, 2019.
81. airlinerwatch, "Ethiopian Airlines considers COMAC C919 instead of Boeing 737 MAX 8," airlinerwatch, 2019.
82. H. Panyue, "A sole international market for China's J-10 fighter jets is taking shape - China Military," China Top Stories, 2019.
83. Ho Kit Hen, "Hey! Come back...," FMT news, 2019.
84. B. Halder, "China, The Next Big Supplier of High-Tech Arms to Emerging Nations | Fast Forward | OZY," OZY, 2019.
85. Airbus, "Final assembly and tests."
86. "Airbus CFO sees 'surge capacity' of 63 A320 planes a month in 2019 - Reuters."
87. T. Hepher, "Airbus confirms plans to raise A320 output to 63 a month - Reuters," REuters, 2018.
88. Airbus Adds A320 Capacity Amid Ramp-up Discussions | MRO Network."
89. A. Zur, "Opening of 4th Airbus Final Assembly Line and Inauguration of extended Delivery Centre at Hamburg Finkenwerder—Aeronautics Online," Aeronautics, 2018.
90. Airbus, "A320 Family, The most successful aircraft family," vol. 2019, no. April, pp. 1–5, 2019.
91. "Airbus' China assembly facility marks 10 years of quality manufacturing for A320 Family jetliners."
92. "2019 Outlook: COMAC's slow but steady progress - Leeham News and Analysis," Leeham news and analysis, 2019.
93. B. PRENTICE, D. COSTANZA, and J. SMILEY, "When Growth Outpaces Capacity - a Labor Shortage and Out-of-Date Technology May Raise Mro Costs for an Expanding Global Fleet," pp. 1–14, 2017.
94. "Exploring the aviation skills shortage in Asia - AeroProfessional," Aero porfesionals, 2018.
95. Civil Aviation Administration of China, "Statistical Bulletin of Civil Aviation Industry Development in 2014," pp. 12–18, 2014.
96. P. Morgan, "Skills Shortages in the Chinese Labor Market: Executive Summary," pp. 1–22, 2016.
97. D. A. Irwin and N. Pavcnik, "Airbus versus Boeing revisited: International competition in the aircraft market," J. Int. Econ., vol. 64, no. 2, pp. 223–245, 2004.
98. R. W. Hess and H. Romanoff, "Aircraft airframe cost estimating relationships: all mission types." p. 146, 1987.
99. B. Perrett, "C919 May Be Largely Limited To Chinese Market," Aviation Week & Space Technology, 2013. [Online]. Available: https://aviationweek.com/awin/c919-may-be-largely-limited-chinese-market.
100. Y. Dawei and L. Dongmei, "COMAC C919 Airliner Project Flying in the Dark," China aviation daily, 2012. [Online]. Available: http://www.chinaaviationdaily.com/news/23/23020.html.
101. Perspectives for the Aeronautical Research in Europe (PARE), "'What if' study 2 - The NMA Case," PARE YR2, 2019.
102. P. Morrell and L. Dray, Environmental aspects of fleet turnover, retirement and life cycle. Omega, Cranfield University., 2009.
103. P. Consortium, PARE what if analysis: The MMA case. 2019.

Chapter 4
The Two Boeing 737 Max Accidents and Consequences for Aircraft Certification

Luiz Manuel Braga da Costa Campos

Contents

L. M. Braga da Costa Campos (✉)
Instituto Superior Técnico, University of Lisbon, Lisboa, Portugal
e-mail: luis.campos@tecnico.ulisboa.pt

L. M. Braga da Costa Campos (ed.), *Accelerating Sustainable Aviation Initiatives*,
Sustainable Aviation, https://doi.org/10.1007/978-3-031-73993-4_4

4.1 Introduction

The two similar successive accidents with the Boeing B737 Max [1–11], first Lion Air in Indonesia [12–21] and then Ethiopian [22–38], led to a worldwide grounding [39–48]. The Federal Aviation Administration (FAA) was one of the last airworthiness authorities to ground the B737Max, but then took a much more proactive stance [49–66] until ungrounding 20 months later, that spread gradually worldwide with longer delays in China [67–91]. The practices of Boeing [52, 92–102] were the subject of multiple often critical investigations [103–130]. The present account starts much earlier with the technical and marketing issues that led to design flaws in B737Max and proceeds to consider the consequences as regards certification of all future aircraft. The main "culprit" implicated in both accidents was the Maneuver Characteristics Augmentation System (MCAS).

The Maneuver Characteristics Augmentation System (MCAS) that Boeing designed specifically for the 737 Max has been implicated in both accidents. Although the 737 Max was certified by the Federal Aviation Administration (FAA) as a modification of the earlier two generations (original B737 and B737NG—New Generation), the MCAS was a new addition to the aircraft systems. Thus, three factors relating to the accidents (Sect. 4.2) are (Sect. 4.2.1) the motivations that lead to the introduction of the MCAS in the third-generation B737Max; (Sect. 4.2.2) the list of issues on questionable design features that could have safety implications; and (Sect. 4.2.3) the information publicly available about the two accidents and the role the MCAS may have played.

The near-term consequences (Sect. 4.3) can be viewed from three aspects: (Sect. 4.3.1) the grounding of the Boeing 737 Max by airworthiness authorities worldwide and the path toward ungrounding; (Sect. 4.3.2) the changes to the MCAS performed by Boeing and the extent to which they meet current safety concerns; (Sect. 4.3.3) the steps taken toward the resumption of normal service as ungrounding of the B737Max was decided by the FAA and followed on different time scales by the various aviation safety authorities around the world from almost immediately by EASA with conditions to longest delays in China.

Some near-term consequences may have longer-term effects (Sect. 4.4) at least in three areas: (Sect. 4.4.1) the loss of credibility of Boeing as an aircraft manufacturer and the FAA as a certification authority, which may affect other aviation stakeholders, even if they did not influence the accidents that rocked the public and the aviation community; (Sect. 4.4.2) the multiple contributors to financial losses by Boeing and the extent to which these are temporary and can be recovered at a later stage, when airline operations restart and new aircraft deliveries resume; (Sect. 4.4.3) the effect on the current and future Boeing 737Max order book in the context of the duopoly competition with the Airbus A320neo and derivatives, and the repeatedly debated case for a 737 replacement as a middle-of-the-market aircraft (MMA) or a direct future single aisle (FSA) replacement or a family of new aircraft.

The ultimate aim of all current activities, by Boeing as an original equipment manufacturer (OEM), the FAA, and other certification authorities worldwide and airline operators around the globe, was a safe return to services as soon as possible, ending the B737Max grounding after 20 months. The safety measures needed (Sect. 4.5) include (Sect. 4.5.1) reducing the probability of failure of the MCAS, making it redundant or taking other fail-safe measures; (Sect. 4.5.2) ensuring that the elevator runaway procedure is robust and sufficiently fast-acting without excessive control forces; and (Sect. 4.5.3) providing a pilot training package including failure scenarios and manageable control forces.

A tentative analysis is made of root causes and their consequences in future scenarios (Sect. 4.6) within the currently available information that includes limited results of a Boeing internal enquiry (Sect. 4.6.2). The root causes may be multiple like most aircraft accidents, combining questionable engineering and management choices driven by an unclear priority between safety, cost, and expediency (Sect. 4.6.1). While some corrective actions resolve some of the 13 issues (Sect. 4.3.2), others may require time and resources whose availability is less clear. The efforts to return the B737Max quickly to service were not helped by market factors and other unrelated Boeing activities that may reduce incomes and consume available resources (Sect. 4.6.3).

The predictions of Boeing (Sect. 4.7) that the grounding of the 737Max would last 3, then 6 and then 9 months, were dispelled when it became clear that recertification by the FAA would not occur in 2019 and would be delayed well into 2020 (Sect. 4.7.1). The emergence of the COVID-19 pandemic in the first half of 2020 created disruption in the airline market with far greater impact than the two Boeing B737Max accidents. Part of the delay may be due to the reluctant and slow response of Boeing to provide a simulator training package that would satisfy an international team of pilots assembled by consensus of the airworthiness authorities worldwide (Sect. 4.7.2). The Boeing approach of piecemeal palliative actions dragging over 1 year may, in the end, prove more costly in compensation and other losses than a fundamental redesign that would restore its credibility (Sect. 4.7.3).

The change of leadership at Boeing, from Dennis Muilenburg to David Calhoun at the beginning of 2020 (Sect. 4.8), marked a policy change from a focus on

financial aspects like (Sect. 4.8.1) using profits to buy back shares and pay sizeable dividends to (Sect. 4.8.2) investment in engineering efforts to improve safety based on (Sect. 4.8.3) closer cooperation with airworthiness authorities as regards the provision of technical information and pilot training materials. The replacement as CEO of David Calhoun by Kelly Ortberg was again motivated by failures in the B737Max program, this time related to persistent production quality separation problems, leading to separation in flight of a cabin door of an Alaska Airlines aircraft, fortunately without casualties.

The two B737Max accidents led the FAA to undertake a thorough review of the whole flight control system, uncovering further issues (Sect. 4.9) not directly related to the MCAS, including (Sect. 4.9.1) rewiring the electrical connections to the tailplane to provide fail-safe operation in the case of short circuits or other electric disruptions; (Sect. 4.9.2) amending the list of minimum operating equipment required to start a flight following an inconsistency found by an operator. The recertification of the B737Max has motivated a more strict and thorough approach that will also apply to all future aircraft, in particular with closer scrutiny of human factors and man–machine interface, and institutional safety management systems that are critical aspects of flight safety (Sect. 4.9.3).

The main modifications made by Boeing to the B737Max to obtain (Sect. 4.10) recertification and return to service are (Sect. 4.10.1) redesign of the flight control system, including software changes to make the two flight control computers fully redundant; (Sect. 4.10.2) a comprehensive simulator flight training package containing new information not only on the B737Max but also on its predecessor B737NG, to train pilots not only on normal flight modes but also putting particular emphasis on all possible non-normal flight conditions. The changes did not include (Sect. 4.10.3) major hardware changes such as adding angle-of-attack sensors to provide fail-sate operations or servo assistance to reduce large manual control forces with electric pitch trim disengaged.

About 1 year into the grounding of the B737Max, the sudden emergence of the COVID-19 pandemic caused a much greater disruption of the airliner market, and the effects on the Airbus–Boeing duopoly (Sect. 4.11) can be seen in chronological order separated by (i) the B737Max grounding and (ii) the COVID-19 pandemic as the two major disruptive events: (Sect. 4.11.1) the competition in the single- and twin-aisle sector before the grounding of the B737Max; (Sect. 4.11.2) the period before the COVID-19 pandemic including the successful takeover of the Bombardier C-series by Airbus and the breakdown of the rival Boeing–Embraer alliance; (Sect. 4.11.3) the sharp recovery from the COVID-19 pandemic providing both Boeing and Airbus with healthy order books with relative further gains of Airbus relative to Boeing.

The conclusion (Sect. 4.12) highlights the consequences of the Boeing "expedient" of using "grandfather rights" to make the old B737 revamped as B737NG a temporary competitor to the all-new A320: (i) when forced to develop the third-generation B737Max against the second-generation A320neo, the latter gained a significant competitive advantage at the upper end of the single-aisle market with stretches beyond the potential of the former; and (ii) to regain a more competitive

position following the distraction of the middle-of-the-market-aircraft (MMA) and the loss of credibility of the B737Max, Boeing has developed new versions of the B737Max that have a loyal and significant customer base, waiting for deliveries to start, when certification delays are overcome. Concerning a future single aisle (FSA), it requires technological advances and resources for development that may not be available soon. A more likely time scale is the next decade, when advances in propulsion, aerodynamics, control, structures, and other areas will be sufficiently mature to provide gains in efficiency and environmental friendliness that justify replacing the current fleets.

4.2 Main Contributing Factors

The Boeing 737 Max was certified (Sect. 4.2.1) by the FAA for pilots without any transition simulator training from the preceding generation 737NG, requiring an MCAS to mask its different flying characteristics. The design of the MCAS incurred a surprising number of basic flaws (Sect. 4.2.2) despite Boeing's experience and competence in the area. Based on the publicly available information on the two accidents (Sect. 4.2.3) that has gradually emerged over time, the MCAS is widely suspected to be implicated in both cases.

4.2.1 The Need for an MCAS in the B737Max

The first-generation Boeing 737 already had automatic pitch trim that was carried over in the next-generation 737NG and could be disabled by pilot action on the control column, reverting to manual control.

The more recent Airbus A320 has comparable systems, and the two competitors shared almost evenly a duopoly of the single-aisle (SA) airliner market, with perhaps a slight lead of the all-new more advanced first-generation challenger over the second generation of the incumbent benefiting from grandfather rights that required it to meet only past rather than current certification standards. The stable "status quo" was disturbed by Bombardier developing the C-series as an all-new aircraft in the 100–130-seat class competing in the low end of the A320/B737 market with efficiency gains mainly due to new engines.

Bombardier, like its rival Embraer, is about one-tenth the size of Airbus and Boeing, and brave or foolish enough to challenge them, although claiming that this was not the case. Embraer, perhaps more wisely, chose not to challenge head-on the "giants" and remained within the upper reaches of the regional aircraft market with the E-series. Boeing was aware of the limited development potential left in the B737 and did not wish to try a compromised third-generation design and steadily worked toward an all-new clean sheet replacement toward the end of the decade; Boeing expected airlines to wait for its superior future product and saw no need to counter

Bombardier scratching the bottom of its SA market. Airbus took the opposite position, realizing that adding to the development potential of the A320 the engine technology in the Bombardier C-series, it could produce in a short time and with modest budget as an A320neo that would prevent the newcomer from gaining its intended market share.

The Bombardier C-series was not helped by development delays and cost overruns that might have led to bankruptcy had the Canadian government not intervened. In the meantime, Airbus's determination to develop the A320neo without a launch customer was amply rewarded: the A320neo was a runaway success beyond the Airbus's wildest expectations with more than 1000 orders in a few months after the launch, validating the Airbus's faith in the concept: airlines could not ignore a 15% improvement in fuel consumption at a time of high fuel costs. The message was brought home to Boeing when American Airlines, an all-Boeing operator, made a record order for the A320neo. Boeing threatened to sue American Airlines for breach of its long-term soon-to-expire exclusive supply agreement but was told to develop a re-engined B737 if it wanted a share of the order. In a few days, Boeing decided it had no choice to keep a valued customer other than to shelve its continuing work on an all-new B737 replacement in favor of the hasty development of a third-generation 737 Max compromised from the beginning by its origins half a century earlier.

The B737Max had little in common with the original B737 that first flew in 1967. A cabin with almost the double seating capacity, a much larger wing, more powerful engines, etc. Yet it was certified as modified aircraft. Boeing wanted pilots certified for the second-generation B737NG to fly the third generation, B737Max, without simulator training for the transition, and the FAA required that the flying characteristics should be the same.

Perhaps the biggest handicap inherited from the original half-a-century-old B737 design was a low wing designed to fit below a first-generation turbofan with a low bypass ratio in a small diameter nacelle. The second-generation B737NG design was challenging in fitting a turbofan with higher bypass ratio: the engine nacelle was no longer circular but had a flattened bottom close to the ground and was also closer to the wing with a shortened pylon, resulting in the formation of shock waves in cruise flight. The resolution of aerodynamic and structural compromises did not prevent the commercial success of the B737NG, but Boeing became acutely aware that development potential was near its limits, and accordingly planned the methodical development of an all-new replacement.

The requirement to develop the third generation, B737Max, with high bypass ratio turbofans posed almost impossible challenges on how to fit the large engine nacelle under the wing. The nose landing gear was extended to increase ground clearance but less could be made about the main landing gear retracting into the wing, short of a major structural redesign. The engine nacelle was pushed up and forward creating a strong pitch-up at high angles of attack and low speed. Despite all this, the critical fan diameter in the B737Max could not match the A320neo, giving the latter a fuel-efficiency advantage that could not be clawed back by aerodynamic or structural measures. Ultimately competitiveness was ensured by reduced

seat pitch and higher passenger capacity in a B737Max cabin that is smaller than that of the A320neo; this is possible by the B737 "grandfather rights," which means that it does not meet the more recent and stringent emergency evacuation standards that apply to the A320.

A problem remained: the 737 Max was required by the FAA to fly like a 737NG but had a more severe pitch-up at low speed and high angle of attack (AoA); the MCAS was developed by Boeing to automatically compensate this pitch-up and make the B737Max respond like B737NG. The execution of the MCAS should fulfill that aim.

4.2.2 Thirteen Issues: Questionable Design Features

The design of the MCAS involved at least 13 issues on questionable design features that could hardly be expected of a manufacturer with the competence and experience of Boeing, and should not escape scrutiny in a certification process with the high standards expected of the FAA:

Issue 1—single-sensor failure: The MCAS was triggered into action by a signal from a single AoA sensor exceeding a threshold indicating a pitch-up. Although the B737Max has two AoA sensors (the Airbus aircraft have three), Boeing chose to use only one per flight, alternatively from the left and right sides. Thus, a single AoA sensor failure signaling a false high AoA, without any warning, could activate the MCAS.

Issue 2—safety as a cost option: A cost option was available to use both AoA in all flights. In case of major disagreement, an indication in the cockpit would serve as a warning of sensor failure, with no possibility to identify which might be the correct or incorrect sensor. It is not clear why this valuable safety feature was made a cost option. This may have contributed to most airlines not taking the option since they could not see its usefulness because the existence of the MCAS was not disclosed (see issue 4 below).

Issue 3—runaway elevator pitch-down: Once an excessive AoA signal triggered the MCAS, it would automatically apply a preset pitch-down, at regular intervals, without limit, until the AoA fell below the specified threshold. In the event of a failed AoA sensor, locked or frozen at an excessive high value, the MCAS would apply a runaway pitch-down trim beyond the control power of pilots, who did not even know the MCAS existed.

Issue 4—the omission of reference to MCAS in the flight manual: Boeing decided that since the MCAS was "transparent" to the pilots, they did not need to know about its existence because the 737 Max would fly like a B737NG if the system operated properly. It was later reported that the FAA representation/derogation of authority at Boeing agreed that the MCAS was mentioned explicitly only in internal documents, with a vaguer reference to pitch trim in external documentation. The argument of "transparent" borders on contradiction because:

- The MCAS is vulnerable to runaway (item 3) due to a single-sensor failure (item 1).
- In such a case, the pilots could not know the cause of the runaway because of the existence of the MCAS was not disclosed.
- The pilots find a B737Max pitch runaway when they expect B737NG-like handling and were not trained for the difference.

Issue 5—assumption of fast pilot reaction: Boeing assumed that, in case of MCAS failure, the pilots would within 4 s implement the elevator runaway procedure. This assumption could be optimistic on several grounds:

- Since the MCAS was unknown to the pilots and it was not mentioned in the flight manual, there was no instruction on how to identify an MCAS failure or to counter it.
- Despite this, the crew was supposed to realize within 4 s that there was a pitch anomaly from an unknown cause and decide within that time the right response was to apply the elevator runaway procedure.
- If the MCAS failure occurred in a critical flight condition involving other factors, leading to a higher workload or stress, the reaction time could be longer.

The pilots should know the checklist for elevator runway checklist by memory, but there were omissions in that checklist.

All this looks more like "lack of transparency." The pilots had to identify a pitch runaway without knowing about its cause.

Issue 6—identification of a pitch runaway: The flight manual identified pitch runaway as "a continuous uncommanded movement of the trim wheel." However, this is not the effect of activation of the MCAS that acts by short bursts of trim wheel rotation with intervals of several seconds. These short interventions of the MCAS can be easily confused with frequent movements of the trim wheel during flight phases such as climb after take-off. The pilots could not be expected to detect and act upon in 4 s on sharp trim wheel movements at longer time intervals due to a system whose existence they did not know of, not acting like the "continuous trim wheel movement" expected as a runaway pitch trim. As part of the B737Max recertification process, the flight manual was amended to define runaway pitch trim as "a continuous trim wheel movement or movements inconsistent with the flight condition." Despite poor clues on how to identify a pitch runaway condition, the pilots were supposed to apply within 4 s the corresponding non-normal checklist that they should know by memory.

Issue 7—omissions in the checklist: In the original B737 and B737NG, the automatic pitch trim could be disengaged by the pilots moving the control column beyond a point leading to manual control. The MCAS in the B737Max was designed to override this feature. It could be disengaged only if the pilots used some toggle switches together with the movement of the control column. This was mentioned in the pitch runaway checklist, although some airlines informed erroneously pilots that control column movement was sufficient to regain manual pitch control. This was not Boeing's fault, but omissions in the pitch runaway checklist might be. When

Boeing demonstrated in a simulator to airline pilots a redesigned MCAS, they noticed that to recover from an MCAS-induced pitch-down, the control forces were so large that both pilots should pull their sticks together. Furthermore, to reduce the control forces, the initial stick movement should be the opposite.

These can be serious omissions in a critical situation requiring large control forces. Boeing always maintained that the way out of an MCAS failure was to apply the elevator runaway procedure and that the checklist should be known by memory by pilots. However, the checklist had the two omissions mentioned above.

Issue 8—effectiveness of the electric pitch trim: The pilots could counter the pitch-down imposed by the MCAS by commanding an opposing pitch-up using the electric pitch trim. However, as long as the angle of attack exceeded the preset limit, the MCAS would apply a pitch-down at regular intervals, and the electric pitch trim might not be able to compensate.

Issue 9—heavy pitch trim wheel: If the MCAS was disengaged using the toggle switches, then the electric pitch trim was deactivated and pitch trim had to be performed manually. The stick forces were high and the pitch trim wheel could not be moved with one hand on the stick. Moving the trim wheel required two hands, and the effort increased with angle of attack and speed until it became almost impossible to move, as stated in the intermediate report of the second accident.

Issue 10—roller-coaster maneuver: In the B737NG, the pilots could reduce the effort to turn the trim wheel using a roller-coaster maneuver: pitching down to start a dive, then pitching up to use the transition to turn the trim wheel more easily. The roller-coaster maneuver could be repeated until the aircraft was fully trimmed; it is clear that the roller-coaster maneuver implies an altitude loss and cannot be used near the ground.

After the two B737Max accidents, in the revised flight manual for recertification, Boeing stopped short of recommending the roller-coaster maneuver. A possible reason is that the B737Max has more powerful engines further forward than the B737NG, generating a stronger pitching moment, making the roller-coaster maneuvers more numerous and hazardous.

Issue 11—an increase of forces with speed: In the revised flight manual for recertification of the B737Max, Boeing stops short of recommending the roller-coaster maneuver, and instead suggests that the two pilots turn the trim wheel together. This was found to be possible in in-flight tests, although Boeing recommends reducing speeds. This raises two issues. If both pilots are needed to turn the trim wheel, presumably with one hand each, then each of them has only one other hand free for other actions; assuming that this is sufficient in all flight conditions, making strong efforts with one hand in turning trim wheel and light and precise actions with the other hand elsewhere might not be easy; second, reducing the speed to reduce the effort to pitch-up with the steering column or to turn the trim wheel may not be feasible if the aircraft enters a dive, which is precisely what the MCAS does after erroneous high AoA signals. The report on the second B737Max accident states that the controls became unmovable after the aircraft entered a dive when the MCAS was reactivated by the pilots, after failing to reverse the dive with manual control.

Issue 12—slow pitch response: After the two accidents, Boeing made software changes to the MCAS and resubmitted to FAA certification. The FAA pilots tested on the simulator the pitch trim runaway procedure that was presented by Boeing as the escape route from MCAS malfunction. They found in certain cases very slow response times of the order of seconds. This was traced down to some computer chips requiring hardware changes. The Boeing claim that these control issues are unrelated to the MCAS is to some extent questionable since the elevator runaway procedure was presented as the escape route to MCAS failure. Boeing agreed with the FAA that control system redesign is essential since such slow responses are not acceptable in an emergency. The requirement for hardware as well as software changes to the pitch control system introduced another step and delay in the recertification process.

Issue 13—standard safety assessment (SSA): The MCAS was designed for use at high speed and subjected to the comprehensive safety assessment covering all combinations of failure modes. Later, Boeing decided to use the MCAS at low speed and did not repeat the SSA, considering that the high-speed case was more critical. Also, Boeing may not have fully informed the FAA representative about the changes to the MCAS or its use in a wider speed envelope.

4.2.3 Sequence of Occurrences in the Accidents

Since the information on the two B737 has been made available gradually over time and is not always completely clear about surrounding circumstances, any judgment of contributing factors must be made with caution. Enough is known to implicate the MCAS in both accidents, although there are different views on how responsibility should be assigned.

The B737Max of Lion Air involved in the first accident already showed dubious readings of the AoA while on the ground before take-off, leading to speculation about some maintenance issue or other damage. After the take-off and about 15 min into the flight, the aircraft started an uncommanded dive into the ground. The crew may not have identified a runaway elevator condition and did not apply the corresponding checklist. They failed to manually reverse the pitch-down, which may have exceeded the control power available to the pilots.

Boeing blames the crew for not applying the runaway elevator checklist they should know by heart. This assumes that the crew:

- Suspects a faulty AoA sensor, although it has no warning of the situation.
- Identifies an elevator runaway condition without knowing about the existence of an MCAS that was implicated.
- Recognizes that manual control cannot cope with the runaway inputs of an unseen system.

Maybe this is to expect more from the pilots who fly the aircraft than from the engineers who designed the MCAS and the authorities who certified or derogated it.

The accident report lists multiple causes. Some relate to poor Boeing design of the MCAS and inadequate certification procedures of the FAA. Also, the AoA sensors of that aircraft had a history of failures that the crew did not know about; the AoA might have a wrong calibration, and the workshop that did that work was closed. Finally, although the captain was an experienced pilot, the co-pilot was not, showing limited skills in earlier training. Perhaps aggravated by the stress of the situation with fault warnings, the communication was poor, and the co-pilot may not have fully understood when the captain handed over control to him.

Boeing was aware that the MCAS could be implicated in the Lion Air accident and:

- Issued a bulletin that did not explicitly mention the MCAS but reminded of the procedures following a suspected elevator runaway condition and advised that the bulletin be appended to the flight manual.
- The elevator runaway checklist had the omissions concerning control forces (issues 9–11 in Sect. 4.3.2) that were identified later in demonstrations to airline pilots of the redesigned MCAS.
- The FAA reviewed with the European Union Aviation Safety Agency (EASA), as part of their cooperation, all past B737Max incidents, and found no relation with the Lion Air accident, so made no flight ban, although it is not clear if the MCAS was considered a possible contributor or if EASA knew or had been informed of its existence.
- Also following the Lion Air accidents, Boeing started an internal effort to redesign the MCAS to issue a modification within 2 months, but that date slipped as more changes were considered, possibly not only to MCAS but also to other systems.

Within 6 months of the Lion Air accident, a similar occurrence led to the crash of an Ethiopian Airlines B737Max about 5 min after take-off. The similarities were obvious: an uncommanded steep dive into the ground that the pilots could not counter with repeated pitch-ups. This time:

- The AoA only malfunctioned after take-off, leading to speculation about a bird strike or similar event, which cannot be confirmed.
- The pilots did recognize an elevator runaway condition and did follow the checklist using both the control column and toggle switches to regain manual control.
- Although the pilots did follow the elevator runaway checklist, they were unable to manually trim the aircraft and stop the dive: The intermediate accident report made clear that the effort needed to reverse the dive was too large and increased as the aircraft accelerated beyond the never exceed speed in a dive.
- It is not clear if delays in pitch control (issue 9 of Sect. 4.3.2) played a role since control forces (issues 9–11) were in any case beyond manual control.
- Perhaps in a last desperate attempt to reverse the dive, and having failed to do so manually, the pilots used again the toggle switches to engage automatic pitch trim, thereby activating the MCAS that only made matters worse.

The excuse of Boeing that the pilots "did not apply the checklist consistently" borders on the cynical. The pilots (i) did recognize the elevator runaway condition without information about the AoA sensor malfunction and knowledge of the existence of the MCAS and its effects; (ii) fully applied the runaway elevator checklist as it was known to them; (iii) could not have guessed the omissions in the checklist about control forces and directions; and (iv) it is not known if the slow response to the pitch trim runaway condition found in subsequent FAA tests (issue 12 of Sect. 4.2.2) occurred in this instance. But clearly, the control forces were too high for manual pitch trim. It appears that the crew did all it could have done but was betrayed by incomplete information and a hidden MCAS system. In both accidents, the crew repeatedly tried to counter the dive with mandated pitch-ups but failed to overcome the pitch-downs commanded by the MCAS that ultimately prevailed and caused a dive into the ground.

4.3 Near-Term Consequences

The almost immediate consequence of the second B737Max accident was a ban on operations, which the FAA was the last to implement on orders from the US president (Sect. 4.3.1). Boeing was already developing several software changes to the MCAS, which it claimed would safeguard against further accidents (Sect. 4.3.2). The return to normal service took place when the aviation safety authorities around the world lifted their bans (Sect. 4.3.3), starting with the FAA after 20 months of grounding.

4.3.1 Ban of Operations from National Authorities

The first to react to the similarities of the two B737Max accidents were the Chinese airworthiness authorities: they questioned the FAA as the original certification authority, were not satisfied with the answers, and within 24 h banned all B737Max flight operations. This swift action might have been seen as another episode in the US–China trade war if Singapore and Australia had not placed similar bans soon after. The ban on B737Max operations quickly spread worldwide, including Canada and starting in Europe with Ireland that hosts low-cost airlines that are major B737Max operators. The ban was quickly followed by France and Germany, and then EASA. Given the efforts at cross-certification between EASA and the FAA, and their most successful cooperation, EASA may have been less inclined to question the FAA certification.

The FAA was the last airworthiness authority in the world to ban the B737Max operations, claiming against all others that this was an unjustified action based on

insufficient evidence. It was President Donald Trump who ordered the FAA to ban B737Max operations. His motivation was that there had been no major jet airline accidents in the United States (as in most Western countries) for some years, and he wanted to avoid a new one. When finally banning B737Max operations, the FAA argued it had received refined satellite data that correlated the two accidents: similar altitude excursions resulting from MCAS-enforced pitch-down opposed by crew-commanded pitch-ups that ultimately could not cope, resulting in a terminal dive. A national ban on aircraft operations has two elements: (i) it forbids national airlines from flying the airplane and (ii) it does not allow the airplane of any foreign airline to enter national airspace.

The B737Max accidents damaged the credibility of the FAA almost as much as that of Boeing. If Boeing designed a flawed MCAS, how come the FAA did certify it? Boeing started the development of the MCAS to satisfy the FAA that pilots could transition from the second-generation B737NG to the third-generation B737Max without any specific simulator training. And Boeing succeeded in making the B737Max handle undistinguishably from the B737NG in pilot tests in the simulator, as long as the MCAS worked as intended. However, the questions about the unintended operation are unavoidable: (i) Why a safety system was left vulnerable to a single sensor failure without warning? (ii) How come an automatic system can react to erroneous signals with runaway repeated commands beyond pilot control? Bearing in mind that the MCAS was not fail-safe, why were pilots not informed of its existence and trained at least in the simulator to identify and counter its failures? Boeing did not hide the existence of the system, at least in presentations to airlines about the differences between the B737NG and B737Max and in the maintenance documentation of the B737Max. But no mention of the MCAS was made in the flight manuals, and most pilots were not aware that the real flight characteristics of the B737Max at high angle of attack were different from those of the 737NG.

These questions apply equally to Boeing, which developed the MCAS, and to the FAA, which certified it. Having allowed pilot transition from the second-generation 737NG to the third-generation 737 Max without simulator training, did the FAA not check the development of the MCAS by Boeing in detail? Did it not notice the single-point sensor failure, the runaway trim, or the lack of documentation in the flight manual? Did the FAA explicitly approve all this, or did it rely on a delegation of the supervisory authority to the manufacturer? Given all these questions, several airworthiness authorities, including EASA, Canada, and Australia, stated that they would no longer accept an eventual FAA recertification of the B737Max and would conduct their independent investigations until they were convinced of the safety of the MCAS and perhaps other systems. The FAA decided to hold regular meetings with other airworthiness authorities on the Boeing 737 Max ban, reported being sometimes hot with the FAA trying to restore its credibility in the face of skepticism of others, in the beginning, leading to smoother cooperation afterward.

4.3.2 Boeing Modifications to the MCAS

Boeing started modifications to the MCAS in the period between the two 737 Max accidents, and since the second accident and the worldwide grounding of the fleet this became its highest priority, to submit the recertification file to the FAA and unground the aircraft. Under intense pressure over its apparent laxity in the certification of the original MCAS, and with the scrutiny of other airworthiness authorities around the world, this time the FAA would not delegate its certification authority and went into greater scrutiny of the details of the changes. Boeing delivered the list of software changes hoping for approval in a few months, but the FAA always maintained it had no timeline for acceptance and would focus only on safety. The discovery of the situation of slow pitch control response (issue 11 of Sect. 4.3.2) led the FAA to mandate and Boeing to agree to hardware changes delaying certification by several months till an uncertain date.

The Boeing first redesign of the MCAS consisted only of software changes and was claimed by Boeing to provide three layers of protection against similar accidents occurring:

1. First, the data from both AoA sensors are used in every flight, and if a significant discrepancy is detected, the MCAS is disconnected, and the pilots warned.
2. Second, the MCAS can make only one pitch-down intervention unless the pilots require or authorize further actions.
3. Third, the pilots always have at least 1.2 g of pitch-up control authority.

These changes correspond to how the MCAS should have been designed in the first place and are certainly an improvement but may not answer all questions.

The redesigned MCAS remains not fail-safe since a single-sensor failure deactivates the system. Errors of AoA sensors, due to freezing in icing conditions, or mechanical and electrical failures, are neither common nor extremely rare and have been involved in other accidents. Is it worthwhile to design a safety system that is disabled after a single-sensor failure? Airbus uses three AoA sensors to be able to vote out one dissident sensor, which Boeing cannot do with two AoA sensors. Even three AoA sensors are not immune from safety issues with recorded instances of two consistent frozen sensors outvoting the correct one. But if the A320 that does not have the B737Max pitch-up issue uses three AoA sensors, should the B737Max use four AoA in a fail-safe quadruplex system? Boeing appeared to be reluctant to implement any hardware changes that would require a costly development, bring new certification requirements, and delay ungrounding and resumption of operations.

However, Boeing had to make other hardware changes after the software changes it submitted proved insufficient. This time the issue was not the MCAS itself, but the pitch trim runaway procedure repeatedly put forward by Boeing as the answer to MCAS failures. The FAA pilots tested the procedure in the simulator and found rare cases of the slow control response of the order of seconds. This may or may not be a new problem since in the original certification some safety situations were found that could only be resolved with pilot input. It was considered at the time that

such situations were sufficiently rare not to require specific corrective action. Much tighter scrutiny has come into play since the two B737Max accidents, and this time the slow pitch control response was traced down to a computer chip. Boeing agreed with the FAA that pitch control response times of the order of seconds are not acceptable in emergencies and had to make hardware changes to the computer system. The need for hardware changes found subsequently may also lead to more stringent requirements for recertification.

4.3.3 Return to Service and Deliveries

More worrying was the Boeing's initial position that following the MCAS changes "computer training" or touchpad experience is sufficient and no simulator training of pilots is required, which Boeing will provide only at airlines' request perhaps at extra cost, as a new business line; along similar lines, Boeing was expecting the FAA to unground the B737Max without including pilot simulator training in the minimum requirements and limiting itself to software and documentation changes. This rather light-hearted position called into question again the Boeing commitment to safety, and if accepted by the FAA could further damage its credibility.

Pilots resisted the notion that they need no training against the failure of a safety system disabled by one sensor malfunction and leading to a pitch-up at high AoA for a real aircraft (B737Max) far removed from the flying characteristics they were trained for (B737NG). Airlines took into consideration the concerns of their pilots as guardians of passenger safety and their own. And several airworthiness authorities around the world clearly stated they require pilot simulator training as part of the recertification process. Boeing and the FAA surely realized that not including pilot simulator training in the minimum requirements could reopen the credibility gap they are trying to close. Convergence of the FAA and other airworthiness authorities worldwide on the conditions for ungrounding the B737Max and allowing the return to service led to the requirement to submit and demonstrate a pilot training package besides system software and hardware changes.

Boeing set aside 100 million dollars as compensation for the deaths of 364 passengers in the two B737Max accidents. Had just one of those accidents been in the United States, the compensation bill could have been much higher in billions. Had both accidents occurred in the United States with its tradition of litigation, criminal charges might have been leveled at Boeing and the FAA. Although safety is always the stated first priority, Boeing was also driven by the aim to unground the B737Max as soon as possible and the FAA tried to shake away the impression of lack of independence in its certification authority. As the International Air Transport Association (IATA) director general noted as his main concern about aviation safety: How can it be explained that the same airplane is safe to fly in one part of the world and not in another? If Boeing convinces the FAA to set minimum recertification requirements that other airworthiness authorities around the world would not accept, it will further damage the reputation of both and provide a sad image of certification

spreading geographically at a pace determined by national criteria. Also, the situation where (i) some airlines can and others cannot fly the same aircraft, and (ii) the same aircraft can fly in some, but not other parts of the world, may leave passengers wondering whether they should be flying at all in that particular aircraft.

Even if all airworthiness authorities around the world agree on a common recertification standard and the same date for lifting the current ban, it does not imply that the whole 737 Max fleet will come back into operation at the same pace. The 385 B737 Max already delivered before the ban have been replaced by other aircraft, mostly leases, and will require inspection and software and hardware changes before returning to flight. It is no longer only software changes requiring a few hours as Boeing claimed at one time.

The B737Max production at a rate of 52 per month was temporarily halted after the ban of operations and then reduced to 42 per month. Boeing then started parking the aircraft it produced and could not deliver using a special FAA permission to ferry the aircraft away from the production site to another airfield with more available space for parking. The parked fleet totaling 470 B737Max before production was halted will require flight testing before handing over to airlines. All aircraft will require software upgrades to the MCAS, which Boeing said can be performed in a few hours before the need for hardware changes was found. These changes will apply to (i) the 385 aircraft delivered to airlines and grounded, and (ii) the 470 aircraft produced and parked but not delivered. The ungrounding of more 855 aircraft would take place at a rate of 70 per month over at least 8 months according to the initial Boeing plan. However, return to service took much longer as aircraft are delivered to an airline distinct from the initial order, requiring modifications. The recertification and modification of existing aircraft in parallel with the production of new aircraft lead to delays because work capacity is exceeded. Also, it may take longer for airlines to reintroduce the 737 Max to the fleet depending on how many aircraft are involved relative to the size of the airline fleet and the time of the seasonal travel market when the activity is intense. For all the reasons stated, clearing the stock of all grounded and undelivered aircraft was not completed several years after ungrounding.

4.4 Long-Term Effects

The financial losses due to the grounding of the 737 Max (Sect. 4.4.1) have mostly short-term effects while the large order book may be safe in the next few years (Sect. 4.4.2). Resolving the 737 Max issues as quickly as possible and returning the aircraft to operation was essential not only to stem short-term losses: Boeing might wish to regain credibility and evolve away from the damaged 737 Max reputation toward all-new aircraft in the short-haul sector with a new future SA, and perhaps the MMA as well (Sect. 4.4.3).

4.4.1 Financial Losses and Recovery

Boeing has roughly a 100 billion yearly turnover, of which 60 billion comes from airliner sales, generating a 9 billion profit. Its strong financial position is backed by a backlog of orders covering more than 5 years of production worth hundreds of billions. In the event, its internal financial resources prove insufficient, there should be no problem in obtaining credit. While Boeing may be able to absorb the losses associated with the grounding of the B737Max, they can certainly affect profits and turn them into losses due to multiple causes.

First, the airlines owning the 385 B737Max already delivered and then grounded claimed compensation from Boeing: for lost flights and passenger revenue, for leasing replacement aircraft and the disruption to their flight plans and schedules. Second, Boeing continuing producing the B737Max at a monthly rate reduced from 52 to 42 for the first months of grounding, and halted production only when it's the expectations of early ungrounding faded. The reduction of production rate was initially a welcome relief to an overstretched supply chain but halting of production had unwelcome effects on suppliers that Boeing tried to support. However, at a nominal value of 100 million each (without discounts), 42 aircraft a month represent a missing revenue of 4.2 billion; Boeing has permission to fly away aircraft to parking areas and is paying fines for late delivery. The missing revenue will be recovered when the ban is lifted and the aircraft are delivered to airlines, but not the fines. In the meantime, Boeing allocated with the highest priority internal resources to solve the B737Max issues and stem as soon as possible the losses.

Boeing quoted 4.9 billion over one quarter associated with B737Max costs, resulting in an unprecedented loss of 2.9 billion over that quarter, wiping out two billion of profit from other programs. This was the biggest quarterly loss ever at Boeing. Boeing had to halt production as the grounding dragged into 2020 depending on the outcome of hardware changes. The software changes were submitted within 3 months with the expectation of recertification within a similar period. The discovery of additional control issues requiring hardware changes sets the clock back to the starting point, all depending on how long Boeing took to perform the hardware changes and how long the FAA took to validate them hopefully with a consensus of other airworthiness authorities worldwide. The FAA had repeatedly stated it has no timeline for ungrounding the B737Max, and the emergence of new hardware issues vindicated that position. Boeing had publicly stated that, after submitting the second set of changes, the aircraft could be certificated before the end of 2019, although ungrounding by the FAA was achieved only in November 2020.

On the other hand, Boeing intended to raise production to the 57 per month it planned before the ban as soon as the B737Max was ungrounded. In order to quickly resume production, Boeing would have to mind not only its airline customers but also its industrial supply chain. The main airframe supplier, Spirit AeroSystems, is sheltered from the B737Max troubles by a favorable contract previously signed with Boeing guaranteeing a monthly production rate of 52. Others in the supply chain may be dependent on the B737Max production and not have the resources to

survive a long interruption without Boeing support. The target of delivering 57 B737Max per month after ungrounding was dependent not only on Boeing's ability to do so but also on customer airlines agreeing on a suitably fast transition from the current "substitution" schemes. For most airlines, the sooner they receive new more efficient aircraft the better as long as feasible for transition plans.

The COVID crisis with a reduction of orders and supply chain capacity (Chap. 6) implied that both Boeing and Airbus struggled to increase production rates to meet the post-COVID surge in demand, requiring a ramp-up period of at least 2 years. The post-covid supply chain limitations affected production ramp-up both to 78 per month for the A320 at Airbus and to 52 per month for the B737 at Boeing In fact the production quality issues found post- MCAS and post-covid at Boeing and its chief airframe supplier Spirit AeroSystems, lead to a Boeing repurchase of its former subsidiary, amid the inability to reach a production limit of 38 per month imposed by the FAA.

4.4.2 Current and Future Order Book

Airbus has a backlog of 6500 A320 orders corresponding to more than 8 years of production at a rate of 60 aircraft per month. Boeing has a backlog of 4500 B737 orders corresponding to more than 7 years of production at a rate of 52 per month. Both Airbus and Boeing were struggling to increase their production rates from overstretched supply chains barely able to provide the required number of engines and other major and minor pieces of equipment; in some cases, the acquisition of suppliers was the way to increase production. The B737Max ban does not change the fact that switching orders with the A320 is not a viable proposition in either direction. Airlines cannot afford to wait 5 years or more for a new more efficient and more environmentally friendly aircraft and keep operating current older models and must stick to their precious earlier production slots.

In this situation where demand for airliners far exceeds supply, and waiting times exceed 5 years, the current order books of Boeing and Airbus are safe, and the B737Max troubles are unlikely to lead to large-scale cancellations. There have been some rumors of airlines talking to Airbus about shifting the order from the B737Max to the A320: Airbus does not confirm this, merely stating that it is in continuous contact with airlines. Airbus has been carefully silent about the B737Max issues, being conscious that inflating safety concerns will harm the whole aeronautical industry. Instead, Airbus has been trying to expand its lead at the upper end of the SA market, with the A321LR and A330neo squeezing the potential MMA market from above and below. The main problem at Boeing is simply catching up on delayed B737Max deliveries due to production quality issues preventing an increase in delivery rate.

While airlines are unlikely to shed their precious B737Max production slots, the situation may change with regard to new orders, for which the long waiting time will be comparable for the B737Max and A320neo. The negative publicity resulting

from the two B737Max accidents may explain why it did not receive new orders in the first few months since the grounding while Airbus sold more than 100 A320neo over the same period, Airbus planned to highlight its strength by announcing its new A321XLR stretch at the Paris air show to collect 150 new orders. Contrary to all expectations, it was Boeing that stole the show with a larger order for 200 B737Max from a single customer: a leasing company. Boeing was relieved that the order "could not have come at a better time." The customer expressed its "confidence in the B737Max" and later gave as a motivation a concern of "excessive dependence on the A320neo." The financial details were not disclosed, as is the usual practice, and there was no shortage of gossip on how big a discount the leasing company as a customer had obtained out of Boeing's need to bolster its credibility. In the longer term Boeing continued to receive orders for the B737Max from its strong customer base, that remains loyal in spite of unwelcome delivery delays.

4.4.3 The B737 Replacement and the MMA

The previous Boeing strategy was to develop an all-new MMA aimed at a higher efficiency that cannot be attained with modifications of existing aircraft like the A321XLR and A330neo. However, those aircraft can dry up some of the market for the MMA in the 5 years it takes to develop. The key to the MMA success could not be almost the same generation of engines but rather a new more automated 4.0 production infrastructure lowering costs and increasing flexibility. This strategic objective is also pursued by Airbus in a different approach of gradual modernization of production facilities, which is not turning out to be easy; it is open to debate what that implies for Boeing more radical global grassroots approach to modernization of production.

Following the Boeing rationale, the MMA would implement a more efficient and flexible production approach essential to reduce costs and overmatch the A321XLR/A330neo in the middle of the market, wiping out the current weak spot of the B737Max family, whose larger Max 9 and Max 10 models enjoy satisfactory sales with reported sizeable discounts and are not yet certified after a long development period. The proven MMA production infrastructure would allow the rapid introduction of an all-new B737 replacement overwhelming the core short-haul market and tipping the current balance away from the A320neo in Boeing's favor. One might imagine that Airbus would not be idle for so long and could prolong its current strategy of developing existing aircraft dovetailing with new designs benefiting from the hindsight of Boeing's known proposals and commitments. Under the same Boeing strategy, the MMA would cover up the development of the 737 replacement, without jeopardizing the sales of the current model.

The B737Max accidents and the resulting tarnished reputation may have weakened that strategy. The development of an all-new 737 replacement dovetailing with the current B737Max order book would provide a way out of the tarnished reputation of the current model: the airline customers would hold to the valuable

production slots of the current aircraft and order the new design as a follow-on in the 5-year plus time frame when new slots are available. An all-new 737 replacement would not undermine the current model, but rather provide the exit route from its troubled recent past unworthy of earlier more glorious times. Thus, the 737 replacement could become the declared priority delegating the MMA to the status of a possible extended family option. Or the 737 replacement could be designed to span a wider payload range, making a separate MMA a redundant or doubtful business case.

Subsequently, Boeing stopped all mention of the MMA and started discussing with airlines plans for an FSA. The FSA cannot be the MMA that was advertised as a twin aisle (TA). The choice of a TA smaller than the large twins (A350 and B777/B787) was questionable because of the higher drag of the large diameter fuselage for the same capacity. The unstated "shelving" of the MMA and airline push for the FSA leaves little doubt about where Boeing is heading.

The replacement as CEO of Boeing of Denis Muellenberg by David Calloun brought a strong focus on resolving the B737Max grounding issues and certificating the new versions that have received a substantial order from loyal airline customers. Thus, the Boeing priority remained to deliver on its large B737Max backlog, leaving the development of a successor to a later date, perhaps in the next decade, when game-changing technologies like more efficient engines and aerodynamics mature. There was some dismay at the statement by former CEO David Calhoun that the development of an B737 successor would costs 20 billion and would not be competitive with current technology, and would have to wait until a new generation of more efficient engines beyond 2030; others claim that an A321 competitor is needed before then. The new CEO Kelly Ortberg has been more concerned with other issues like improving production quality and ramping up B737 deliveries to stem current financial losses and reduce debt instead of borrowing more.

4.5 Safety Measures to be Implemented

The B737Max accidents have shaken the aeronautical community and brought a level of unfavorable publicity unknown for decades and at total variance with the trend of years without a major airline accident. The damaged credibility affects not only Boeing and the FAA but is also reflected in reduced trust in the whole aviation community. The FAA did scrutinize the B737Max beyond the MCAS until the ungrounding of the B737 and continued its scrutiny as production quality problems were found subsequently and not yet resolved. Other airworthiness authorities made their independent assessments, Boeing may be more cautious with other aircraft developments like the 777X, Airbus can only worry about the damage to the aviation sector it has not contributed to, and the supply chain might be holding its breath for the storm to pass but is instead now struggling to cope with increasing orders. Boeing has the competence, the resources, and the determination to overcome the 737 Max problems, and a successful return to service has led to a resumption of

orders so that the challenge has become to raise production rates to meet them. History shows that Boeing can rise to the challenge as it did when launching the B747 to overcome the challenge of the Douglas DC-8-60 series that could not be matched by further development of the Boeing 707–320; the 747 was an aircraft that needed a new factory and did not fit any existing airport; it could bankrupt the company if it failed and its success instead lifted Boeing above its competitors.

Airbus has its reputation for technical excellence unblemished, a strong SA market position to balance the Boeing advantage in the twin-aisle (TA) market and will not miss a good challenge. The B737Max accident must never happen again, and the lessons must be learned by all, whether involved or not. Boeing earlier CEO Dennis Muillenberg maintained that its design methods "were safe but can be improved." although that claim faced some skepticism after the MCAS failures. After the second B737Max accident, Boeing started two internal reviews: (i) one specifically about the B737Max design and (ii) another about its design methods in general.

The aim of the recertification effort was to make sure no accident occurs again with any similarity to the preceding two. On the technical side, there were three issues: (Sect. 4.6.1) Should the MCAS be made fail-safe? (Sect. 4.6.2) If not, at least the elevator runaway procedure must be robust and manageable by pilots beyond doubt? (Sect. 4.6.3) Is pilot simulator training in manual control of the high AoA pitch-up necessary in any case?

4.5.1 Should the MCAS be Made Fail-Safe?

Failed AoA sensors have an undesirable long record of being implicated in aircraft incidents, often not as the only cause and mostly an initial trigger. icing, ground damage during maintenance, bird strikes in flight, water ingress on the ground, and freezing at altitude have all been involved at various times and circumstances. Freezing of AoA sensors can be particularly perverse, with cases of two consistent false readings from frozen sensors outvoting the correct reading of another AoA sensor. Maybe it is time to have de-icing of AoA sensors and indication of physical damage.

Besides the general issue of reliability of AoA sensors, their use in the MCAS is not reassuring. Using two instead of one in every flight is an unquestionable improvement in allowing comparison: it is surprising it was not always this way from the outset and the "cost option" was made "free standard" only after the two accidents. But two frozen AoA sensors can agree on a wrong reading. And if two working AoA sensors disagree beyond a certain threshold, there is a warning, but the MCAS disengages. Should a flight-critical system disable itself after a single-sensor failure? All the effort put into the development of the MCAS might be better matched by a triplex or quadruplex AoA system allowing one or two failures. Airbus uses three AoA sensors in its aircraft that do not have high AoA/MCAS issues of the B737Max that uses only two.

Without being too prescriptive about the ultimate solution, whether it includes or not more AoA sensors, it is clear that the probability of failure of the MCAS must be very low, and that safeguards should exist in case sensor failure does occur. EASA has required that there must be a safeguard after an AoA sensor failure in terms of a third independent sourcr of AoA data. There is no expectation, even at EASA, that Boeing will fit more AoA sensors: the time and cost of modification are excessive. Instead, Boeing is looking at alternative sources of AoA information after an AoA sensor failure. However, the AoA vane was originally chosen as a sensor because other alternatives had a too slow response. Thus, Boeing must be looking for new sources of fast and reliable AoA data. The issues of availability, accuracy, and reliability of AoA data increase the focus on the elevator runaway protection.

4.5.2 Robustness of Elevator Runaway Protection

Having a non-fail-safe MCAS places an increasing burden of safety on the pitch control at high AoA. The procedures must be clear, the physical effort of the pilots manageable, and the control response time short enough. The omissions in the elevator runaway procedure highlight the first issue and hopefully there are no more. The need for both pilots to pull the stick up together, after an initial pull-down or the use of two hands to turn the steering wheel, points to large control forces bordering on the excessive. Even assuming that control system delays are eliminated, the scenario is far from an "easy-to-fly" aircraft at high AoA and pilot training is crucial. In contrast to the unfortunate crews of the two-accident aircraft, pilots must be aware of the normal operation and failure modes of the MCAS, learn to identify the latter, and take corrective action, with acceptable control forces and short time delays. The issue of excessive time delays is being addressed by hardware changes to the computer system, but it is less clear whether a specific action is needed to reduce control forces and provide more actuator assistance that could require another kind of hardware changes. Again, Boeing may exclude major changes and simply rely on manual control, in spite of high control forces, rather than provide emergency servo assistance. The availability of servo assistance when the MCAS is deactivated is also a relevant situation bearing in mind the large manual control forces involved. Some kind of dedicated servo assistance could counter or relieve excessive forces in manual flight, like both pilots having to pull the stick together or two hands needed to turn the trim wheel.

4.5.3 Pilot Training Package

Boeing succeeded in making the MCAS completely mask the pitch-up tendency at high AoA of the third-generation B737Max to the extent that pilots in the simulator could not tell the difference from the second-generation B737NG or know which of

the two they were flying. This success at "transparency" led Boeing to omit all information about the MCAS in the flight manual, inevitably mentioning the system in the maintenance documentation. The *Oxford English Dictionary* defined transparency, in a figurative sense not on physical light, as "open, candid, ingenious, easily seen through, recognised or detected; manifest, obvious." By this definition, the MCAS system is the opposite of transparent to pilots since they could not be aware of its existence. "Transparency" would require that the pilots be made aware of a system so clever that they could not detect its presence.

The reverse side of the coin of "transparency" or lack of it is that when the MCAS system failed the pilots could not know the cause because they were unaware of the existence of the system. Also, if the MCAS fails or disables itself, the pilots trained for the B737NG find themselves flying a different machine; the 737 Max that is far less tractable at high AoA. A non-fail-safe MCAS gives a sense of security that vanishes at the first AoA sensor failure, and thus pilots must be trained to fly without it. The training package took into account the probability of failure of the MCAS, the skill and force levels required for the successful implementation of the elevator runaway procedure, and gave a clear and transparent demonstration to pilots of the differences between (i) the B737Max with MCAS operating like a "fake" 737NG and (ii) the real flying characteristics of the B737Max when the MCAS is not available as a "mask."

4.6 Analysis of Causes and Consequences

While lying blame is not a contribution to safety, understanding the ultimate root cause of major accidents is essential to avoid recurrence and learn the hard-earned lessons. The tentative analysis of possible root causes and contributors (Sect. 4.6.1) that ultimately led to the two B737Max accidents may be compared with the limited available information on the results of the Boeing internal enquiry (Sect. 4.6.2). This may allow a hypothetical assessment of the evolution of the "B737Max crisis" and how it might have been influenced by internal and external factors (Sect. 4.6.3).

4.6.1 Conditions that Could Lead to 10 Issues

The remarks of a former CEO of Boeing David Muillenberg "about unexperienced pilots" adding that it "would not have happened with American pilots" were adding insult to injury, contradicted by American pilots who flew the B737Max and were an unfair excuse to hide Boeing's internal failures. Most, though not all, of the pilots involved in the accidents were experienced and to be fair were betrayed by the failures of the system whose existence was unknown to them and would need to implement an emergency procedure with omissions, requiring large control forces in a critical high angle-of-attack regime for which they were never trained or had even

been warned of. The ultimate root cause cannot be placed with the pilots who were the victims of a losing fight against a hidden fatal system.

The main root cause may lie in a phenomenon that is almost certainly not unique to Boeing, but rather a trend in the aerospace industry for at least two decades. Engineers in informal exchanges comment on the negative influence of other groups lacking in the technological background and thus less aware of safety concerns and implications: (i) the finance department trying to reduce costs, cutting 'unnecessary' developments for the 'fun' of engineers; (ii) the sales 'department' looking for profit and suspicious of developments with less return on investment; (iii) the legal department considering sufficient to meet minimum requirements or satisfying the letter of the rules rather than the spirit of safety; and (iv) the public relations department eager for innovations to show and less keen on 'invisible' improvements that consume resources.

There is indirect evidence that the development of the MCAS was affected by nonengineering priorities, for example, as concerns the AoA sensors. It was decided that as a standard feature only one of the two available AoA sensors would be used per flight. A cost option was to use both in every flight, compare data, and warn the pilots of the significant discrepancy. This cost option was made standard in the redesign of the MCAS after the two B737Max accidents, confirming that it is an important safety feature. It is hard to imagine that an engineer would turn a safety feature into a cost option. It is more likely that such a questionable idea might come from a sales priority overriding safety.

Although the engineers may have lost the safety argument against other interests, there is also evidence of plain design error. Issue 3 of runaway elevator pitch-downs beyond pilot control at regular intervals as long an erroneous sensor signal persists must be seen as an engineering oversight, unlikely to have other causes than hasty and ill-considered development. The MCAS was developed in response to an FAA certification requirement and fast response could override a careful closer analysis, although B737Max development time was not short, maybe more because of modest resources allocated.

Issues 9–11 of large control forces could have been overcome by a servo system, which would have involved further investment in time and money. Disabling a faulty electric pitch trim is necessary, but it is questionable, leaving as a consequence large manual control force. Issue 5 of omissions in the elevator runaway procedure again suggests that work may have been done under time pressure rather without sufficient scrutiny and checks. Issue 11 of excessive time delays in the pitch control system points to problems found during certification and deemed sufficiently rare to require no action or assumed to be within the control abilities of the flight crew. It is difficult to substantiate the assumption that within 4 s the pilots would detect a pitch runaway without clues when caused by a system unknown to them and implement a procedure with flaws requiring hardly manageable control forces.

Thus, an analysis of the 13 main issues (Sect. 4.2.2) without any or much inside information points toward several possible root causes. Aircraft accidents seldom have a single cause and often result from a combination of factors. Similarly, the 13

issues associated with the two B737Max incidents point to a range of organizational issues: (i) safety concerns of engineers overridden by other interests; (ii) flawed engineering design perhaps due to haste and certainly not subject to proper checks; and (iii) a tendency to let issues that are noticed remain unaddressed based on convenient assumptions. These conjectures based on the analysis of the 13 main issues related to the MCAS may be compared with the limited publicly available information about the results of a Boeing internal enquiry (Sect. 4.6.2).

4.6.2 Results of the Boeing Internal Enquiry

The internal review following the two B737Max accidents recommended that the engineering function be separated from other groups. That each project has a lead engineer, reporting directly to the vice-president for engineering who reports directly to the president. This direct engineering line to the top existed in the past, but has been eroded over time, not only at Boeing but also elsewhere, so its restoration is a good Boeing example that others would do well to follow before they also fall into a Boeing-style crisis.

The separation of the engineering function required a major internal restructuring at Boeing. The previous three major divisions (commercial aircraft, defense, and space) included engineering together with other functions. The new structure creates four major divisions, with engineering, serving the other three divisions, but with no hierarchical dependence on them and with a separate direct line to top management.

The internal Boeing enquiry was led by a former vice-chief of staff, with a long career in the US Navy and experience with nuclear submarines, thus well versed in safety measures. The four-person team was given almost unlimited freedom to interview anyone they wished inside or outside Boeing. They were not tasked to investigate the accidents but inevitably touched on related issues. They interviewed current and former staff at Boeing and in other organizations that had faced a comparable major crisis to assess how they had faced the challenge. The restoration of engineering pre-eminence and line to the top together with greater accountability were only two of the new safety drivers.

The leader of the enquiry was appointed by the Board of Boeing as the chairman of a new Airspace Safety Committee (ASC) that will oversee the organization, including design, testing, and certification. The ASC will consider "anonymous reports" on safety issues and should ensure that no "undue pressure" is placed on technical staff. It is perhaps no coincidence that the highest-level dismissal was at the lead of the commercial division that will no longer have authority over engineering. The new head of the commercial business unit was formerly in charge of cutting supply chain costs, leading to complaints from suppliers that felt "squeezed." The new focus on accountability reached level 2 before reaching the top level.

The president of Boeing, Dennis Muilenburg, faced tough questioning on several public occasions and not always provided fully satisfactory answers. There is no

simple answer other than the admission of the error to obvious questions such as: Why a safety system was designed to be vulnerable to single-sensor failure and lead to a runaway condition? The answer that "the B737Max is safe but can be improved" is a public relations exercise of not fully assuming responsibility in difficult circumstances. Dennis Muilenburg resisted calls for resignation. As a first step, he lost the position of chairman of the Board but remained a member and retained the title of president and CEO (Chief Executive Officer). The changes were presented as allowing him to concentrate on resolving the 737 Max issues, and Muilenburg did not disagree with that choice.

The Board may have considered more drastic action to be undesirable since it could be interpreted as a tacit admission of more responsibility than Boeing is willing to accept. Staying on as president of Boeing, even deprived of some other duties, may allow changing the perceived status from a "potential culprit in the crisis" to the outcome of the "hero who saved the company" if the current challenges were eventually overcome. The Board of Boeing is not shy of dismissing a president: Harry C. Stonecipher had to resign not for any professional failure but rather for some personal involvement with a female subordinate. As external image prevails in both cases, there may be more tolerance to management failures than to personal drift. The internal reorganization may take some time to be implemented and could help preempt some of the criticism that Boeing has received from enquiries led by Congressional committees, government departments, and agencies. Following hearings in congress, Boeing has already been accused of "putting profit before safety," which is a simplistic but perhaps not unfounded assessment of a complex situation. The appointment of a new CEO, with a long experience at Boeing and a strong engineering background, signaled the intention to focus on resolving technical issues to achieve ungrounding of the fleet and resume deliveries against the large order backlog.

A major new activity is a total overhaul of flight deck design covering not only technology and certification but also pilot training, which can differ between airlines. Boeing chose to retain a more traditional appearance of the flight controls, understating automation features, and giving the pilots the feeling they are still in control in the age of complex systems. Airbus has adopted a more engineering-led approach, letting through the fundamental role played by automation, while trying to keep pilots informed of all options available as well as limits enforced. In reality, the level of automation is not very different between Airbus and Boeing aircraft, with Airbus playing the role of the innovator (two crew cockpit, fly by wire) and Boeing adopting similar technologies without much delay without giving as much external evidence of change to retain the preference of pilots. Thus, in the future, Boeing will extend its activities in design and certification to offer support in pilot training packages for airlines that is seen as a major profitable new business activity. The engineers have won, but the sales no less.

4.6.3 Internal and External Factors and Evolution

Of direct concern to the ungrounding of the B737Max were at least the following partially open issues:

- Are the software changes sufficient to make the probability of failure of the MCAS sufficiently low?
- Is the elevator runaway procedure sufficiently fast in response with acceptable control forces?
- Will the FAA accept the Boeing preference for software and hardware changes, revised documentation, and computer training, but not simulator training for pilots?
- Will Boeing have to provide simulator training for pilots, which it says it will do at the request of airlines, but perhaps at extra cost?
- Will airworthiness authorities worldwide agree with an FAA ungrounding and there will be sufficient coordination to have a common standard that satisfies the independent analysis of each authority?
- If a non-simultaneous ungrounding of the B737Max allows flights in some parts of the world and not others, how will passengers react?
- When ungrounding does occur, will airlines accept the B737Max at a rate of 70 per month from direct production and storage, or will other considerations (refurbishing, retesting, leases of other aircraft, market and seasonal conditions) limit the achievable rate?

Besides direct issues, other aspects could affect the financial health of Boeing and its reputation and ability to address the issues:

- The initial estimated losses of $4.6 billion were exceeded as grounding extended into 2020 and production was stopped, although compensation to airlines (and perhaps suppliers) will continue to drain resources.
- The sales of wide-body airliners were slowing, and the reduced production rates may limit the cash flow that could help compensate B737Max-related losses.
- Besides the commitment of resources to the solution of B737Max problems as the highest priority, other current developments such as B777X face delays due to engine problems. The delays in the development of the B777X imply increased losses, and Boeing needs to take extra care with testing and questions about FAA independence may mean that the certification of the B777X will face closer scrutiny than the B737Max did in the past.
- There are signs that, after several years of booming aircraft sales, there was a slowing down due to weaker worldwide economic growth and perhaps airline overcapacity and over-ordering.
- The continuing trade disputes, between the United States and China, tend to exclude Boeing from one of the world's largest markets.

Boeing had been quite aggressive in winning military aircraft orders, including supersonic trainers for the Air Force and unmanned refueling aircraft for the Navy

and in competing in rotorcraft for the Army. The Boeing bids were in all cases significantly lower than all other competitors and in some cases below customer expectations. Since Boeing had been drifting out of the military aircraft business for several years, this strong revival has been seen by some as securing contracts at a loss, to be recovered later in production and sales, with the gap filled by expected continuing profits in commercial aircraft. The reversal of the latter assumption may imply that Boeing may not have commercial aircraft profits to bridge the gap to lucrative military production.

This ensemble of circumstances did not make it any easier for Boeing to overcome the B737Max crisis, with the COVID pandemic a major complicating factor, until the post-COVID sharp rebound of demand changed the situation again. Even after ungrounding, the production quality problems and slow production rate of the B737Max and B787, the delays in the delays in certification of the new B737 Max variants and B777X, and the losses in military and space programs, eroded further Boeing finances and increased debt and borrowing.

4.7 Grounding Extended from 9 to 20 Months

The Boeing publicly stated expectation that the B737X would be recertified by the FAA at the end of 2019, after 9 months of grounding, was dispelled when it became clear that the operation would not resume for another 11 months (Sect. 4.7). The extension of grounding to 20 months had serious implications for suppliers and airlines, and the consequences extended for a longer period after return to service (Sect. 4.7.1). A major factor in the additional delay was the provision of a pilot training program that meets the demands of airworthiness authorities (Sect. 4.7.2). There is some contrast between the ideal clean sheet design (Sect. 4.7.3) and the remedial actions that can be applied at a late recertification stage with limits on time and cost.

4.7.1 Consequences of Extended Grounding

After losing the position of chairman of the Board of Boeing on October 15, 2019, and while still retaining a seat on the board and the title of CEO, the change was intended to allow Dennis Muilenburg to concentrate on bringing the B737Max back to service as early as possible. He publicly stated that FAA certification was expected by the end of 2019, a claim viewed with skepticism by most observers. He also recognized that recertification of the B737Max should be simultaneous for all airworthiness authorities around the world, a very desirable goal even less likely to be achieved in such a short time. The FAA administrator called Dennis Muilenburg to tell him who set the schedule for certification and that Boeing could do more to

bring that date closer. It became clear that the B737Max would remain grounded probably for at least the first half of 2020.

The Boeing Board expressed lack of confidence in his ability to deal with the B737Max and stripped him of all duties including CEO. Dennis Muilenburg received the news from David Calhoun, a veteran of the Boeing, who had replaced him as chairman of the Board in October 2019 and took over as CEO as well in January 2020. David Calhoun has a strong engineering background and is a veteran of the Board, more than once in the past considered a candidate for CEO. Before the 737 Max crisis, the expected successor of Dennis Muilenburg as CEO of Boeing was the head of the commercial aircraft division, who was the first to be dismissed after the two accidents. Since David Calhoun is 62 and Boeing enforces compulsory retirement at 65, his task may be seen as bringing Boeing out of the Boeing 737 Max crisis and handing over a better situation to his successor. In fact, his term has been extended beyond formal retirement age, undoubtedly in recognition of the good work done and in progress and needing to be continued and completed. The change of leadership may also relate to the unwelcome news of extended grounding of the 737 Max. David Calhoun was replaced as Boeing CEO later by Kelly Ortberg as a consequence of other B737Max issues, namely production quality issues that lead to in-flight separation of a cabin door.

The airlines had already struggled to meet demand in the peak of 2019; despite resorting to leases and rescheduling, there was no way to avoid massive cancellations of tickets. The prospect of a similar situation in 2020 was most unwelcome as leasing companies were short of aircraft and crews. Faced with mounting compensation for airlines, for canceled flights, leases, and undelivered aircraft, Boeing was not in the best position to support its supply chain as well. With production halted at the end of 2019 and unlikely to resume before the end of 2020, some suppliers, in particular the smaller ones, struggled to survive until production resumed and had to lay off a large part of the workforce that may not return later. Also, the return to service and stable production at pre-crisis rates of 52 aircraft per month would be gradual after recertification and could take more time than predicted to be achieved and in fact has not yet been reached 4 years later.

Boeing initially claimed a swift return to service after recertification since only software changes to MCAS would be needed. Later it became apparent that the replacement of some computers would imply hardware changes. As the grounding extended past 6 months with the aircraft parked for longer periods, the FAA made clear that return to service will require inspection after recertification; the priority must be bringing back to service the 385 aircraft already delivered to airlines so that compensation is no longer justified. Then the 470 aircraft produced and parked must be delivered, netting some revenue instead of fines for late delivery and mothballing or upkeep costs. In addition, resumption of all production should deliver new aircraft against existing orders. The COVID-19 pandemic in the first half of 2020 worsened all the preceding situation until the sharp post-COVID recovery replaced low by high demand.

4.7.2 Provision of Pilot Training Material

The failures and deficiencies of the MCAS have highlighted the issues of pitch trim and the relation with pilot training. Boeing was initially overly reluctant to offer pilot training packages to cover the failure of MCAS, and this has contributed to delays in recertification from the outset of the B737Max program. Boeing concluded that a major economic and sales advantage would be to require only computer-based training (CBT), with no simulator hours, when transitioning pilots from the previous generation B737NG to the new 737 Max. A case in point was Lion Air that would later have the first 737 Max accident. Lion Air wanted to add to CBT also 3 h of simulator training. The Boeing representative insisted this was unnecessary and wasteful and succeeded in convincing the airline to give up simulator training for pilots. Thus, no Lion Air pilots had simulator training on the B737Max, including those involved in the accident. Boeing had convinced most airlines to bypass simulator training and pressured Lion Air to prevent it from becoming a precedent that other airlines might follow, thus damaging sales prospects. During development, Boeing convinced the authorized representative (AR) of the FAA that the MCAS designation could be used internally but not externally, being in the latter case referred to as a pitch trim modification. Also, the MCAS was not mentioned in the flight manual, the pilots did not know about its existence, and even after the first B737Max accident, it was not mentioned in the service bulletin that addressed the runaway pitch trim procedure.

After the first B737Max accident, the FAA did a study about the probability of recurrence over the lifetime of the aircraft. It assumed a fleet of 4800 aircraft operating over 20 years and that only in 1% of cases of MCAS failure the crew failed to apply successfully the runaway trim procedure: the result of 14 accidents was an unacceptably high value. The assumption that the crew would overcome an MCAS failure in 99% of the cases looks optimistic bearing in mind that they did not know the existence of the system and the recovery procedure description had omissions and involved large control forces. The two B737Max accidents in retrospect do not suggest a 99% success rate in overcoming MCAS failures. In any case, the prediction of 14 accidents due to MCAS failures over the lifetime of the B737Max fleet was alarming enough, making a second B737Max accident less than surprising. What is surprising is that nothing was done, and the study was reported much later in an external hearing after the accidents.

Even after the second accident, Boeing maintained that computer training was enough, with the option of providing training packages tailored to the requests of individual airlines. This was seen in internal reports as a new business opportunity to extend the activities of the Boeing services division and might offset some of the losses associated with the B737Max crisis. Boeing hopes that the FAA would accept CBT only, and other airworthiness authorities around the world would follow, were wrong, and only served to prolong the grounding since Boeing was eventually forced to provide a training package including simulator time. Under pressure from other airworthiness authorities around the world, the FAA assembled an international team of pilots with different levels of experience. The tests by both FAA and other pilots

highlighted the large control forces required to apply the runaway pitch trim procedure: both pilots pulling on the stick to pitch-up, and two hands to turn the pitch trim wheel. This raises the issue of fundamental design versus palliative measures.

4.7.3 Clean Sheet Design Versus Remedial Measures

Assuming that the B737Max was a redesign of the B737NG with more powerful engines further forward, a safe clean sheet design could include:

1. An MCAS with limited authority, allowing a pilot to override, using four AoA sensors to be fail-safe allowing two failures. This contrasts with two AoA sensors, leading to disengagement if they disagree, in the redesigned B737Max MCAS that is not fail-safe. This would go further than the three AoA sensors used by Airbus allowing one failure. The undesirable high AoA flight characteristics might justify four AoA in a double fail-safe configuration.
2. The large control forces in high AoA could be compensated by trim tabs (ii-1) or servo hydraulic assistance (ii-2). The trim tabs were used in large World War II bombers with manual control and high control forces. For example, a small trim tab up would push the elevator up for a pitch-down. The control forces on the small trim tab are much smaller than on the control surface as long as flow separation does not occur. The other alternative is a hydraulic servo system to power the pitch control with reduced pilot effort. Instead of duplicating the electric pitch actuation with hydraulic or other system, another alternative after disengaging a faulty electric pitch trim would be servo assistance to allow the pilots to turn the trim wheel and pull the stick with moderate effort with one hand.

Both designs (1) and (2) add complexity and cost, but not much at the design stage, with significant safety benefits they allow. The implementation of these changes to a certified aircraft would be a considerable effort in design, testing, recertification, production, and retrofitting but would restore confidence in a fail-safe system.

The remedial measures like software changes and training documentation may both provide the same level of safety and credibility. As their acceptance dragged recertification over more than 20 months, there were large commercial and collateral losses perhaps, beyond those of a serious engineering effort. Palliative measures that struggle to gain credibility may, in the end, cost more than correcting errors with a full redesign.

The short conclusion may be summarized in a single question and answer: How was it possible that having a large cadre of very experienced and highly qualified engineers, the B737Max came into service with an MCAS with multiple designs and implementation flaws? Large organizations let financial, legal, publicity, and profit issues override technical competence and safety concerns at their own risk. Incomplete development processes, stringent cost-cutting, and excessive profit-seeking increase risks instead of promoting a thorough and safe design that is the key to long-term

success sustaining a good reputation. The issues of maintaining the right balance of efficiency, safety, profit, and expediency may remain as relevant in the future as they are now and have been since the start of the B737Max crisis. That enduring lesson may last beyond any particular case study or assessment like the present one.

4.8 Change of Leadership and Work Toward a Solution

The change of leadership at Boeing marked a new course, from the pursuit of profit leading to the technical and financial crisis (Sect. 4.8.1) to the return to engineering quality as the best approach to safety (Sect. 4.8.2) based on close cooperation with airworthiness authorities (Sect. 4.8.3).

4.8.1 The Management of Profit, Dividends, and Publicity

David Muilenburg led Boeing in the period 2015–2020 following the policies established over more than a decade 2005–2015 by his predecessor James McNerney. The remark "no more moon shots" by Alan Mullaly, head of the commercial aircraft division, stated the aversion to advanced technology and preference for plain profitable solutions. The CEO before McNerney, Harry Stonecipher, who came from McDonnell Douglas and managed its incorporation into Boeing, once answered "yes" to the question: "You care only about money?" The leadership of Harry Stonecipher was short, 2003–2005, due to a personal affair with an employee, rather than any policy issues. The priorities of the leadership at Boeing for two decades were (i) externally to increase profits, share values, and dividends; and (ii) internally to squeeze supplier prices, lower salaries, and weaken labor unions. Boeing had profits of the order of $10 billion on a revenue of $100 billion based on its reputation for engineering excellence since its foundation in 1907. For the first time in a century, Boeing was led by finance not engineering, and for nearly two decades.

The profits of Boeing were mostly used to buy back shares and increase dividends. Boeing made bank loans to buy back its shares and pay higher dividends: (i) buying shares increases their prices, so paying back the bank loan should not be a problem; and (ii) larger dividends give more of the profit to the shareholders, and two-figure percentages likely exceed bank lending rates. These benefits might apply to top executives who may have preferential share options, besides bonuses and salaries. Not that the latter are unconsiderable: (i) the last yearly salary of James McNerney was $28 million and (ii) Dennis Muilenburg did not take his $13 million bonus for the year 2019 of the B737Max accidents. He left Boeing with a compensation of more than $60 million, which was described as not a benefit since it was his net value before taking the position of Boeing CEO.

During the leadership of the last three CEOs before David Calhoun, Boeing invested less than one-fifth of its profits into research and development (R&D).

Even in the year 2019 of the start of the B737Max crisis, Boeing borrowed $12.1 billion but used only $2.2 billion for R&D in spite of having to address B737Max technical problems, B777X development, and other programs. The remaining $10 billion in bank loans were used to buy back shares and pay dividends. In 5 years, Being paid $78 billion in dividends. All this financial manipulation did not prevent share price to drop from an all-time high of $446.62 on March 1, 2019, before the B737Max crisis to less than a third $128.28 one year later on April 24, 2020. By that time, the B737Max may have cost Boeing more than $20 billion, and the $60 billion government rescue package for covid may have covered some of the losses associated with the B737Max crisis. Nearly two decades of pursuit of profit and dividends were most damaging for the underinvestment in engineering. The financial exploitation of Boeing's engineering past, in the end, damaged both its share value and engineering reputation.

4.8.2 Balancing Safety, Engineering, and Finance

The external focus on profits and dividends had its counterpart in an internal focus on cutting costs. The suppliers were squeezed to cut costs by two-figure percentages, leading Airbus to claim that it treated suppliers better than Boeing. The architect of the cut supply chain costs action was Kevin McAllister, who became head of the commercial aircraft division, which made more than half of Boeing revenue and profits before the B737Max crisis; he was expected to succeed Dennis Muilenburg as CEO and was the first to be dismissed after the two B737Max accidents. The management policy was also to reduce salaries and break the considerable power of labor unions, for example, transferring work from the giant Everett factory in Washington State whose legislation protects workers to other states that attract investment at the expense of worker's rights. The working atmosphere was described as "toxic," not only for financial reasons; engineers had no access to management and technical issues and test pilot reports were ignored. The 6-year development time of the B737Max was not short bearing in mind it is a derivative aircraft and may have been more a result of weak R&D investment and focus. Reporting problems was considered a bad career move, and McNerney crowed about his ability to cower subordinates.

The reaction of Dennis Muilenburg to the two Boeing 737Max was a public relations exercise that left most skeptical. He blamed "unexperienced pilots" for the accidents and claimed that the B737Max was "safe but could be improved," where most independent observers saw the reverse. He publicized technical solutions that were partial palliatives and resisted the need for essential elements like a pilot simulator training package. He repeatedly claimed certification within 3 months, when the FAA repeatedly asserted "it had no timeline" and failed after 9 months to satisfy the safety requirements of the FAA, now more closely aligned with the other 10 main airworthiness authorities around the world. When Dennis Muilenburg was dismissed, the Boeing board stated that it has lost confidence in him.

4.8.3 Engineering Quality Ensuring Airworthiness

David Calhoun, who brought the news to Dennis Muilenburg, is a veteran engineer rather than a fast-rising manager. He immediately took major decisions in a quiet but determined way. He canceled the middle-of-the-market aircraft (MMA) and recognized that the technology needed for a future single aisle (FSA) as a direct replacement for the B737 would be available only in the next decade. He stated that Boeing would give the highest priority to a complete redesign of its flight control systems in close cooperation with the FAA to ensure no certification issues would arise in the future. The main objective of the new flight control system is safety, including protection from flight excursions comparable to those used by Airbus. He stated that flight control system design will proceed ahead, or at most in parallel with the design of the FSA, though later excluded the development of such an aircraft before 2030.

The emphasis on cooperation with the FAA became clear. In the first 3 months after taking over, David Calhoun did not announce any technical solution or claim any "certification date" as Dennis Muilenburg repeatedly did to his disgrace. Clearly, Boeing was quietly and diligently working with the FAA behind the scenes and was going to let the airworthiness authorities report on their findings and conclusions.

David Calhoun has admitted that Boeing may not be able to pay any dividends for several years. This is a consequence of (i) lack of liquidity due to using past profits to buy back shares and pay dividends and (ii) debt and R&D investment needed to remedy current problems. The architects of the Boeing "financial boom" may have "suffered" losses in the value of their shares but have kept the fat dividends and benefits.

In contrast, to signal his commitment David Calhoun renounced any payment for his work in 2020, for a crisis he had not led. Dennis Muilenburg had waived his bonus only in 2019 for a situation he presided over in deference to the 346 victims of the two B737Max accidents. In spite of his commitment to improving quality at Boeing, David Calhoun was unable to reverse in two years the decline in quality and safety culture affecting the shop floor at least as much or more than engineering. The production quality issues continued to surface post- MCAS and post-covid, with inability to raise production, continuing financial losses, and finally an in-flight incident, finally leading to his replacement as CEO by Kelly Ortberg.

4.9 Additional Unrelated Issues

The B737Max accidents damaged the credibility of Boeing as an aircraft manufacturer and of the FAA as a certification authority, leading as a reaction to a more thorough and stringent review of all aspects of the flight control system (Sect. 4.9.3). This is turn highlighted additional issues remotely or totally unrelated to the MCAS for example: (Sect. 4.9.1) the provision of an additional wiring path to the tailplane to provide redundancy and fail-safe operation in case of short-circuit or other

electrical problem with the existing wiring loom; and (Sect. 4.9.2) identification of an inconsistency between an emergency procedure and the list of minimum equipment functioning for flight dispatch, requiring additional functions in the flight control system.

4.9.1 Rewiring for Short-Circuit Protection

The electrical wiring interconnect system (EWIS) regulations came into force in 2009 to prevent short-circuits or other electrical hazards by requiring at least two separate redundant wiring paths for the actuation of critical systems. The EWIS was not applied to the wiring to the stabilizer electric trim motor of the B737Max on the basis that it was an evolution of the B737 and B737NG that never had that kind of electric failure in their long operational history. The FAA nevertheless asked Boeing to check whether any electric wiring failure could cause an uncommanded elevator failure. Boeing identified about a dozen places in the existing wiring looms where electrical separation could not be assured. Thus, a new wiring path had to be installed separately from that already existing to provide redundancy and fail-safe operation. The modification was introduced in all future production aircraft and those already on the production line. The FAA also required that all existing 470 parked aircraft be modified before delivery to airlines. Concerning the 385 aircraft already delivered to airlines and currently grounded, it had to be decided whether the modification is required prior to return to flight or can be postponed for a limited time.

4.9.2 Revising the Dispatch Conditions

The B737Max, like the B737NG and the B737, has a Master Minimum Equipment List (MMEL) of the equipment that must be functioning correctly in order for a flight departure to be allowed. Although the MMELs of the B737, B737NG, and B737Max are different, they all allow a flight to be dispatched with both flight control computers (FCCs) not operating, provided a triple set of rather stringent conditions are met: (i) limited duration of the flight; (ii) conditions throughout the flight for which autopilot is not essential; and (iii) specific acceptance by the pilots of the additional workload that may be associated with the absence of autopilot assistance modes, such as manual landing without autothrottle. The 737Max differs from its predecessor 737NG in having a fly-by-wire spoiler system. When the elevator-jam landing assist system is operating with spoilers extended, the non-normal checklist instructs the pilots to use the autopilot to retract the spoilers, and then to continue to use the autopilot as needed. This contradicts the MMEL that allows flight departure with neither of the two FCC operating. This discrepancy was found by an B737Max operator Fly Dubai 18 months after B737Max's entry into service, and escaped Boeing design reviews and FAA certification scrutiny, even after the B737Max grounding. Adding to all the issues associated with the MCAS, having an

operator finding an incompatibility in non-normal flight procedures, does not help the Boeing and FAA credibility.

4.9.3 Reconsideration of Certification Procedures

The preceding occurrences remind us that the priorities in aerospace engineering should be:

- Safety first because there are people's lives at stake.
- Engineering quality as the insurance of airworthiness.
- Finance to make whole effort sustainable.

The Boeing 737Max lesson is that persistent short-term focus on financial profit that neglects quality engineering and undermines safety may also lead to large financial losses and loss of confidence that may take longer to recover than money.

An investigation by the combined airworthiness authorities on the certification of the B737Max points to some of the improvements expected. The airworthiness authorities have much more limited resources than the aircraft manufacturer and must delegate technical details while providing overall supervision. Also, there is little point in duplicating the supervisory function, which is led by the airworthiness authority of the country of the aircraft manufacturer. However, closer information exchange and cross-checking among airworthiness authorities are desirable in the seven regions hosting the main aircraft manufacturers (United States, Europe, Brazil, Canada, Japan, Russia, and China).

An area of certification that needs to be improved is human factors: most of the recent accidents and incidents are neither a fault of the pilots nor a failure of the systems but rather a misunderstanding between partial technical failures and mismatched pilot reactions. The B737Max was a case in point when Boeing assumed that following an MCAS failure the pilots would recognize the need to apply the elevator runaway procedure and implement it in 4 s. This sounds rather optimistic even in an ideal situation of zero workload let alone in the midst of an emergency with several other concurrent issues to mind about. A better understanding of human factors is essential to design a symbiotic relation with automated systems and their protective measures and failures.

Besides a much stronger focus on human factors, both the FAA and EASA are preparing new rules concerning safety management systems (SMS). The SMS are recommended by ICAO, are mandatory for airlines in the United States, and have been adopted voluntarily by major industrial actors like Bell and General Electric as useful tools to keep an oversight of complex development and detect issues before they become problems. Both the FAA and EASA are developing rules mandating aircraft manufacturers to implement an SMS. This supplements the transactional information for certification about details with an overall safety picture. In the case of the B737Max, the FAA admits that the disconnected, partial, and incomplete

information it received did not give an overall picture of the flight control system and related issues.

4.10 Modifications for Return to Service

The main modifications made to the B737Max to regain certification are (Sect. 4.10.1) a thorough revision of the flight control system with full redundancy of the two flight control computers (FCCs) and (Sect. 4.10.2) provision of a much more comprehensive flight simulator training package, with much more emphasis on an improved non-normal checklist (NNCs). Although some issues remain (Sect. 4.10.3), this led to ungrounding at the end of 2020.

4.10.1 Redundant Flight Control

The B737Max flight control system has been thoroughly revised using mostly existing hardware, thus stopping short of a larger effort if, for example, additional AoA tensors were used to provide fail-safe operations. The data from the two AoA sensors is used by both flight control computers in a fully redundant way including a comparison to detect discrepancies. The case of consistent but erroneous AoA information, for example, both AoA sensors frozen, is no longer as critical since the MCAS can command only one pitch-down, not a runaway sequence. A further MCAS pitch-down must be allowed by the pilots that retain a 1.2 g authority at all times. Excessive time delays in the system have been eliminated through changes in computer hardware. The remaining issues include that following a single AoA sensor failure, it is not clear how reliable angle-of-attack data can be obtained as an alternative to disengaging the MCAS. For this reason, EASA has required Boeing to provide a third independent source of AoA data, without prescribing the technical solution to be chosen by Boeing, but setting a plan for demonstration of compliance.

4.10.2 Simulator Training Package

In the past, up to the resignation/dismissal of Denis Muilenburg, the Boeing policy was to avoid overwhelming pilots with data, to the point of choosing to omit the existence of the MCAS from the flight manual, possibly leaving some pilot questions unanswered. The takeover by David Calhoun at the beginning of 2020 signaled a radical change of policy toward indulging the pilots with data in a 190-page report detailing the features of the flight control system not only of the B737Max but also of the B737NG. Some of the information had not been available before, and

much of it was accessible only to instructors, and never passed to pilots. The new approach put much more emphasis on non-normal checklists (NNCs) with more clear revised instructions of seven NNGs, namely (i) speed trim fail; (ii) horizontal stabilizer out of trim; (iii) runaway stabilizer; (iv) stabilizer trim inoperative; (v) airspeed unreliable; (vi) altitude disagree; and (vii) AoA disagree. As an example, to the previous definition of elevator runaway as "continuous uncommanded movement" of the trim wheel, which does not cover erroneous MCAS activation, is added "or uncommanded motion inappropriate to the flight condition." The pilots are also warned of large control forces following the disengagement of electric pitch trim and advised to reduce airspeed if necessary. Issues remain like the need to use two hands to turn the trim wheel and whether the effort can be reduced by decreasing speed in a dive.

4.10.3 Remaining Issues and Assumptions

Although some issues remain like fail-safe MCAS design and large control forces, Boeing was rightly confident that the changes to the flight control system (Sect. 4.10.1) and much more comprehensive pilot training package (Sect. 4.10.2) would pave the way for ungrounding the B737Max that was decided first by the FAA in November 2020. The expectations of a return to service were delayed by additional issues with wiring (Sect. 4.9.1) and the Master Minimum Equipment List (Sect. 4.9.2). Boeing performed 9 h of testing in three flights from June 29 to July 1, 2020, covering a wide range of airspeeds up to 500kt altitudes up to 37,000 ft. and demanding maneuvers including steep dives and climbs and banked turns. The path to recertification included flight simulations for the Joint Operations Evaluation Board (JOEB) consisting of pilots nominated by airworthiness authorities around the world, and possibly with different skill levels. The travel restrictions due to the COVID-19 pandemic delayed the JOEB report that the FAA did consider together with other airworthiness authorities. Together with technical evidence supplied by Boeing, the FAA did unground the B737Max, after a further public consultation period. The predictions of Denis Muilenburg after grounding that the B737Max would be flying in 3, 6, or 9 months were not credible and proved wrong; the diligent work under David Calhoun, without public announcements, aimed at recertification after 15 months in mid-2020, did bear fruit toward the end of 2020, after 20 months of grounding.

4.11 Effect on the Boeing–Airbus Duopoly

The effect of the B737Max accidents on the Boeing–Airbus duopoly can be analyzed in three periods of separated major crisis: (i) the two B737Max accidents and grounding that pale by comparison with (ii) the impact of the COVID-19 pandemic.

The three periods are thus (Sect. 4.11.1) competition between the A320 and B737 families before the grounding of the B737Max; (Sect. 4.11.2) an intermediate period during which the Airbus–Bombardier deal was consolidated, and the Boeing–Embraer arrangement collapsed; and (Sect. 4.11.3) the overwhelming and hardly predictable effects of COVID-19 as the biggest crisis in aviation history.

4.11.1 Competition before Grounding

The B737 was the most successful twin-engine single-aisle aircraft of the first generation, in spite of coming into service after its European competitors, the Sud Aviation Caravelle and British Aircraft Corporation One-Eleven. The Boeing 737/737NG also eventually prevailed over its only remaining American rival, the Douglas DC-9/McDonnell Douglas MD-80, after declining sales led to the acquisition of McDonnell Douglas by Boeing.

In spite of the modest success of the Airbus A300/A310, Boeing was still diffident of European 'government' airplanes when the A320 proved to be technically a significant advance over the B737. The Boeing expedient to use grandfather rights to seat more passengers in a second-generation B737NG was effective in matching the sales of the more advanced A320 but left it with an underdog. The sales success of the second-generation A320neo did not give Boeing the time to develop an all-new B737 replacement, forcing a compromised third-generation B737Max.

The lack of development potential of the Boeing B737Max was highlighted by the development by Airbus of the extended A321/LR/XLR that Boeing could not match. The introduction of a new generation of engines, the GE/Snecma Leap and geared fan Pratt & Whitney PW 1200, gave single-aisle aircraft a transatlantic range, which could not be expected with the earlier less efficient CFM56 and V2500. The A321LR/XLR was able to fly the medium-range routes with higher efficiency than long-range twin-aisle wide bodies. Thus, the A321 became a dominant sales force, stealing sales both from wide bodies for thin medium-range routes and from narrow bodies with a more versatile payload-range combination. The A321 accounted for most of the difference between the Airbus backlog of 6500 A320neo versus the Boeing backlog of 4500 B737Max. The Boeing advantage in wide bodies with the B777/B787 over the Airbus A350/A330neo was eroded by the A321LR/XLR that also put the A320 family at advantage with regard to the B737 family. The A321LR/XLR became the best seller of all airliners.

4.11.2 Between Grounding and COVID-19

The grounding of the B737Max after two related accidents was a major blow to the finances and reputation of Boeing and harmed the credibility of the FAA. The boom in the aviation sector, with large order books and long waiting times, shielded

Boeing from cancellations since airlines had to stick to their precious delivery slots. Of greater consequence was the success of the Airbus–Bombardier deal and failure of the competing Boeing–Embraer arrangement, both triggered by Boeing miscalculations. When the Bombardier C-series won a large order in the United States against a Boeing-backed Embraer E-series competitor, Boeing resorted to a legal filing, leading to a large import duty on the C-series assembled in Canada. This controversial trade ruling was subsequently reversed, but by then its consequences were iron-clad. Bombardier offered for a nominal fee its C-series program to Airbus, allowing final assembly at the Airbus factory in Mobile, Alabama, and avoiding US import taxes altogether. The clumsy actions of Boeing led Bombardier into the arms of Airbus, and it would have been worse for the duopoly if Bombardier technology had gone to China to help certify their long-delayed airliner programs. The Bombardier–Airbus deal was a win-win for both. Airbus gained a modern and efficient C-series rebranded A220, to revitalize the modest sales of the A319 at the low end of the single-aisle market; the C-series/A220 gained the support of the vast Airbus network and had the negotiating clout to obtain better conditions from suppliers with the backing of an Airbus 10 times larger than Bombardier.

Boeing tried to correct the initial mistake with the only option left, a deal with Embraer, that at best could not match the Bombardier–Airbus deal. The Embraer E-series is smaller than the Bombardier C-series; the E-series does not fit as well at the bottom of the B737 family as the C-series/A220 fits at the bottom of the A320 family. Airbus incorporated the C-series for a nominal fee, whereas Boeing should pay $4.2 billion for the E-series and the engineering workforce behind it. With Boeing seeking a $60 billion bail-out from the US government, it could not possibly pay $4.2 billion to buy a foreign company. The collapse of the Boeing—Embraer deal was a loss-loss for both: Embraer was left with the E-series competing against a C-series backed by Airbus an order magnitude larger, a situation it had avoided in the past. In the single-aisle market, Boeing was left behind in all three sectors with (i) a grounded B737Max in the middle versus a flying A320neo; (ii) a growing gap at the upper end between B737–9/10 and the A321/LR/XLR; and (iii) a small B737Max derivative versus the new A220 and old A319.

4.11.3 Post-COVID-19 Recovery Scenarios

In March 2020, when the COVID-19 pandemic reached the United States, Boeing was already in dire financial straits: (i) it had ended 2019 with a debt of $27 billion; (ii) the commercial aircraft division alone was consuming $4.3 billion per month; and (iii) a loan of $15 billion had been depleted in less than 3 months. It became clear that if left on its own Boeing could be bankrupt in a few months; not that would ever be allowed to happen to the United States' largest exporter and second-largest defense contractor. The US government bail-out was a sum of $60 billion. This was the time the US Congress was approving the CARES (Coronavirus Aid Relief Economic Stability) act. The Boeing $60 billion bill raised some eyebrows in

Congress even in the CARES context and might have raised more if it had been submitted on its own. Boeing argued that the $60 billion also covered the supply chain, not disclosing how much was keeping for itself. The CARES allocation of $60 billion to Boeing may have ensured its survival in the coming months, probably long enough to recertify the B737Max toward the end of 2020. Although the coronavirus pandemic is an unmitigated disaster for the whole of the aviation sector, it arrived at the right time to help justify a Boeing $60 billion bail-out that was needed before. This was not the end of Boeing financial troubles, prolonged by issues unrelated to the covid-pandemic, namely production quality issues with the B737/b787, delays in certification of the B737Mas-9/10 and B777X and losses in military and space programs, which have lead to debt and borrowing with no US government assistance.

4.12 Conclusion

Viewed with hindsight Boeing has been the victim of its expedient of countering a new technologically superior A320 with an outdated B737NG competing through unequal grandfather rights. The trick would no longer work in the next generation when Boeing was forced to put the third-generation B737Max against the second-generation Airbus A320neo. The profit-before-quality and sales-before-engineering culture at Boeing only made matters worse with the flawed development of a B737Max starting at disadvantage. The coronavirus pandemic had massive negative consequences in the airliner market:

- The long-range twin-aisle wide-body market, in which Boeing is stronger than Airbus, took the longest to recover since international travel restrictions were mainly uncoordinated.
- For the medium-range routes, with reduced market demand, the A321LR/XLR offers superior economics relative to wide bodies and has feeble competition from the B737–9/10; the reduction of the A320 backlog from 6500 to 5000, and B737Max backlog from 4500 to 3000, was accompanied by a comparable decrease of 30% (or more) in production rate, keeping the Airbus advantage at the middle of the SA market, even assuming that the B737Max problems fade away.
- The A220 leaves Airbus in a stronger position at the bottom of the SA market, and with more development potential.
- The lack of orders, cancellations, and delayed deliveries at least until 2022–2023 affect both Airbus and Boeing, but Boeing has the onus of bringing more than 800 grounded B737Max into service.
- Keeping the supply chain in operation was a challenge to both Airbus and Boeing, with many common suppliers and a possibly increasing Airbus market share.

The sharp rise in demand post-COVID concerned

- The single-aisle (SA) aircraft first as COVID restrictions were lifted at national and regional levels.
- The twin-aisle (TA) aircraft as COVID restrictions were lifted worldwide and international travel resumed.

Both Boeing and Airbus have large order books and are struggling to ramp up production with a supply chain that suffered during the COVID crisis and will take time to recover. The outlook in the various sectors of the aircraft market may be briefly summarized as follows:

- Lower end of the SA market: The former Bombardier C-series renamed Airbus A220 is larger than the Embraer E-series and at advantage relative to the smaller B737Max versions; if Airbus chooses to develop an extended range A220 LR, it should be a strong contender but might undermine A320 sales.
- Middle of the SA market: The B737Max 10 with its clever undercarriage design allowing a longer fuselage is a strong competitor to the A320neo, provided Boeing overcomes delays in certification, which keep its loyal customer base waiting.
- Top of the SA market: The Airbus A321 LR/XLR is beyond the B737Max stretch potential and has no competitor, thereby giving Airbus most of its increasing lead in the SA market. With the Boeing future single aisle (FSA) not likely before the next decade, the sizeable Airbus lead in "long thin routes" is only likely to increase.
- Medium TA market: Although Airbus sells more SA aircraft than Boeing, they have lower value than the TA aircraft where Boeing leads the market. Thus a 60:40 Airbus lead in aircraft numbers equates to about equal 50:50 value. The Boeing 787 has been from the outset a success in the TA market, able to face competition from the Airbus A350/A330neo, mainly limited by Boeing's ability to resolve recurring quality production issues.
- Large TA market: The Boeing 777X may be beyond the Airbus A350 stretch potential and should cement Boeing lead in TA market when it overcomes certification delays.

In conclusion:

- Airbus leads the SA market, with the A320neo matching the B757Max 10, the A220 leading below, and the A321 LR/XLR unchallenged above.
- Boeing leads the TA market, with the B787 still benefiting from the late launch of the competing A350/A330neo and the B777X likely to strengthen its lead at the top.
- Both Airbus and Boeing struggle to increase production rates due to supply chain constraints on engines and other equipment.
- The Airbus programs A320/A321/A330/A350 are mostly free from major issues in contrast with Boeing programs: (i) new production issues with B737Max; (ii) delayed certification of B737Max 10; (iii) continuing production issues with B787; and (iv) delayed certification of B777X.

It appears that two decades of profit-first management at Boeing have left it with problems in all its four major airliner programs, which it is diligently addressing while facing the same supply chain limitation as Airbus; Airbus is benefiting from two decades of quality-first management and besides grappling with the same supply chain issues can put a stronger focus on new technologies for the future. Boeing is focusing on the transonic truss-braced wing for the FSA, whereas Airbus is considering a flapping wing as an alternate way to increase aspect ratio and aerodynamic efficiency. Both plan to use a new generation of engines with synthetic aviation fuels (SAF) but Airbus is also putting much creator effort into hydrogen propulsion and could gain a lead that would be hard for Boeing to recover later.

Besides common supply chain and production ramp-up challenges, the Boeing's quality and certification problems and financial situation contrast with the Airbus's relative lack of quality issues and stronger focus on future technologies.

References[1]

1. CAO fact sheet: Accident investigation (PDF). ICAO. 2016. Archived (PDF) from the original on September 30, 2019. Retrieved January 4, 2020.
2. Langewiesche, William (September 18, 2019). "What Really Brought Down the Boeing 737 Max?". The New York Times Magazine. Archived from the original on May 10, 2020. Retrieved September 18, 2019.
3. Aircraft Accident Investigation Report – Update. Final Report (Aviation Division) (PDF). October 25, 2019. Archived (PDF) from the original on October 25, 2019. Retrieved October 25, 2019. {{cite book}}: |work= ignored (help).
4. "Piece Found at Boeing 737 Crash Site Shows Jet Was Set to Dive". Bloomberg News. March 14, 2019. Archived from the original on March 15, 2019. Retrieved March 18, 2019.
5. Glanz, James; Lai, K.K. Rebecca; Wu, Jin (March 13, 2019). "Why Investigators Fear the Two Boeing 737s Crashed for Similar Reasons". The New York Times. Archived from the original on March 15, 2019. Retrieved March 16, 2019.
6. "Data on Ethiopia crash: 'Clear similarity' to Indonesia crash". Politico. March 17, 2019. Archived from the original on March 27, 2019. Retrieved January 24, 2020.
7. Schemm, Paul (March 17, 2019). "'Black box' data show 'clear similarities' between Boeing jet crashes, official says". The Washington Post. Archived from the original on March 18, 2019. Retrieved March 18, 2019—via Los Angeles Times.
8. Campbell, Darryl (May 2, 2019). "The many human errors that brought down the Boeing 737 Max". The Verge. Archived from the original on January 21, 2021. Retrieved January 6, 2021.
9. "How much was pilot error a factor in the Boeing 737 MAX crashes?". The Seattle Times. May 15, 2019. Archived from the original on July 27, 2020. Retrieved July 14, 2020.
10. "Blaming Dead Pilots Brought to You by Boeing". Corporate Crime Reporter. May 22, 2019.
11. Kenigsberg, Ben (February 17, 2022). "'Downfall: The Case Against Boeing' Review: Behind Two Fatal Crashes". The New York Times. Retrieved February 19, 2022.
12. "Boeing Statement on Lion Air Flight 610 Preliminary Report" (Press release). Boeing. November 27, 2018. Archived from the original on June 20, 2019. Retrieved March 13, 2019.

[1]The following reference list documents, reports, and books related to the Boeing B737MAX grounding.

13. Pasztor, Andy; Tangel, Andrew (September 22, 2019). "Indonesia to Fault 737 MAX Design, U.S. Oversight in Lion Air Crash Report". The Wall Street Journal. Archived from the original on September 23, 2019. Retrieved September 23, 2019.
14. Liebermann, Oren (October 25, 2019). "Investigators spread blame in Lion Air crash, but mostly fault Boeing and FAA". CNN.
15. Pasztor, Andy; Otto, Ben; Tangel, Andrew (October 25, 2019). "Boeing, FAA and Lion Air Failures Laid Bare in 737 MAX Crash Report". The Wall Street Journal. Archived from the original on January 26, 2021. Retrieved October 25, 2019.
16. "Flight control feature of Boeing 737 MAX under scrutiny after Lion Air accident". Aviation Safety Network. November 13, 2018. Archived from the original on January 17, 2021. Retrieved January 6, 2021.
17. Beech, Hannah; Bradsher, Keith (November 7, 2018). "In Indonesia Plane Crash Inquiry, New Focus on Possible Aircraft Problems". The New York Times. ISSN 0362-4331. Archived from the original on January 9, 2021. Retrieved January 4, 2021.
18. The message, MOM-MOM-18-0655-01B, is included in the Lion Air Accident Report.
19. Hradecky, Simon (January 14, 2019). "Crash: Lion B38M near Jakarta on Oct 29th 2018, aircraft lost height and crashed into Java Sea, wrong AoA data". The Aviation Herald. Archived from the original on March 9, 2021. Retrieved January 26, 2022. On Nov 10th 2018 Boeing sent out multi-operator messages informing operators about the MCAS.
20. "After Lion Air crash, Boeing doubled down on faulty 737 MAX assumptions". The Seattle Times. November 8, 2019. Archived from the original on November 9, 2019. Retrieved November 9, 2019.
21. Hradecky, Simon (January 14, 2019). "Crash: Lion B38M near Jakarta on Oct 29th 2018, aircraft lost height and crashed into Java Sea, wrong AoA data". The Aviation Herald. Archived from the original on March 9, 2021. Retrieved March 2, 2020.
22. Paris, Francesca; Romo, Vanessa (April 4, 2019). "Preliminary Crash Report Says Ethiopian Airlines Crew Complied With Procedures". NPR. Archived from the original on March 3, 2021. Retrieved May 28, 2021.
23. Picheta, Rob (March 10, 2019). "Ethiopian Airlines crash is the second disaster involving Boeing 737 MAX 8 in months". CNN. Archived from the original on September 11, 2019. Retrieved March 11, 2019.
24. Nicas, Jack; Kaplan, Thomas; Glanz, James (March 15, 2019). "New Evidence in Ethiopian 737 Crash Points to Connection to Earlier Disaster". The New York Times. Archived from the original on March 16, 2019. Retrieved March 16, 2019.
25. Lazo, Luz; Schemm, Paul, Aratani, Lori. "Investigators find 2nd piece of key evidence in crash of Boeing 737 Max 8 in Ethiopia". The Washington Post. Archived from the original on 17 March 2019. Retrieved 16 March 2019.
26. Schemm, Paul (March 13, 2019). "Ethiopian pilots received new training for 737 Max after Indonesian crash". The Washington Post. Archived from the original on March 19, 2019. Retrieved March 20, 2019.
27. Siddiqui, Faiz (March 10, 2019). "U. S. authorities to assist in investigation of Ethiopian Airlines crash that killed 157". The Washington Post. Archived from the original on March 11, 2019. Retrieved January 24, 2020.
28. McKirdy, Euan; Berlinger, Joshua; Levenson, Eric (March 10, 2019). "Ethiopian Airlines plane crash". CNN. Archived from the original on March 11, 2019. Retrieved January 24, 2020.
29. Hepher, Tim (March 13, 2019). "France to analyze Ethiopian Airlines flight recorders: spokesman". Reuters. Archived from the original on March 14, 2019. Retrieved January 26, 2022.
30. Kiernan, Kristy. "The Black Boxes From Ethiopian Flight 302: What's On Them And What Investigators Will Look For". Forbes. Archived from the original on March 15, 2019. Retrieved January 26, 2022.
31. Mohammad Azka, Rinaldi (13 March 2019). "Alasan KNKT Minta Dilibatkan Menginvestigasi Tragedi Ethiopian Airlines" [Reasons NTSC Asked To Be Involved In Investigating Ethiopian Airlines Tragedy] (in Indonesian). Bisnis. Archived from the original on 15 March 2019. Retrieved 24 September 2019.

32. Rosana, Christy (14 March 2019). "KNKT Tawarkan Bantuan Investigasi Kecelakaan Ethiopian Airlines" [NTSC Offers Ethiopian Airlines Accident Investigation Assistance] (in Indonesian). Tempo. Archived from the original on 15 March 2019. Retrieved 24 September 2019.
33. "Aircraft Accident Investigation Bureau Interim Report" (PDF). Ethiopian Civil Aviation Authority, Ministry of Transport (Ethiopia). March 9, 2020. Archived (PDF) from the original on March 10, 2020. Retrieved March 26, 2020.
34. Levin, Alan (March 7, 2020). "Boeing Set to Get Blame in Ethiopian Report on Crash of 737 Max". Bloomberg.
35. Marks, Simon; Dahir, Abdi Latif (March 9, 2020). "Ethiopian Report on 737 Max Crash Blames Boeing". The New York Times. ISSN 0362-4331. Archived from the original on March 11, 2020. Retrieved March 11, 2020.
36. Ranter, Harro. "ASN Aircraft accident Boeing 737 MAX 8 ET-AVJ Addis Ababa-Bole Airport (ADD)". aviation-safety.net. Retrieved August 9, 2022.
37. "Tensions Brewing Between Ethiopia, U.S. in 737 Max Crash Probe". Bloomberg. March 10, 2021. Archived from the original on March 11, 2021. Retrieved March 28, 2021.
38. "Aircraft Accident Investigation Report B737- MAX 8, ET-AVJ" (PDF). Ethiopian Civil Aviation Authority, Ministry of Transport (Ethiopia). December 23, 2022. Retrieved December 28, 2022—via Swedish Accident Investigation Authority.
39. *Boeing 737 MAX groundings*. Wikipedia.
40. Shih, Gerry (March 12, 2019). "China's ban on the Boeing 737 Max inspires others, ramps up pressure on U.S. regulator". The Washington Post.
41. Frost, Natasha (March 13, 2019). "The US is increasingly alone in not grounding the Boeing 737 Max". Quartz. Retrieved February 24, 2020.
42. Kaplan, Thomas; Austen, Ian; Gebrekidan, Selam (March 13, 2019). "U.S. Grounds Boeing Planes, After Days of Pressure". The New York Times. Retrieved March 14, 2019.
43. "Boeing 737 Max 8 planes grounded after Ethiopian crash". CNN. March 13, 2019. Retrieved March 15, 2019.
44. Lu, Denise; Mccann, Allison; Wu, Jin; Lai, K.K. Rebecca (March 13, 2019). "From 8,600 Flights to Zero: Grounding the Boeing 737 Max 8". The New York Times. Retrieved December 24, 2019.
45. "U.S. Joins Other Nations in Grounding Boeing Plane". The New York Times. March 13, 2019. Archived from the original on March 16, 2019. Retrieved March 16, 2019.
46. Schemm, Paul; Aratani, Lori; Laris, Michael (March 12, 2019). "Boeing Max 8 will continue to fly in U.S. and Europe even as China, Ethiopia ground it". The Washington Post. Archived from the original on September 15, 2019. Retrieved March 14, 2019.
47. "Report: Europe closing in on decision to let 737 Max fly". AP News. October 16, 2020.
48. Lahiri, Tripti (March 11, 2019). "China is the first country to ground the Boeing 737 Max after its two crashes". Quartz. Retrieved August 8, 2020.
49. "Editorial: Why was the FAA so late to deplane from Boeing's 737 Max?". Los Angeles Times. March 14, 2019. Archived from the original on July 20, 2021. Retrieved July 20, 2021.
50. "Boeing expects 737 Max to fly again by New Year". BBC News Online. October 23, 2019. Archived from the original on October 23, 2019. Retrieved October 23, 2019.
51. "Boeing Statement on Operations Manual Bulletin" (Press release). Boeing. November 6, 2018. Archived from the original on January 8, 2021. Retrieved January 6, 2021.
52. "Boeing nearing 737 Max fleet bulletin on AoA warning after Lion Air crash". The Air Current. November 7, 2018. Archived from the original on October 13, 2019. Retrieved July 10, 2019.
53. "Emergency Airworthiness Directive (AD) 2018-23-51" (PDF). FAA. November 7, 2018. Archived (PDF) from the original on March 24, 2020. Retrieved March 23, 2020.
54. "FAA Airworthiness Directive" (PDF). Federal Aviation Administration. Archived (PDF) from the original on January 24, 2021. Retrieved January 27, 2021.

55. Jon Hemmerdinger, Boston (December 12, 2019). "FAA 2018 analysis warned of 15 fatal Max crashes months before second accident". Flightglobal. Archived from the original on February 13, 2020. Retrieved March 14, 2020.
56. Summary of the FAA's review of the Boeing 737 MAX (PDF) (Report). FAA. November 18, 2020. Archived (PDF) from the original on January 23, 2021. Retrieved January 25, 2021.
57. Gregory Polek (November 27, 2019). "FAA To Take Full Control of Max Airworthiness Certification". AIN online. Archived from the original on July 27, 2020. Retrieved April 3, 2020.
58. "Boeing Statement on 737 MAX Simulator Training" (Press release). Boeing. January 7, 2020. Archived from the original on March 16, 2020. Retrieved March 28, 2020.
59. Wolfsteller, Pilar (July 1, 2020). "FAA completes three days of 737 Max flight testing". Flight Global. Archived from the original on July 3, 2020. Retrieved July 3, 2020.
60. Hemmerdinger, Jon (September 30, 2020). "FAA's Dickson flies Max, declares 'I like what I saw'". Flight Global. Archived from the original on October 3, 2020. Retrieved October 1, 2020.
61. "Continued Airworthiness Notification to the International Community" (PDF). Federal Aviation Administration. November 18, 2020. Archived (PDF) from the original on November 24, 2020. Retrieved January 14, 2021.
62. "FAA Probing Boeing's Alleged Pressure on Designated Inspectors". BNN Bloomberg. July 9, 2020. Retrieved July 30, 2020.
63. Gates, Dominic (November 18, 2020). "Boeing 737 MAX can return to the skies, FAA says".
64. "FAA issues emergency AD regarding potential erroneous AOA input on Boeing 737 MAX". ASN News. November 7, 2018. Archived from the original on June 7, 2019. Retrieved June 7, 2019.
65. "Continued Airworthiness Notification to the International Community" (PDF). FAA. March 11, 2019. Archived (PDF) from the original on November 3, 2019. Retrieved March 19, 2019.
66. "Airworthiness Directives; The Boeing Company Airplanes". rgl.faa.gov. November 20, 2020. Archived from the original on April 18, 2022. Retrieved December 12, 2020.
67. "FAA clears Boeing 737 Max to fly again 20 months after grounding over deadly crashes". November 18, 2020. Archived from the original on November 18, 2020. Retrieved November 18, 2020.
68. Gates, Dominic (November 18, 2020). "Boeing 737 MAX can return to the skies, FAA says". Archived from the original on November 30, 2020. Retrieved November 18, 2020.
69. "EU to Clear 737 MAX to Fly in January". The American Machinist. November 23, 2020.
70. David Koenig (March 20, 2019). "European, Canadian regulators to do own review of Boeing jet". Associated Press—via Business Insider.
71. *Yu, Sophie; Freed, Jamie (January 13, 2023).* "Boeing 737 MAX makes first passenger flight in China since March 2019". *Reuters*. Retrieved January 13, 2023.
72. Chokshi, Niraj (November 18, 2020). "Boeing's 737 Max Will Be Flying Again. What Do Travelers Need to Know?". The New York Times. Archived from the original on November 19, 2020. Retrieved November 18, 2020.
73. "Boeing 737 Max: Brazilian airline resumes passenger flights". BBC News. December 9, 2020. Archived from the original on December 9, 2020. Retrieved December 9, 2020.
74. "Boeing Max Returns to US Skies With First Passenger Flight". U.S. News. December 29, 2020.
75. "Removal of Notice of Flight Prohibition Number 001/2021" (PDF) (Press release). Trinidad and Tobago Civil Aviation Authority. January 13, 2021.[permanent dead link]
76. "Transport Canada introduces additional requirements to allow for the return to service of the Boeing 737 MAX" (Press release). Transport Canada. January 18, 2021. Archived from the original on January 28, 2021. Retrieved January 28, 2021.
77. "Transport Canada Validates 737 MAX Design Changes, Expects Return to Service in January". Aviation Today. December 18, 2020. Archived from the original on January 28, 2021. Retrieved January 28, 2021.

78. "Boeing 737 Max cleared to fly in Europe after crashes". BBC News. January 27, 2021. Archived from the original on January 28, 2021. Retrieved January 28, 2021.
79. Kaminski-Morrow, David (January 27, 2021). "EASA yet to approve 737 Max for certain precision approaches". Flight Global. Archived from the original on January 30, 2021. Retrieved January 28, 2021.
80. "Australia lifts ban on Boeing 737 MAX, a first in Asia-Pacific region". Reuters. February 26, 2021. Archived from the original on April 22, 2021. Retrieved February 26, 2021.
81. "DGCA 26 August 2021". August 27, 2021. Archived from the original on August 27, 2021. Retrieved August 27, 2021.
82. "Boeing 737 Max Aircraft: India allows Boeing 737 Max to fly again| India Business News". The Times of India. August 27, 2021. Archived from the original on August 27, 2021. Retrieved August 27, 2021.
83. Gebre, Samuel (September 2, 2021). "Ethiopian Aims to Fly Max Jet by January After Boeing Deal". Bloomberg. Archived from the original on September 3, 2021. Retrieved September 3, 2021.
84. Ngui, Yantoultra (September 2, 2021). "Malaysia Lifts Ban on Boeing 737 Max After Two-Year Grounding". Bloomberg. Archived from the original on September 3, 2021. Retrieved September 3, 2021.
85. "Malaysia lifts ban on Boeing 737 MAX". Bangkok Post. Reuters. September 2, 2021. Retrieved September 3, 2021.
86. Wolfsteller, Pilar (December 2, 2021). "China approves changes to Boeing 737 Max and clears way for jet's return". Flightglobal. Archived from the original on December 3, 2021. Retrieved December 3, 2021.
87. Christina, Bernadette (December 28, 2021). "Indonesia allows Boeing's 737 MAX to fly again after deadly crash". Reuters. Archived from the original on December 28, 2021. Retrieved December 28, 2021.
88. Endeshaw, Dawit (February 2, 2022). "Ethiopian Airlines flies 737 MAX with passengers for first time since deadly crash". Reuters. Retrieved February 2, 2022.
89. "Transport Canada introduces additional requirements to allow for the return to service of the Boeing 737 MAX" (Press release). Transport Canada. January 18, 2021.
90. "Boeing 737 Max cleared to fly in Europe after crashes". BBC News. January 27, 2021.
91. Transportation (November 13, 2019). "CAE builds extra Boeing 737 Max simulators, expecting pent-up demand | Financial Post". Financialpost. Retrieved November 14, 2019.
92. *Isidore, Chris (November 17, 2020).* "Boeing's 737 Max debacle could be the most expensive corporate blunder ever". *CNN.* Archived *from the original on December 21, 2020.* Retrieved December 21, 2020.
93. "Boeing fires commercial planes boss as 737 MAX crisis deepens: SKY". sky. October 23, 2019. Archived from the original on October 23, 2019. Retrieved October 23, 2019.
94. Gates, Dominic (April 3, 2019). "Why Boeing's emergency directions may have failed to save 737 MAX". The Seattle Times. Archived from the original on March 13, 2021. Retrieved January 26, 2022.
95. "Boeing 737 MAX crisis: a timeline (Part I)". aerotime.aero. July 22, 2019.
96. Graham Dunn (March 13, 2020). "Timeline of the twists and turns in the grounding of the Boeing 737 Max". Flightglobal. Archived from the original on July 27, 2020. Retrieved March 14, 2020.
97. Jon Hemmerdinger (May 27, 2020). "Boeing restarts 737 Max production". Flightglobal. Archived from the original on June 6, 2020. Retrieved July 4, 2020.
98. "Boeing Tells Airlines to Stop Flying Some 737 Max Planes". The New York Times. April 9, 2021. Archived from the original on April 12, 2021. Retrieved April 12, 2021.
99. "Boeing's 737 MAX Crisis: Coverage by The Seattle Times". The Seattle Times. December 15, 2019. Retrieved March 5, 2020.

100. Gates, Dominic; Miletich, Steve; Kamb, Lewis (October 2, 2019). "Boeing pushed FAA to relax 737 MAX certification requirements for crew alerts". The Seattle Times. Retrieved October 4, 2019.
101. Chris Isidore (November 17, 2020). "Boeing's 737 Max debacle could be the most expensive corporate blunder ever". CNN. Archived from the original on December 21, 2020. Retrieved November 21, 2020.
102. Boeing's Fatal Flaw (full documentary) | FRONTLINE, retrieved July 1, 2022.
103. Rucinski, David Shepardson, Tracy (October 30, 2019). "U.S. lawmakers question Boeing's $1 mln rebate clause for Southwest 737 MAX orders". Reuters. Archived from the original on February 7, 2021. Retrieved January 28, 2021.
104. Zeleny, Jeff; Schouten, Fredreka (March 12, 2019). "Trump speaks to Boeing CEO following tweets on airline technology". CNN. Retrieved December 19, 2019.
105. "U.S. Senate to hold crash hearing as lawmakers urge grounding Boeing 737 MAX 8". Reuters. March 12, 2019.
106. Final Committee Report on the Design, Development, and Certification of the Boeing 737 MAX (Report). The House Committee on Transportation and Infrastructure. September 15, 2020. p. 20, Executive Summary. [permanent dead link] This article incorporates public domain material from websites or documents of the United States House of Representatives.
107. "The Boeing 737 MAX: examining the design, development, and marketing of the aircraft" (PDF). Hearing before the committee on transportation and infrastructure house of representatives. U.S. government publishing office. August 30, 2019. p. 134. Archived (PDF) from the original on January 17, 2021. Retrieved January 6, 2021. the Maneuver Characteristics Augmentation System (MCAS) flight control law was not originally included In the 737 NG to 737 MAX differences tables nor was a specific reference included in the FCOM/QRH.
108. Brady, Chris. "737 MAX – MCAS". The Boeing 737 Technical Site. Retrieved January 6, 2021. Flight Standardisation Board (FSB): The purpose of this revision is to add the B-737-7, B-737-8200, and ... (MCAS). ... the Design Differences Table from the Boeing 737-800 to the Boeing 737-8 is revised to include ATA 27 Flight Controls addition of MCAS.
109. per Mr Bahrami's testimony, reported in the House FINAL COMMITTEE REPORT: BOEING 737 MAX, section 9, Post-Accident Response.
110. "Final Committee Report on the Design, Development, and Certification of the Boeing 737 MAX". The House Committee on Transportation and Infrastructure. September 2020. pp. 193–194, 210. Archived from the original on January 8, 2021. Retrieved January 6, 2021.
111. "Pilot Union President to Call for Accountability from Boeing and Reassessment of FAA Certification Process" (Press release). Allied Pilots Association. June 17, 2019.
112. "After 18-Month Investigation, Chairs DeFazio and Larsen Release Final Committee Report on Boeing 737 MAX" (Press release). House Committee on Transportation and Infrastructure. September 16, 2020. Archived from the original on September 20, 2020. Retrieved September 20, 2020.
113. "Boeing Charged with 737 Max Fraud Conspiracy and Agrees to Pay over $2.5 Billion" (Press release). The United States Department of Justice. January 7, 2021. Archived from the original on January 8, 2021. Retrieved January 8, 2021.
114. "Boeing to pay $2.5bn over 737 Max conspiracy". BBC. January 7, 2021. Archived from the original on January 8, 2021. Retrieved January 8, 2021.
115. Chokshi, Niraj (October 14, 2021). "Former Boeing Pilot Is Indicted in 737 Max Inquiry". The New York Times. ISSN 0362-4331. Archived from the original on November 1, 2021. Retrieved November 1, 2021.
116. Chokshi, Niraj (March 23, 2022). "Jury Finds Former Boeing Pilot Not Guilty of Fraud in 737 Max Case". The New York Times. ISSN 0362-4331. Retrieved April 9, 2023.
117. Slotnick, David. "The DOJ is reportedly probing whether Boeing's chief pilot misled regulators over the 737 Max". Business Insider. Retrieved February 22, 2020.
118. "Inspector General report details how Boeing played down MCAS in original 737 MAX certification – and FAA missed it". The Seattle Times. June 30, 2020. Retrieved July 2, 2020.

119. "Final Committee Report on the Design, Development, and Certification of the Boeing 737 MAX". The House Committee on Transportation and Infrastructure. September 15, 2020. p. 141. [permanent dead link]
120. Laris, Michael (June 19, 2019). "Changes to flawed Boeing 737 Max were kept from pilots, DeFazio says". The Washington Post. Archived from the original on December 5, 2020. Retrieved February 29, 2020.
121. "The decision by both the FAA and Boeing to omit information about MCAS from communications with the public, 737 MAX customers, and MAX pilots did not last long.—House report".
122. Andy Pasztor; Andrew Tangel; Alison Sider (May 6, 2019). "Boeing Knew of Problem for a Year". The Wall Street Journal. p. A1.
123. Tangel, Andrew; Pasztor, Andy (July 31, 2019). "Regulators Found High Risk of Emergency After First Boeing MAX Crash". The Wall Street Journal. Archived from the original on July 31, 2019. Retrieved July 31, 2019.
124. "Multi Operator Message" (PDF). Skybrary. Archived (PDF) from the original on June 28, 2021. Retrieved January 26, 2021.
125. "Boeing: US regulator admits 'mistake' over aircraft crashes". BBC. December 11, 2019. Archived from the original on December 12, 2019. Retrieved July 21, 2021.
126. "Exclusive: Boeing kept FAA in the dark on key 737 MAX design changes – U.S. IG report". Reuters. July 1, 2020. Archived from the original on May 6, 2021. Retrieved July 21, 2021.
127. Josephs, Leslie (October 29, 2019). "In brutal Senate hearing, Boeing admits its safety assessments of 737 Max fell short". CNBC.
128. "How Boeing went from appealing for government aid to snubbing it". Reuters. May 2, 2020. Archived from the original on February 22, 2021. Retrieved July 22, 2020.
129. "Review: Netflix doc 'Downfall: The Case Against Boeing' explores cost of greed over safety". Los Angeles Times. February 18, 2022. Retrieved February 19, 2022.
130. "Boeing refuses to play ball as Dutch MPs reopen 2009 crash involving 737". euronews. February 7, 2020. Archived from the original on February 9, 2020. Retrieved February 8, 2020.

Chapter 5
Comparison of Pre- and Post-COVID Pandemic Airline Operations

Sonia Avelar

Contents

5.1 Introduction

Hospitality businesses have been affected by different crises and disasters (natural disasters or epidemics) with limited or short-term impacts [15]. But the effects of COVID-19 will be evaluated as unprecedented [3, 10, 13, 15]. Travel restrictions and lockdowns limited free mobility, affecting the occupancy and revenue of hotels [10] and airlines [1, 5, 13].

The hospitality industry has been consolidating its global economic importance, but the pandemic exposed its vulnerabilities. The so-called prevalence of a service economy makes studying crisis management in hospitality mandatory to avoid firms' financial difficulties or unemployment [17].

The COVID-19 pandemic brought health and economic insecurity, resulting in people redefining their needs and priorities—the basic psychological and safety needs became the primary concern of customers and businesses, highly impacting the customer experience structure [1, 3].

S. Avelar (✉)
SATA Azores Airlines, Ponta Delgada, Portugal
e-mail: sonia.avelar@sata.pt

L. M. Braga da Costa Campos (ed.), *Accelerating Sustainable Aviation Initiatives*, Sustainable Aviation, https://doi.org/10.1007/978-3-031-73993-4_5

There was an expectation on how hospitality businesses would adapt to the pandemic scenario in terms of developing their businesses and ensuring health and safety for employees and customers, along with motivating customers to return to previous travel patterns [9]; the change and adaption to new conditions were mandatory in search of a new normal [3].

During past epidemics and pandemics, recommendations for the hospitality industry included contingency plans, highlighting the need to consider hygiene, cleanliness, and safety to recover activities and businesses [15]. Hotels designed some measures in response to the COVID-19 pandemic, including hygienic and protective measures, processes' redesign and organization, and the adoption of technology and digital innovations, intending to effectively communicate a new safety-conscious operation [3]. In the case of airlines, medical recommendations included changing the air filters, renewing the ventilation system, regularly disinfecting seats or the entire aircraft, redesigning the boarding procedures, taking the body temperature of passengers (before boarding), and letting one seat empty between two passengers [8].

Government restrictions on a global scale highly affected the transport sector, especially the operation of airlines, with forced quarantines, curfews, and travel restrictions with grounded aircraft [1, 13].

Recognizing the extraordinary situation of the COVID-19 pandemic, the main purpose of this study is to understand the impact of this pandemic on airline operations, comparing pre- and post-scenarios based on passengers' comments generated online through TripAdvisor.

5.2 Background

Pre- and Post-pandemic Airline Operations

Airlines represent an essential means of transportation, facilitating the movement of people and goods worldwide. Its activities should be managed with a focus on efficiency, clearly defining key long-term competitive advantages and development goals; they are also required to combine flexibility and stability in terms of performance to assure a competitive position in the market [12].

Before the COVID-19 pandemic, passenger demand was increasing, with many people opting to travel by air, whether it was for business or leisure. Airlines offer various services and options to cater to different customer needs, ensuring a comfortable and enjoyable travel experience. Air carriers operated broad networks, connecting major cities and smaller destinations through direct or connecting flights; major hub airports served as crucial transfer points for passengers traveling between different regions.

Flight frequencies were abundant, with airlines offering numerous daily flights on popular routes. This dynamic flow gave passengers flexibility in departure and arrival times, making it easier to plan their trips. High-demand routes often had

multiple flights throughout the day to accommodate the ever-increasing passenger demand [1, 7, 8, 13].

To meet the demand for air travel, airlines maintained large aircraft fleets; the different aircraft types included narrow-body jets (short-haul flights) and wide-body jets (long-haul, international routes). This diverse fleet allowed airlines to cater to various market segments and offer suitable services to destinations. Airlines can also differentiate their value propositions by offering multiple services and amenities to passengers, including meals, entertainment systems, Wi-Fi connectivity, and upgraded seating options (business and first class).

Passenger safety and security were of utmost importance. Airlines implemented stringent safety measures, including systematic security screenings and baggage checks, to ensure the well-being of travelers. Moreover, the adherence to aviation regulations was strictly enforced to maintain the highest standards of safety and security [6, 7].

International cooperation was a common practice in the airline industry, where airlines had partnerships and alliances with other carriers, forming global networks that provided smooth travel experiences; codeshare agreements allowed passengers to book a single ticket for a journey involving multiple airlines, enhancing convenience and connectivity [6, 18].

Besides ticket sales, airlines generated additional revenue through ancillary services, which included baggage and seat selection fees, onboard retail, and loyalty programs. These ancillary services contributed to airlines' overall financial performance and offered competitive pricing for the base fares [6, 18].

However, the COVID-19 pandemic profoundly impacted the airline industry, leading to travel restrictions, reduced demand, and significant operational changes. The industry had to adapt and implement new measures to ensure passenger safety and rebuild traveler confidence. Post-COVID-19 airline operations had to enhance health and safety protocols, modify flight schedules, and adapt to customer expectations in the new travel landscape.

Some key factors embody the successful performance of airlines as the relationships established with passengers, business allies, and supporting organizations; strategic management allows airlines' competitiveness and leadership [12].

Some characteristics collectively shape airline operations' dynamic and complex nature, with carriers continually adapting to meet customer demands, regulatory requirements, and market challenges. The main features of airline operations can be summarized as follows in Table 5.1 ([1, 6, 8, 13, 18, 19]).

In responding to the pandemic restrictions, intimately related to a drastic drop in demand and reduction in revenues, airlines focused their actions on avoiding and reducing direct and indirect operating costs; many airlines decided to change the flight operations, rationalize their fleet, reduce staff numbers, and reconfigure the networks and capacity [5].

The COVID-19 pandemic had a significant impact on the airline industry globally, being possible to compare the pre- and post-COVID-19 airline operations in different aspects ([1, 5, 7, 8, 11, 13]) (Table 5.2).

Table 5.1 Characteristics of airline operations

Main characteristics	Description
Global network	Operations on a global scale, connecting different countries and regions through networks of routes and destinations, domestic and international travel
Hub-and-spoke model	Establishing the central hub airports and the connection of these with various spoke airports allow for an efficient transfer of passengers between different flights (maximizing connectivity)
Scheduled operations	Operations on a scheduled basis with predetermined departure and arrival times for flights
Safety and regulatory compliance	Strict safety regulations and standards set by national and international aviation authorities (rigorous safety protocols, regular inspections, and maintenance and training programs to ensure safe operations)
Seasonal demand	Seasonal variations influence operations in travel demand
Competitive market	Competition on price, service quality, frequent flyer programs, and route network
Customer service	Quality customer service to enhance the passenger experience (e.g., efficient check-in procedures, comfortable cabin amenities, onboard services, responsive customer support, and loyalty programs)
Complex operations	Flight planning, crew scheduling, aircraft maintenance, ground handling, and cargo logistics
Technological integration	Computerized reservation systems, flight planning software, electronic ticketing, online check-in, automated baggage handling, and real-time communication systems for pilots and crew
Economic sensitivity	Fuel prices, exchange rates, and overall market conditions

Table 5.2 Comparison of pre- and post-COVID-19 airline operations

	Pre-COVID-19	Post-COVID-19
Flight schedules	Airlines worldwide operated numerous flights on various routes, providing passengers an extensive range of options	Airlines' flight schedules were significantly reduced, canceling or consolidating routes due to reduced demand and travel restrictions
Passenger demand	The airline industry experienced a solid growth in passenger demand	Travel demand dropped due to travel restrictions, lockdown measures, and passenger concerns about health and safety
Health and safety protocols	Health and safety protocols primarily focused on standard procedures such as security screenings and onboard safety	Airlines implemented more strict measures to ensure passenger safety, including mandatory mask-wearing, enhanced cleaning and disinfection of aircraft, social distancing measures at airports, and health screenings
Passenger experience	The passenger experience emphasized convenience, comfort, and a wide range of amenities onboard	Airlines have made adjustments to prioritize safety over comfort, including reduced in-flight services, limited food and beverage options, and modified boarding procedures to minimize contact and maintain social distancing

(continued)

Table 5.2 (continued)

	Pre-COVID-19	Post-COVID-19
Financial impact	Airlines were generally profitable and expanding their operations	Many airlines faced significant losses, reduced revenues, and even bankruptcies. Governments provided financial aid and support to help the industry recover
Travel restrictions	Travel was relatively unrestricted for most destinations, with travelers requiring passports and visas as necessary	Travel restrictions were implemented globally, including border closures, quarantine requirements, and testing protocols. These restrictions varied across countries and changed frequently, impacting the ability to travel
Industry regulations	Policies and regulations adjusted to the airline industry	The pandemic resulted in introducing new regulations and guidelines for the airline industry. These included health and safety protocols, passenger screening measures, and guidelines for international travel regulatory bodies and health authorities closely monitored and adapted these regulations as the situation evolved
Flight schedules	Airlines worldwide operated numerous flights on various routes, providing passengers an extensive range of options	Airlines' flight schedules were significantly reduced, canceling or consolidating routes due to reduced demand and travel restrictions.
Passenger demand	The airline industry experienced a solid growth in passenger demand	Travel demand dropped due to travel restrictions, lockdown measures, and passenger concerns about health and safety
Health and safety protocols	Health and safety protocols primarily focused on standard procedures such as security screenings and onboard safety	Airlines implemented more strict measures to ensure passenger safety, including mandatory mask-wearing, enhanced cleaning and disinfection of aircraft, social distancing measures at airports, and health screenings

It is important to note that the specific impact and changes in airline operations have varied across regions and airlines. Airlines' actions included grounding aircraft, reconfiguring networks, requesting national governments' financial support, or placing staff on furlough [5]. The recovery and restoration of pre-pandemic operations depended on factors such as vaccination rates, virus control, and lifting of travel restrictions.

During the pandemic, the state of the aviation industry was the same as the entire world—an unimaginable lockdown. The scenario consisted of businesses without businesses, no planes or passengers, and scarce flight movements. There were some emergency priorities to help the industry, generating cash and cutting costs, besides financial relief measures to ensure companies' sustainability. Consumers were willing to travel again, although under certain conditions—price and virus safety measures turned out to be on the same path of consideration [1, 7, 13].

Airline industry stakeholders have followed the regulatory standards and international pandemic recommendations set by local authorities and international industry associations to continue providing safe services. All cost base structures were severely impacted due to both the operational decrease and the regulatory requirements. Airlines faced challenging times trying to determine how to survive in the short term, how to plan for the medium term, and how to position for the future. This pandemic changed how most people worked daily; many business professionals became reliant on technology and remote working, and business environments had to establish logical plans to allow for flexibility and quick adaptation to changes.

5.3 Methodology

This study adopts a case study approach as the research method to unveil customer perceptions of airline operations in the pre- and post-COVID-19 scenario. The Portuguese airline SATA Internacional—Azores Airlines was selected as it represents the main mobility mean of transportation of the Azores archipelago, a European ultra-peripheral region. The Archipelago has political and administrative autonomy, and for these reasons, local policies were somewhat different from Portugal Mainland during the pandemic, influencing the air mobility patterns. In the initial period, air mobility was limited to those returning home; then a complete lockdown was imposed; and afterward, a soft reopening occurred in late 2020. As a result, the airline had to change operational and communication strategies to comply with health safety measures and overcome financial losses that may occur due to low demand. Thus, the present study attempts to unveil how passengers react and communicate their experience with the airline, highlighting the dimensions mainly valued before, during, and after the pandemic, and how these influence the overall rating of the company from a customer perspective.

The data was collected from the TripAdvisor site, which contains information and opinions generated by users. All comments and ratings until the present date (June 2023) were analyzed to understand the phenomenon under study comprehensively. The data was divided into three periods: before 2020, during 2020, and after 2020. A mixed-method approach was applied, combining multilinear regression models and content analysis, all supported by QDAMiner software: WORDSTAT and SIMSTAT.

5.4 Case Study

SATA: Eight Decades of History and Experience

The Azores archipelago has an undeniable geostrategic position and value. The first air crossing of the North Atlantic occurred during the First World War, passing through Ponta Delgada, São Miguel Island, on May 20, 1919. Several landmarks in the history of aviation intersect with the contemporary history of the Azores, demonstrating the importance of the air transport sector at the global and regional levels

[16]. It is also in this context that the role of SATA in the development of the Azores and its autonomy is highlighted in promoting the unity of the Azoreans (islands and diaspora); the social, economic, and cultural progress of the Archipelago; and the Azorean identity.

The Sociedade Açoriana de Estudos Aéreos was created on August 21, 1941, to study the feasibility of exploring air routes over the Azores and from the Archipelago to other destinations. The designation was changed to Sociedade Açoriana de Transportes Aéreos Ltd. (SATA) on February 17, 1947. The commercial activity and baptism of the aircraft "Açor" (a Beechcraft with registration CS-TAA) began on June 15, 1947, with the first flight operated between Ponta Delgada (PDL) and Santa Maria (SMA) with seven passengers on board [14].

The company's motto was the connection to the world at a time when air transport was only for the elites: it connected the islands of São Miguel and Terceira (which represented 75% of the Archipelago's population) to the island of Santa Maria, where the European and American airlines crossing the Atlantic used to make stopovers (Imperial Airways, Pan American Airways, TWA, Canadian Pacific, Air France, and TAP) [14].

The expansion to all the Azorean islands was gradual following the respective airport infrastructure constructions. First was the island of Faial in 1971, followed by the island of Flores in 1976; the airports of Graciosa, Pico, and São Jorge were opened between 1981 and 1983. Corvo would be the last island to have regular connections operated by SATA starting in 1993. From the Beechcraft aircraft, the SATA fleet evolved to DOVE, then Dakota, Avro, ATP, and the Bombardier [14, 16].

The reality of an air operation in an ultra-peripheral region such as the Azores includes challenges that arise from the need for self-sufficiency, which leads to vertically integrate all activities related to air transport, namely, flight operations, maintenance and engineering, assistance services in each station, aerodrome management, and specific training of human resources.

Since air transport is an international business, SATA had to maintain rigorous industry standards, both in its operation and commercialization processes, as well as in ground handling services (SATA has developed this activity due to the commercial and technical stops of the large international companies operating on the island of Santa Maria).

When stopovers in Santa Maria began to decrease, SATA established a direct bridge with immigrant communities in North America, creating two tour operators in 1985, one in Canada and the other in the United States, which chartered local planes to transport passengers to the Azores. The company acquired the first Boeing 737–300 a decade later and started flying outside the Archipelago [14, 16].

From the end of the 1990s, SATA evolved into the Business Group that it is today, composed of five companies: SATA Air Açores, the parent company that ensures the connections between the islands of the Azores with the Dash 8 aircraft, Q200 and Q400; SATA Internacional—Azores Airlines, the company that flies outside the region, with jet aircraft, Airbus A320 and A321; SATA Gestão de Aeródromos, which manages the airport infrastructures of the islands of Graciosa, São Jorge, Pico, and Corvo; and two tour operators, SATA Vacations Canada and SATA Vacations America (Figs. 5.1 and 5.2).

Fig. 5.1 SATA Air Açores fleet. (Source: Azores Airlines)

Fig. 5.2 SATA Internacional—Azores Airlines fleet. (Source: Azores Airlines)

The liberalization of air transport in Europe (open skies) occurred only after the third legislative package (1993). In 1995, SATA acquired a medium-haul aircraft and, while unable to access the market between Portugal Mainland and the Azores, gained experience operating on Madeira Island. SATA Internacional was created in 1998; in 1999, the company won the tender to explore the public service routes between Lisbon, Oporto, Funchal, and Ponta Delgada. In 2000, it started operating in Boston and Toronto, where a sizeable Azorean community resides. Creating a hub in the Azores is consolidated with increased frequencies and connectivity through a new generation of more efficient aircraft, the Airbus A321neo (new engine option) and LR (long range).

The company continues to have ground handling services on all islands, has aircraft maintenance and engineering services, both basic and line (assisting third parties in line), and has the Atlantic Aeronautical Training Centre (in Santa Maria). SATA's history has been built to service the Azores, an internationally recognized service demonstrated by the ICAO award of a medal of honor on its 50th anniversary, due to its relevant contribution to civil aviation.

SATA emerged in a context of expectations regarding the advantages of air transport for the material and cultural progress of the islands' population. It represents a regional purpose in its condition as a public company (since 1980). The contribution to the unity of the Azoreans, the social and economic development of the Azores, and the cultural promotion of the different islands is undeniable. It also makes commitments to serve the diaspora, attract tourism flows to the region, and boost traffic in Macaronesia—materializing its internationalization and inter-regionalization strategies.

In 2021, SATA celebrated 80 years of foundation and 74 years of its first flight. There are more than eight decades of history and activity, a remarkable case of longevity in world civil aviation. It presents as its current mission (and vocation) "to unite the Azoreans and to break their isolation from the world, through the activity of a group of companies and complementary business areas whose importance for the Azores constitutes an indisputable reality" [14].

SATA Internacional—Azores Airlines

SATA Internacional—Azores Airlines is a full-service airline operating within a broad geographical area; its network includes the Azores, Madeira, Portugal Mainland, besides different European airports, Cape Verde, the United States, and Canada. The company's value proposition to its customers includes the transportation of passengers, cargo, and mail; passenger service is available in business and leisure classes on domestic and international routes. The company's mission is to bring the world to the Azores and take the Azores to the world centered on the values of friendliness, reliability, and innovation.

Today SATA is a small regional airline with large strategic and operational ambitions. It is no longer focused on uniting each of the nine islands of the Azores but, instead, taking each island out to the world and bringing the world to each of the nine islands, collectively and individually. The availability of such air transportation services in a unique regional context substantially positively impacts the region's economic growth; airport activities generate jobs, influence firms' decisions regarding exports and imports, and nourish the tourism industry. Flows of passengers and cargo enable the flow of goods, services, knowledge, opportunities, investment, and employment and can affect the region's GDP, demand, and business environments.

Strategic management of an airline should be based on the identification and establishment of critical long-term competitive advantages, which ensure the creation of new opportunities for development and growth; the company's performance would be a successful one when focusing on the relationships with passengers, alliance partners, and supporting organizations [12]. An airline needs to focus on

improving transportation processes and constructing a chain of added value formation in the development of all operating stages; this will enable the establishment of competitive advantages to achieve long-term objectives and attain the desired results at an optimal cost.

Chronology: SATA and the Pandemic

The COVID-19 pandemic has disrupted normal business operations on a global scale. The impact of this unprecedented crisis resulted in many industries being forced to reduce or cease their business. Worldwide travel restrictions have been imposed due to the necessary containment of the virus, and these have disturbed the aviation industry. SATA Group airlines were not an exception; the following chronology presents some of the main initiatives and measures adopted by the company in response to the initial pandemic scenario.

Surveillance Phase

The first information available about COVID-19 was monitored in December 2019. In January 2020, SATA established daily contact with the International Air Transport Association (IATA) Medical Advisor and a daily information update from the World Health Organization (WHO). Informational content for internal use and information to customers and passengers was created, and internal procedures for notification and promotion of more frequent hygiene habits were implemented. The distribution of disinfectant gel and information leaflets on board aircraft was adopted at this time, and all procedures on board were reviewed; the stocks of protection kits, masks, and gloves were reinforced. The airlines also established a network of contacts with the Regional Health Directorate, General Health Directorate, and Civil Aviation Authority. The Safety, Health, and Hygiene at Work procedures were reviewed based on the SARS 2003 case and Flu Outbreaks in 2005, 2009, 2013, and 2015.

Alert Phase

The beginning of March has revealed increased cases in all countries and the subsequent retraction of economies. SATA used all communication channels to inform passengers about preventive measures taken by the Group's airlines. Also, a working group was created to implement actions related to the new coronavirus COVID-19 crisis. As tight circulation containment measures were in force globally, the commercial rules were changed to allow the change of trips later, without penalty or increase of fare. This phase corresponded to daily monitoring of the general situation and the communication to business partners and passengers about the US flight restrictions; in addition, the reinforcement of operational procedures on board, the placement of COVID-19 information on all public access places (terminals, aircraft, and workplaces), and the replenishment of disinfectant gel in all exposed locations and communication of mitigation principles of contagion.

Contingency Phase

The Contingency Phase began with the World Pandemic Declaration from WHO. The rapid spread of COVID-19 and the deaths caused by the disease made it very clear that actions were mandatory, so preventive measures were reinforced

among all SATA employees. Specific plans for cleaning and decontaminating all aircraft and contingency plans for the Handling and Airport areas were defined. Preventive kits and protective materials were replaced, and a mitigation plan was triggered to prevent material shortages. Risk assessment was performed for exposed employees and business travel for training. The Contingency Plan was published.

The announcement of the suspension of entry of nonresident passengers in the United States indicated demand contraction on international flights, specifically in the leisure and tourism segments. By mid-March, SATA announced the suspension of flights to three of the five gateways in the Azores, concentrating its operations in Ponta Delgada (PDL) and Terceira (TER) airports. Meal service was suspended on board (alternatives offered), and SATA sales offices were closed to mitigate the contagion risk and strengthen the Contact Centre (telephone and online service). SATA's commercial flight network was redesigned, and internal agents, partners, passengers, and the general public were informed. New commercial solutions were created to streamline passenger accommodation processes (e.g., the creation of a nonflown ticket replacement voucher).

March 15, 2020, was the date of the first confirmed case of COVID-19 in the Azores. Also, on this date, SATA Internacional operated a repatriation flight between Lisbon (LIS) and Punta Cana (PUJ). On March 16, the restriction of flights to Terceira was imposed, and SATA maintained the Ponta Delgada gateway only—SATA Air Açores' operation was reorganized without passing through Terceira Island (passenger flights); a daily cargo flight between Ponta Delgada and Terceira was scheduled. SATA Internacional also operated a repatriation flight between Terceira (TER) and Toronto (YYZ).

Emergency Phase

March 18 was the day of the declaration of a state of emergency by the president of the Portuguese Republic. All safety procedures were reviewed regarding the need to transport cases of force majeure and adapt to new circumstances. Due to the reduction in air operations, services and structures were duly adapted, and the Contingency Plan was updated during the emergency period. A total suspension of flights to Boston and Toronto was prepared.

The Emergency Phase brought the communication and implementation of new preventive and more restrictive measures on board as the restriction of seats (social distance) and the limitation of cabin baggage allowed on board; the implementation of new measures to mitigate contagion in the operational areas; the transfer to the teleworking regime of 100% of the administrative human resources; the suspension of all SATA Group commercial flights; the reorganization of the network to allow for force majeure and cargo transportation flights; and the creation of an application to be integrated into the reservation and check-in system to allow prior screening of passengers bound for Ponta Delgada.

Repatriation flights were organized in cooperation with the Government of the Azores, the Government of Cape Verde, and Tour Operators (Shanghai—PVG; São Vicente—VXE; Praia—RAI; Fortaleza—FOR) and operated by SATA Internacional. SATA Air Açores Q200 aircraft was converted to a combi version to ensure the

transport of additional cargo within the Azores archipelago, and SATA Internacional A321neo cabin was adapted to transport hospital supplies between Lisbon and Ponta Delgada. Specific procedures were created to be adopted by the areas still active (maintenance, handling, and crews), and the establishment of medical support contact lines and advice for SATA employees was activated.

By the end of March, IATA forecasts indicated a 44% impact on revenue; SATA revenue projected a drop of 49%. The airline group adhered to the tax and contributory measures announced by the Government of the Portuguese Republic to grant the payment of social security and withholding IRS, and requested a moratorium on all credit lines contracted. Mandatory training was reorganized in an online training regime; all nonmandatory training, internal mobility, and recruitment processes were suspended. The company met with all Unions and Workers' Commissions to agree on Lay-Off measures, with the Lay-Off period (total or partial) beginning on April 7.

SATA Internacional operated two necessary cargo flights in early April, one in China and another in Cape Verde. In mid-April, the airline operated a flight to transport Portuguese workers returning to Austria. A cargo transportation record was achieved from operating with A321neo: 11 tons of food items between Ponta Delgada and Lisbon (mainly fresh fish and pineapples).

Following the communication on extending the suspension of flights until May 15, 2020, and preparing for the second period of Lay-Off, the Social Partners were consulted to identify possible improvements to be implemented. On April 30, the president of the Regional Government of the Azores communicated how the suspension of the state of emergency in the Archipelago would be implemented, and the suspension of flights until May 31, 2020. The change from a state of emergency to a state of public calamity for the entire national territory occurred on May 1, 2020.

Calamity Phase

The Calamity Phase marked the month of May and registered the suspension of flights until the end of May (inter-islands) and until mid-June for international routes duly communicated to partners and passengers. All employees were clarified about the current situation. Some new cargo and repatriation flights were operated to Cape Verde.

Flights resumed within the Azores Archipelago on May 29, operated by SATA Air Açores. Flights operated by SATA Internacional to Portugal Mainland and Madeira resumed on June 15; international flights to Boston (BOS) and Toronto (YYZ) on July 1 (although with traffic restrictions imposed by the European Union), and to Frankfurt (FRA) on July 2. During June and July, besides its regular operation, SATA Internacional—Azores Airlines also operated charter flights for repatriation (Bermuda—BDA/Ponta Delgada—PDL and Fortaleza—FOR/Praia—RAI) and for humanitarian purposes (Havana—HAV/Bissau—OXB and Accra—ACC/Barbados—BGI).

5.5 Data Analysis

SATA Internacional—Azores Airlines is identified on TripAdvisor as SATA—Azores Airlines, and its overview is rated as 3.0 (out of 5). Following a similar approach to the one developed by Ban & Kim [2], the users' comments were analyzed to understand customers' evaluations and opinions about the airline service, and their experience as passengers during the pandemic and in the pre-and post-pandemic settings. Ban and Kim [2] found that practical implications for developing sustainable airline industry strategies can emerge from a customer-centric business operations perspective by understanding online reviews.

Some word clouds were composed to compare the comments and the analyzed periods. From the word clouds generated by the titles (Fig. 5.3) and the text content (Fig. 5.4) of the comments in 2020 and 2021, it is possible to understand that customers mainly complained about poor customer service, flight cancellations, and difficulties in obtaining ticket refunds from the flights canceled or rescheduled due to the travel restrictions in place in the pandemic period.

The word cloud resulting from the titles of the comments in 2022 and 2023 (Fig. 5.5) denotes a different focus, specifically on the experience and service. It is possible to find positive and negative adjectives that qualify the customers' opinions; however, the negative comments are more visible.

The number of comments increased in 2020 in comparison with the comments posted before this year (Fig. 5.6).

The distinction between comments posted before and in 2020 can also be compared in terms of service, flight experience, and the airline (Figs. 5.7, 5.8 and 5.9).

The comments increased in 2020 regarding the flight experience and the airline compared to comments generated before this year. It is possible to understand that the pandemic influenced passengers' pre-disposition to share their comments with online communities.

The frequency analysis of the distribution of the keywords allows a clear vision of the preponderance of negative comments in the overall comments available online about SATA (Fig. 5.10).

The users of TripAdvisor can attribute a general rate about an organization, write a comment (free text), and rate a group of specific dimensions related to the type of service performed by the organization under evaluation. In the case of an airline, the dimensions available for rating (from 1 to 5) are legroom, seat comfort, in-flight entertainment, customer service, value for money, cleanliness, check-in and boarding, and food and beverage. It is also possible to add information about the type of flight, domestic or international; business or economy class; and the route and date of travel.

Passengers traveling on international flights and in economy class are the ones who evaluate the most (Figs. 5.11 and 5.12).

Before the pandemic, the overall rating of the airline was negatively influenced by the ratings of the specific dimensions of cleanliness, legroom, and in-flight entertainment (results from the multi-regression model):

Fig. 5.3 Word clouds from titles in 2020 and 2021

Fig. 5.4 Word clouds from text content in 2020 and 2021

Fig. 5.5 Word cloud from titles in 2022 and 2023

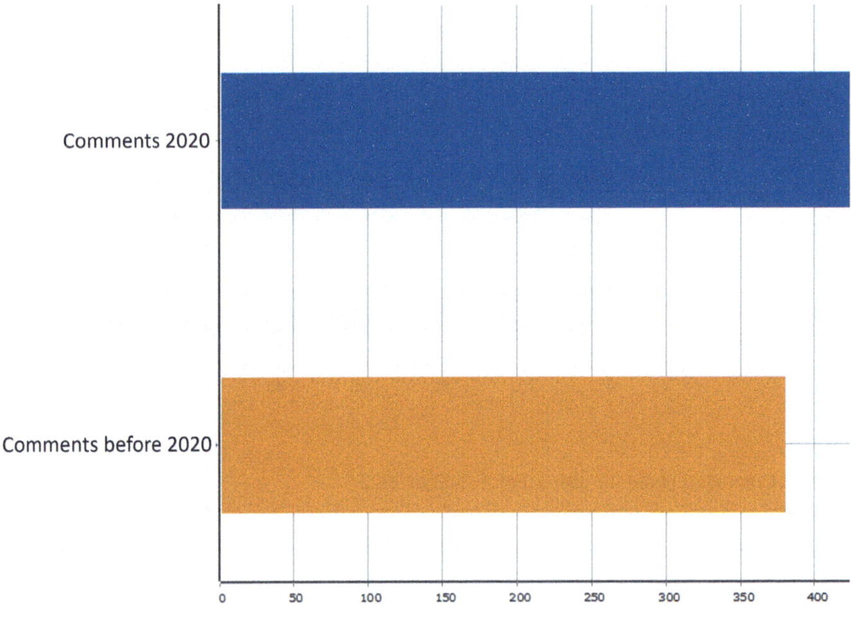

Fig. 5.6 Number of comments before and in 2020

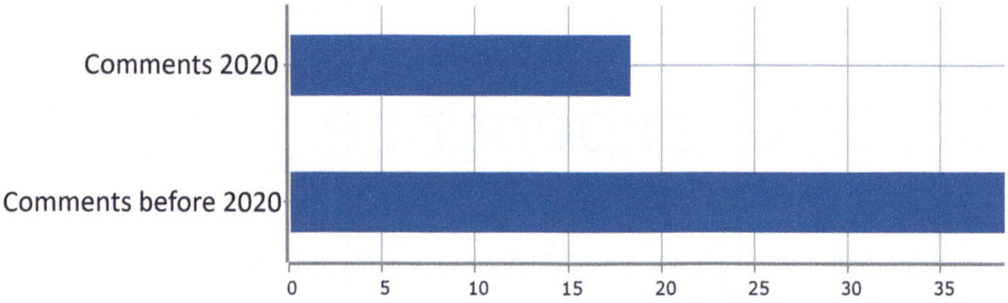

Fig. 5.7 Comments about the service before and in 2020

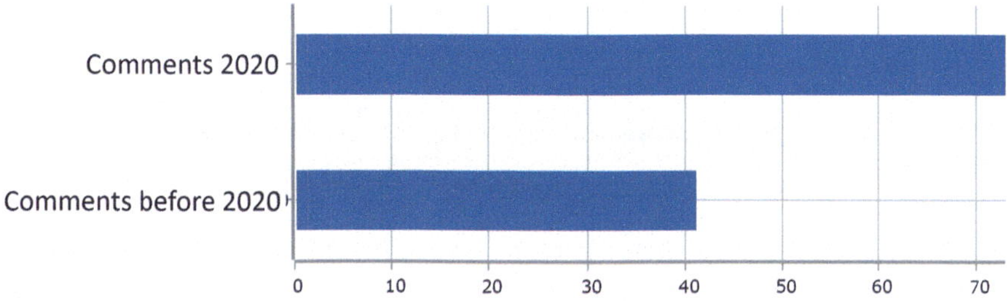

Fig. 5.8 Comments about the flight before and in 2020

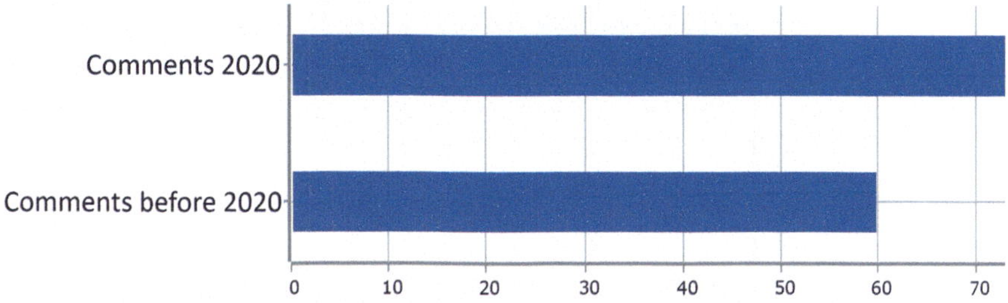

Fig. 5.9 Comments about the airline before and in 2020

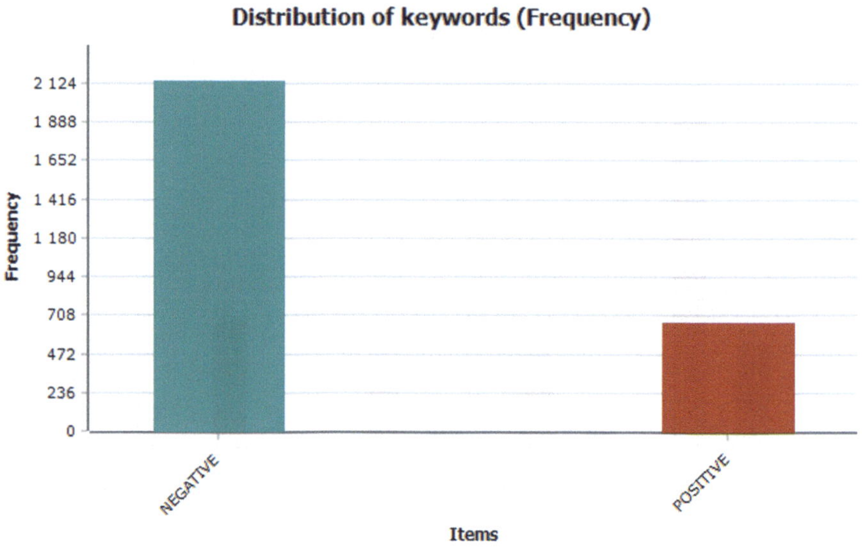

	FREQUENCY	% SHOWN	% PROCESSED
NEGATIVE	2143	76,18%	12,28%
POSITIVE	670	23,82%	3,84%

Fig. 5.10 Frequency of the distribution of keywords in the total period

$$\begin{aligned}\text{Rating} = {} & 1,7224 + \left(-,1326^{*}\text{CLEANLINES}\right) + \left(-,2115^{*}\text{LEGROOM}\right) + \left(,1070^{*}\text{SEATCOMFOR}\right) \\ & + \left(-,1182^{*}\text{INFLIGHTEN}\right) \\ & + \left(,4766^{*}\text{CUSTOMERSE}\right) + \left(,1865^{*}\text{VALUEFORMO}\right) + \left(,07747^{*}\text{CHECKINAND}\right) \\ & + \left(,01903^{*}\text{FOODANDBEV}\right)\end{aligned}$$

During the pandemic, the negative ratings of the dimensions seat comfort, in-flight entertainment, check-in and boarding, and food and beverage contributed the most to the overall rating, as observed in the results of the multi-regression model applied:

$$\begin{aligned}\text{Rating} = {} & ,3182 + \left(-,1449^{*}\text{CLEANLINES}\right) + \left(-,7380^{*}\text{LEGROOM}\right) + \left(-,5645^{*}\text{SEATCOMFOR}\right) \\ & + \left(-,4030^{*}\text{INFLIGHTEN}\right) \\ & + \left(,7244^{*}\text{CUSTOMERSE}\right) + \left(,1417^{*}\text{VALUEFORMO}\right) + \left(-,1070^{*}\text{CHECKINAND}\right) \\ & + \left(-,01379^{*}\text{FOODANDBEV}\right)\end{aligned}$$

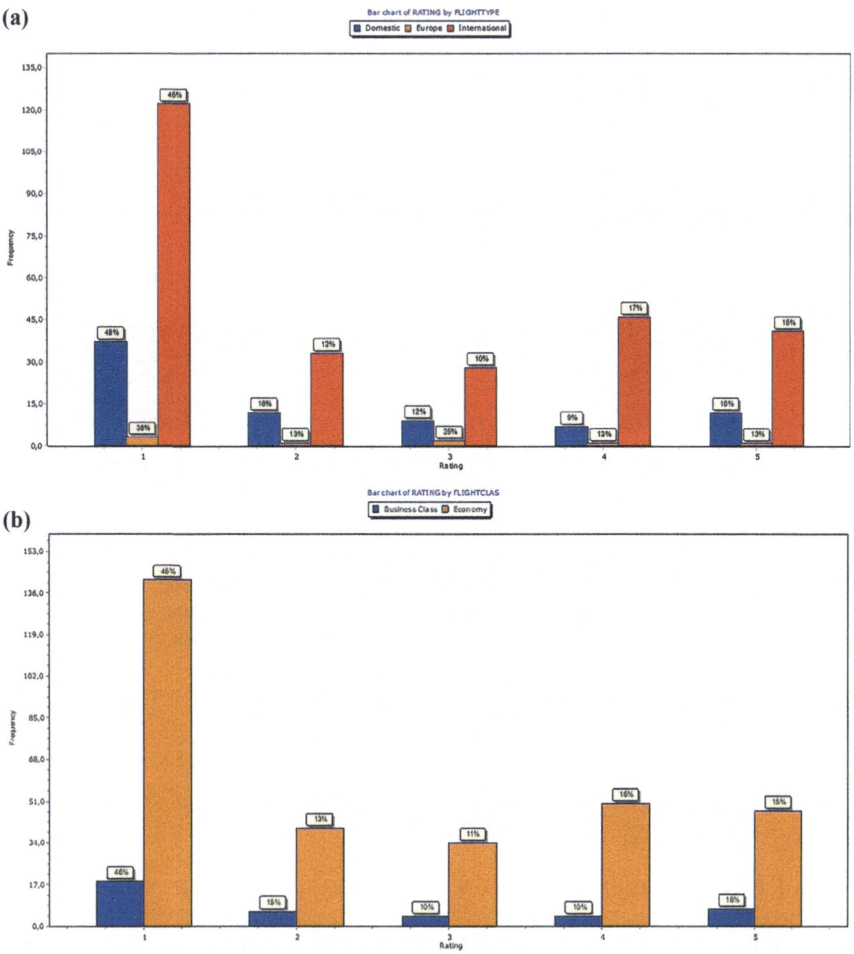

Fig. 5.11 Ratings per flight type and class pre-pandemic (**a**, **b**)

Reassessing the multi-regression model for the post-pandemic scenario, the dimensions negatively impacting the overall rating were legroom and in-flight entertainment:

$$\begin{aligned}\text{Rating} = &-{,}5424 + ({,}005470^{*}\text{CLEANLINES}) + (-{,}3281^{*}\text{LEGROOM}) + ({,}5570^{*}\text{SEATCOMFOR}) \\ &+ (-{,}006312^{*}\text{INFLIGHTEN}) \\ &+ ({,}1747^{*}\text{CUSTOMERSE}) + ({,}3763^{*}\text{VALUEFORMO}) + ({,}2654^{*}\text{CHECKINAND}) \\ &+ ({,}1231^{*}\text{FOODANDBEV})\end{aligned}$$

(a)

(b)

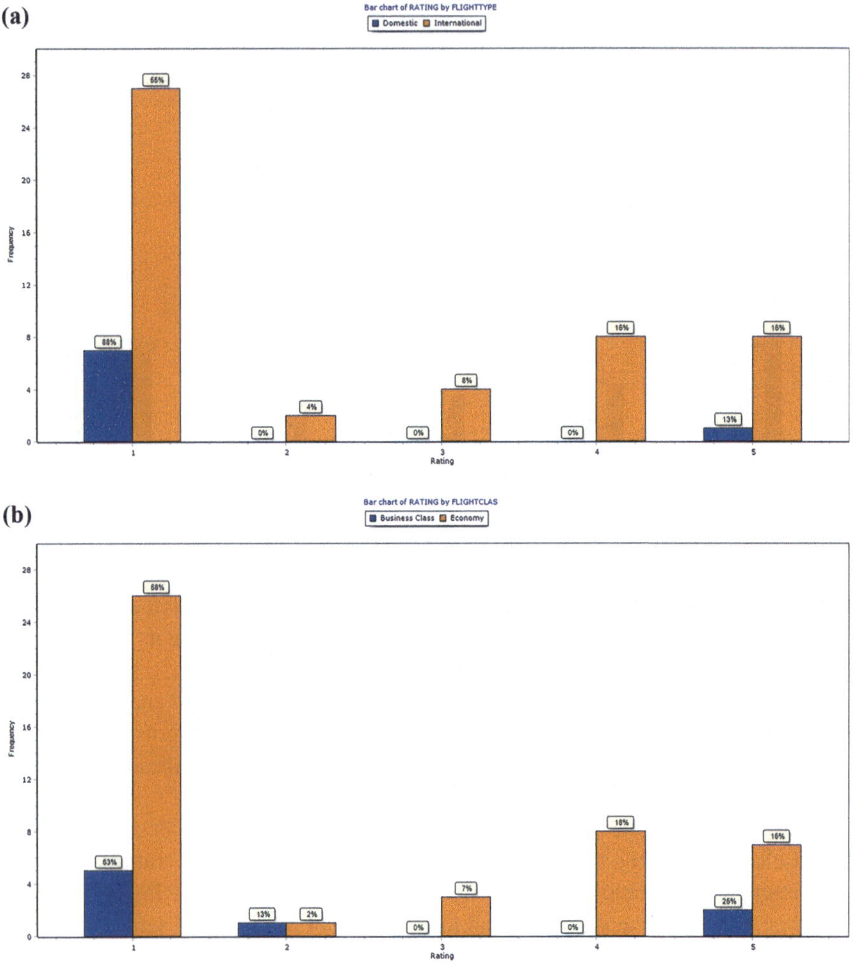

Fig. 5.12 Ratings per flight type and class post-pandemic (**a, b**)

In-flight entertainment is an issue that must be analyzed and improved to better cater to passengers' expectations and needs when traveling with SATA because it is the dimension that contributes negatively to the overall rating in the three periods under analysis. Legroom is negatively rated in the pre- and post-pandemic scenarios; as most comments are related to international flights (long haul), legroom is an important aspect considered by passengers in terms of comfort on board. Cleanliness was an issue with negative weight before the pandemic; during and after the pandemic, this dimension obtained positive ratings, which can be directly related to the strict hygiene protocols implemented to contain the spread of the virus. In the pandemic period, the dimensions negatively rated, namely check-in and boarding and food and beverage, are associated with specific protocols that were in place for the

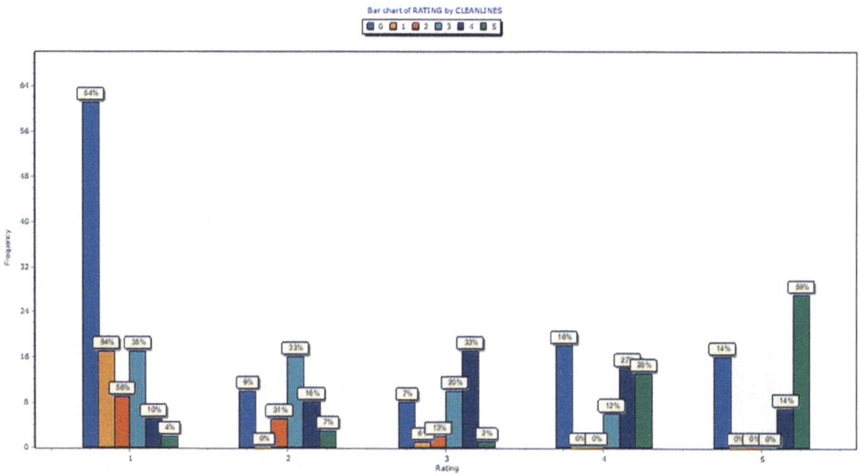

Fig. 5.13 Rating by cleanliness, pre-pandemic period

containment of the pandemic; these included social distancing, disinfection, and the suspension of the offer of meals and drinks on board (long-haul flights were offered a lunch box meal).

The cleanliness issue was further analyzed by concentrating on analyzing some of the most critical dimensions intimately related to the main concerns arising from the pandemic. In the pre-pandemic period, when passengers rated the airline, cleanliness was the least rated dimension (Fig. 5.13).

In the post-pandemic scenario, an expressive positive rating evidences the component cleanliness (Fig. 5.14). This change in terms of the users' rating indicates that passengers' awareness and sensitivity to cleanliness issues are higher after the pandemic, mostly because people had to follow strict hygiene and safety protocols imposed as virus containment measures, incorporating the importance of the procedure, and valuing much more this aspect in the airline offers.

Customer service is one critical aspect of any organization's performance and success. Regarding the ratings attributed to this item, it is possible to note that in both pre-and post-pandemic settings, users rate in the scale extremes—1 corresponding to a bad service or 5 as an excellent service. Figures 5.15 and 5.16 allow this interpretation.

Clients usually value food and beverage; for airlines, meals on board often represent a moment of enjoyment and relaxation, especially on long-haul flights. Analyzing SATA's ratings on this component, it is possible to understand that pre-pandemic passengers were not much concerned with valuing this aspect of the flight experience (Fig. 5.17). However, post-pandemic passengers were more incisive in rating food and beverage available on board (Fig. 5.18), denoting that the clients gained a new perspective of the airline meal offer as the customer experience is

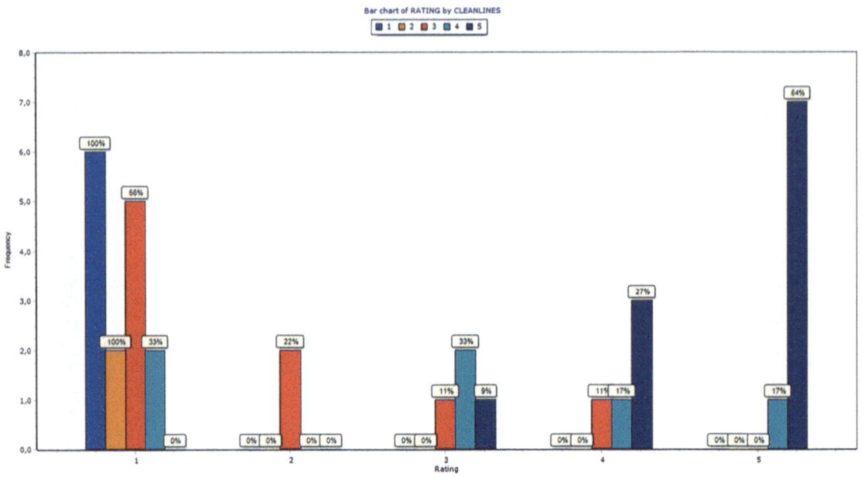

Fig. 5.14 Rating by cleanliness, post-pandemic period

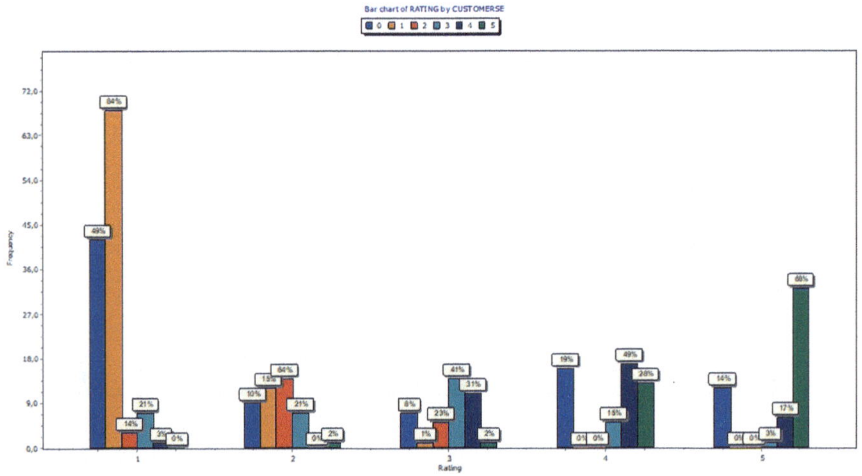

Fig. 5.15 Rating by customer service, pre-pandemic period

fundamental; in the case of tourism travels, the flight experience incorporates the whole tourism travel.

The data analysis shows that passengers are more aware of the service experience and more predisposed to share their experiences online. This attitude of rating and sharing experiences and opinions with other customers is a fundamental aspect that organizations must consider monitoring customer experience and quality of service, improving the aspects negatively rated by the consumers.

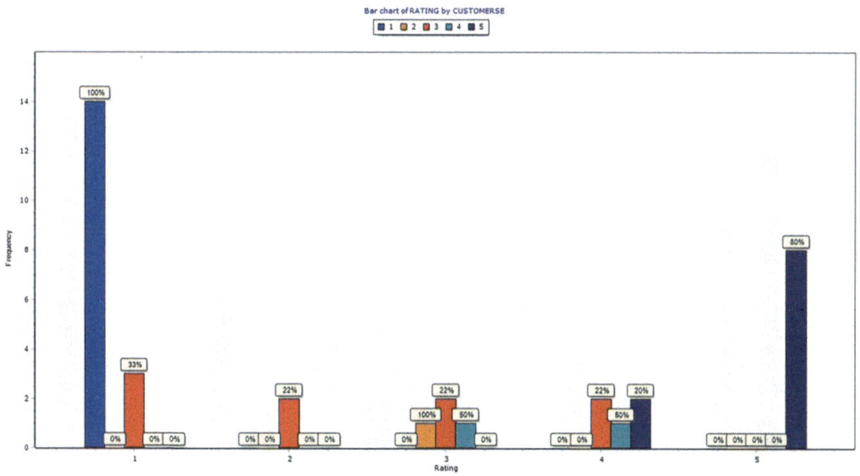

Fig. 5.16 Rating by customer service, post-pandemic period

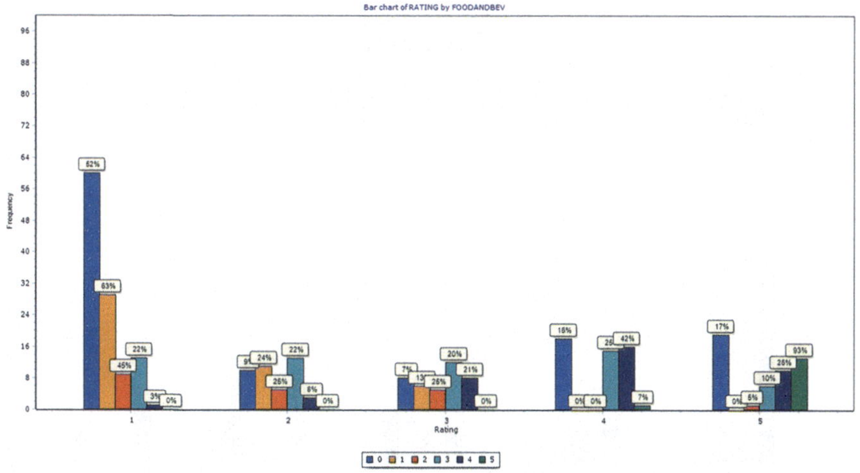

Fig. 5.17 Rating by food and beverage, pre-pandemic period

5.6 Conclusion

Airline passenger volume and traffic decreased during the first months of the pandemic, foreseeing a continued flat demand for air travel for as long as COVID-19 remained a public health issue—economic and financial fallout called for budget reviews and cost adjustments to deliver high-quality and efficient services. The scenario was real: airlines were forced to pause for too long and in a context of

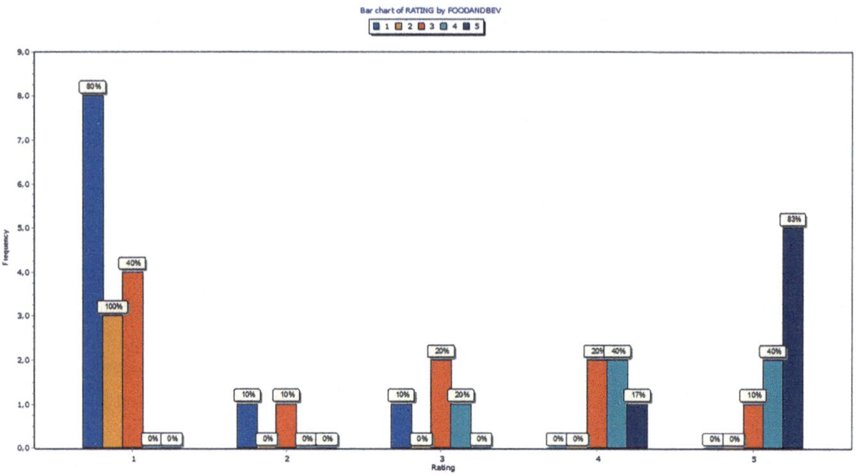

Fig. 5.18 Rating by food and beverage, post-pandemic period

uncertain demand considering the economic and financial impacts of the pandemic—a demand likely to return slowly and differently from before with direct implications for the whole aviation industry.

When the world started to reopen carefully after the lockdown, the aviation industry started to get back to business, reconnecting with customers and creating campaigns to raise awareness of their offerings. To this end, it was necessary to know how consumers were thinking about travel and tourism, one of the most affected sectors by the pandemic, and what the journey back to pre-COVID-19 activity levels would look like—times urged to look much further ahead and envision how the crisis could be surpassed, concentrating all efforts on managing the coronavirus impact, with rapid responses and strategic shifts.

The travel and tourism industry creates jobs, drives exports, and generates prosperity worldwide. However, it must be responsive to change; hotels, airlines, and other travel segments are particularly vulnerable to economic conditions [17]. The novel coronavirus paradigm presented new management challenges to organizations as work resumed; companies had to concentrate on high-performance standards, management, and health safety. As SATA Group companies emerged from lockdown, they were determined to ensure their people's and passengers' safety and the continuity of their operations, making it cost-effective and looking for transformation and renewal.

Like many other airlines, SATA faced a passenger demand reduction and a consequent revenue loss. Besides all the arrangements to control the transmission of COVID-19, the company needed to react quickly to the market dynamics, rationalizing transformations and rethinking the business based on strategic compromises. Real-time data, short-term planning, and optimization of processes and procedures were essential for good performance. The "new now" post-pandemic resulted in

new opportunities and business value creation in a significantly disrupted travel environment; passengers' experience had to be reshaped, with safety and trust as core values [17].

To guarantee strategic long-term success, airlines must have control and know-how of their operational capacities to maintain a balanced combination of flexibility and stability [12]. The companies' aim must be to gain and maintain a high strategic and operating management system, which comprises the development of the airline route network; optimization of aircraft fleet; development of an efficient tariff policy; a highly efficient cost and revenue management that enables the company to set the right price, to the right passenger, on the right time and in the right platform; development and implementation of airline strategy according to long-term objectives; employment of resource potential; creation of airline's positive image; enhancement of airline's product quality; and implementation of modern IT to optimize airlines' business processes [1, 6, 8, 13, 18, 19].

Moreover, as pointed out in the literature [2, 4], airline companies can gain by adopting a customer orientation. Prioritizing customer satisfaction cultivates loyalty, positive word-of-mouth, and a competitive edge. Through the analysis of the reviews found on TripAdvisor, this study clearly shows which dimensions are mainly valued by passengers and how they vary in time, reflecting operations modifications. As the model implies, reduced customer churn and employee satisfaction contribute to a better overall travel experience since in-flight services highly impact the rating. Continuous feedback-driven improvements foster innovation, while regulatory compliance is streamlined, and higher revenue and profitability result from repeat business and enhanced brand reputation. Ultimately, this customer-focused approach ensures sustainable growth and success in the fiercely competitive aviation industry.

SATA Group Airlines' strategic mindset should be based on a consistent value proposition, sustained by the commitment, quality, and safety of its operations and service, and a solid profit structure (fares, low costs, cost control, and profitable growth). Due to its peripheral geographical location and the performance of public service obligations, a well-succeeding venture can be accomplished by combining public service commitment with profit growth.

A small regional airline faces numerous challenges while competing with broader-based giants. Limited financial resources may hinder the ability to invest in advanced technology, modern aircraft, and expansive route networks. These operational conditions, in turn, affect the capacity to offer competitive pricing and extensive services. Brand recognition and reputation can be harder to establish, making it challenging to attract and retain customers amidst the dominance of well-known airlines. The smaller scale may also lead to operational inefficiencies and reduced supplier bargaining power. Regulatory compliance costs and industry-specific challenges further strain their resources, making it crucial for small regional airlines to strategically differentiate themselves by focusing on exceptional customer service, niche markets, and unique routes to carve out their space in the competitive landscape.

References

1. Arena, M., & Aprea, C. (2021). Impact of COVID-19 pandemic on air transport: Overview and implications. Advances in Environmental and Engineering Research, 2(1), 1–14.
2. Ban, H. J., & Kim, H. S. (2019). Understanding customer experience and satisfaction through airline passengers' online review. Sustainability, 11(15), 4066.
3. Bonfanti, A., Vigolo, V., & Yfantidou, G. (2021). The impact of the Covid-19 pandemic on customer experience design: The hotel managers' perspective. International Journal of Hospitality Management, 94, 102871.
4. Boudia, M., Mohamed, S., Bondoux, N., & Delahaye, T. (2021). Traveler centric airline offer design and optimization. Journal of Revenue and Pricing Management, 20(6), 634–645.
5. Budd, L., Ison, S., & Adrienne, N. (2020). European airline response to the COVID-19 pandemic–Contraction, consolidation and future considerations for airline business and management. Research in Transportation Business & Management, 37, 100578.
6. Cook, G. N., & Billig, B. G. (2023). Airline operations and management: a management textbook. Taylor & Francis.
7. Garrow, L., & Lurkin, V. (2021). How COVID-19 is impacting and reshaping the airline industry. Journal of Revenue and Pricing Management, 20, 3–9.
8. Gole, I., Dobrea, R. C., & Gombos, C. C. (2021). Aviation industry-challenges and uncertainties after the Covid-19 pandemic. In SHS Web of Conferences (Vol. 92, p. 01010). EDP Sciences.
9. Gössling, S., Scott, D., & Hall, C. M. (2020). Pandemics, tourism and global change: a rapid assessment of COVID-19. Journal of sustainable tourism, 29(1), 1–20.
10. Gursoy, D., & Chi, C. G. (2020). Effects of COVID-19 pandemic on hospitality industry: review of the current situations and a research agenda. Journal of Hospitality Marketing & Management, 29(5), 527–529.
11. Hassan, L. K., Santos, B. F., & Vink, J. (2021). Airline disruption management: A literature review and practical challenges. Computers & Operations Research, 127, 105137.
12. Lytvynenko, S. L., & Lytvynenko, L. L. (2018). Basic aspects of airline strategic management under joint arrangement of cargo-passenger flights. In International Scientific and Practical Conference World science (Vol. 3, No. 6, pp. 43–50). ROST.
13. Melas, D., & Melasova, K. (2020). The early impact of COVID-19 pandemic on the aviation industry. Acta Avionica, 22(1), 38–44.
14. Peixoto, E. (2012). SATA – De Sociedade de Estudos a Sociedade de Participações Sociais, Génese e Desenvolvimento Operacional. SATA – Junho 2012.
15. Pillai, S. G., Haldorai, K., Seo, W. S., & Kim, W. G. (2021). COVID-19 and hospitality 5.0: Redefining hospitality operations. International Journal of Hospitality Management, 94, 102869.
16. Riley, C. (coord.) (2019). 100 Anos de Aviação nos Açores. Instituto Açoriano de Cultura, Angra do Heroísmo, Maio de 2019.
17. Shapoval, V., Hägglund, P., Pizam, A., Abraham, V., Carlbäck, M., Nygren, T., & Smith, R. M. (2021). The COVID-19 pandemic effects on the hospitality industry using social systems theory: A multi-country comparison. International Journal of Hospitality Management, 94, 102813.
18. Wu, C. L. (2016). Airline operations and delay management: Insights from airline economics, networks and strategic schedule planning. Routledge.
19. Zhou, L., Liang, Z., Chou, C. A., & Chaovalitwongse, W. A. (2020). Airline planning and scheduling: Models and solution methodologies. Frontiers of Engineering Management, 7(1), 1–26.

Chapter 6
Effects of the COVID-19 Pandemic and Recovery in Aviation

Luiz Manuel Braga da Costa Campos

Contents

L. M. Braga da Costa Campos (✉)
Instituto Superior Técnico, University of Lisbon, Lisboa, Portugal
e-mail: luis.campos@tecnico.ulisboa.pt

L. M. Braga da Costa Campos (ed.), *Accelerating Sustainable Aviation Initiatives*,
Sustainable Aviation, https://doi.org/10.1007/978-3-031-73993-4_6

6.1 Introduction

The COVID-19 pandemic [1–5] has caused the biggest crisis in the history of aviation, affecting operations [6–9], most notably airlines and air transport [10–17], including sectors such as industry [18–20], airports [21], and insurance [22], with some mitigation policies [23–26] providing some relief toward recovery [27–30]. Several years of continuous growth were turned into an almost total grounding in the short time span of a few months (Sect. 6.2). The official Chinese version of the origin of COVID-19 starts with the pangolin in remote forests as the carrier with transmission through bats to people (Sect. 6.2.1) who eat them; the Western press has reported a rather different story (Sect. 6.2.2) of origin in a laboratory working on biological weapons of mass destruction near Wuhan, where the outbreak stated. Several analogies have been made with the COVID-19 pandemic (Sect. 6.3), including (Sect. 6.3.1) earlier historic pandemics with much higher death rates; (Sect. 6.3.2) world wars, which caused a much larger number of deaths with attempts to occupy territory; (Sect. 6.3.3) natural disasters, like earthquakes, volcanic eruptions, floods, fires, and tsunamis of much more limited geographical or temporal extent; and (Sect. 6.3.4) biological weapons of mass destruction with no known cure, leading to diseases whose spread is extremely difficult to contain.

The basic medical aspects to be taken into account in the context of the effects on aviation (Sect. 6.4) include (Sect. 6.4.1) the means of protection against the transmission of the coronavirus via water droplets; (Sect. 6.4.2) the treatment and healing, leading to deaths mainly of elderly people with previous ailments; and (Sect. 6.4.3) the prospects for developing a vaccine that should have high effectiveness and a lasting effect, without undesirable secondary consequences and be available in larger quantities at a moderate cost. Until the vaccine became available, the main way to counter the spread of the infection and avoid overwhelming the available medical resources (Sect. 6.5) was confinement: (Sect. 6.5.1) quarantining in infected cases and work at home to avoid contagion on the outside: (Sect. 6.5.2) avoiding large gatherings in confined spaces through teleconferences. These precautions cannot be easily applied (Sect. 6.5.3) to essential services such as the supply of medicines and food at pharmacies and supermarkets.

The social and economic effects of COVID-19 include some of the most undesirable records (Sect. 6.6) such as (Sect. 6.6.1) the deepest economic recession for a long time due to the decline of industrial production and most services; (Sect. 6.6.2) the highest unemployment level due to the paralyzation of the economy; and (Sect. 6.6.3) increased budget deficits and debts attempting to mitigate business and job losses. Only a few sectors like informatics have benefited from the lockdown (Sect. 6.6.4), with a larger number of people staying longer at home and meetings replaced by teleconferencing.

Aviation is one of the worst-hit sectors (Sect. 6.7) in almost every area: (Sect. 6.7.1) airlines have high fixed costs and no revenue from suspended flights, exhausting existing liquidity and leading to bankruptcy in a few months unless a rescue package is provided; (Sect. 6.7.2) the air cargo sector becomes essential in the fast

delivery of urgent medical supplies and protection equipment, but this hardly compensates the general decline due to reduced economic activity; (Sect. 6.7.3) aircraft leasing, which has grown from 4% to 35% of the market in two decades faces a sharp decline between high demand from airlines needing additional aircraft to less demand from the same airlines having most of their fleet grounded; (Sect. 6.7.4) airports lose revenues from canceled flights and lack of passengers, and cannot balance fixed costs, just as airlines; (Sect. 6.7.5) a similar situation applies to air navigation service providers (ANSPs) with no flight charges to cover costs with personnel like air traffic controllers (ATCs) and navigation ground infrastructure. The only niche sector with growth prospects was business aviation (Sect. 6.7.6), as the only alternative to canceled airline flights with reduced health risks.

The decline in airline traffic (Sect. 6.7) had an equally massive effect on the industry (Sect. 6.8), with losses everywhere without a single exceptional niche: (Sect. 6.8.1) airlines finding themselves unexpectedly with grounded or oversize fleets may not make new orders and be forced into cancellations or delayed deliveries, all of which mean less revenues for aircraft manufacturers to add to production slowdown or shut-down; (Sect. 6.8.2) first-tier suppliers, like aero engine manufacturers, fare even worse, adding to (i) the loss of production of engines for new aircraft and also (ii) the reduction in maintenance and spares revenue due to reduced flying; (Sect. 6.8.3) down the supply chain, the effect (i) of reduced aircraft production may be amplified for smaller companies with less resources and high dependence on the aeronautical market; (Sect. 6.8.4) the maintenance, repair, and overhaul (MRO) sector is hit as first-tier suppliers by (ii) reduced flying, leading to less regular servicing and postponement of major work; and (Sect. 6.8.5) used aircraft prices have a sharp drop as airlines dispose of surplus aircraft, starting with the oldest, less efficient and more polluting models. The prospects for survival (Sect. 6.8.6) are (i) worse for small suppliers of services that can be taken up by higher tiers with excessive workforce and (ii) better for unique small suppliers needed for recovery and large companies that cannot be allowed to fail.

The dire situation caused by COVID-19 was addressed in the United States by the Coronavirus Air Relief and Economic Stability (CARES) bill passed by Congress, which included aviation (Sect. 6.9) besides other sectors. Substantial sums were made available setting conditions lasting only for 6 months such as (Sect. 6.9.1) no dismissal or pay cuts for employees, (Sect. 6.9.2) no reduction in domestic flights, and (Sect. 6.9.3) support to airports. In spite of or because of the large sums involved, the rules of allocation lead to widespread controversy about unequal treatment in all cases: airlines, airports, and employment. The CARES act was less controversial on (Sect. 6.9.4) bailouts for industry, although the size of the $60 billion Boeing fund did raise some eyebrows. However, the biggest issue was after spending $80 billion with airlines and airports for 6 months, what would happen (Sect. 6.9.5) in the future: worker lay-offs and route cancellations or another large subsidy without long-term prospects or economic recovery after crossing a difficult period? The latter proved to be the case with sharp post-Covid recovery of air transport.

Unlike the United States that applied blanket temporary rules lasting 6 months, other countries took a longer-term view, with a case-by-case analysis, leading for

airlines (Sect. 6.10) to (Sect. 6.10.1) a rescue package in case of flag carriers with strong or not so strong records, sometimes with environmental and other conditions, mostly in developed countries with strong economies; and (Sect. 6.10.2) bankruptcies of major airlines and flag carriers with poor performance records mainly in developing countries with weak indebted economies, already in financial strain before aggravation by the COVID-19 pandemic.

In Europe, government policy toward rescuing airlines differed by country (Sect. 6.11), with significant contrasts: (Sect. 6.11.1) the $9 billion rescue of Lufthansa by the German government involved strict conditions on the reduction of fleet size and loss of airport slots, leaving an open road to increased or decreased state control toward nationalization or privatization; (Sect. 6.11.2) France and the Netherlands agreed on an 80:20 split of a $7 billion rescue for Air France-KLM; (Sect. 6.11.3) Italy provided Alitalia with an ambitious rescue plan assuming far better performance in the future than that achieved in the past; and (Sect. 6.11.4) British Airways-Iberia have protested against government travel restrictions, besides receiving limited support. The low-cost carriers (LCCs) may survive on their own, with layoffs, salary reductions, and limited support (Sect. 6.11.5) trying to ramp up services and recover revenues as quickly as possible.

A good example of a comprehensive rescue package for the aerospace sector is provided (Sect. 6.12) by France: (Sect. 6.12.1) the collapse of the civil market was be partially compensated for the industry also active in the defense sector by keeping or expanding military contracts that are stable in the long term as long as economic conditions allow; (Sect. 6.12.2) the industry support included not only the large original equipment manufacturers (OEMs) but also smaller companies down the supply chain whose products and skills were essential for post-COVID-19 recovery; and (Sect. 6.12.3) supporting research and technology for 20–30% more efficient airliners in the 2030–2035 time frame kept the design and development active and ready to support future competitiveness.

Comprehensive and well-coordinated support programs for the aerospace sector, at a national level like in France and Germany, become most effective, together with initiatives at the level of the European Union: (Sect. 6.13), for example, (Sect. 6.13.1) the overall COVID-19 cohesion fund strengthened national economies, enabling the support of several sectors including aeronautics; (Sect. 6.13.2) the aeronautical and space programs of the EU support continued competitiveness and initiatives like electrification, a hydrogen economy, and cleaner environment; and (Sect. 6.13.3) EU-wide measures on deconfinement and health rules, even with local exceptions, were essential to accelerate recovery of the aviation sector from the COVID-19 crisis.

The prospects for recovery of the aeronautical sector from the COVID-19 crisis were (Sect. 6.14) different for (Sect. 6.14.1) a faster V-shaped ramp-up for regional and short-haul flights among regions with almost simultaneous deconfinement; (Sect. 6.14.2) the uncoordinated deconfinement measures at worldwide-level harmed long-haul travel, which remained at low level longer until global deconfinement lead to an abrupt U-shaped recovery. Besides the uncertainty of when short-haul and long-haul traffic recovered, there was the major question (Sect. 6.14.3) of ultimate

recovery level in the range from full (100%) to partial (down to 70%?) due to changed habits, like more teleconferencing instead of professional travel. The post-Covid recovery to near pre-Covid levels was rapid with higher fares, costs and revenues.

The post-COVID-19 recovery of aviation was not only enabled by coordinated deconfinement between travel destinations but also by the commitment to on non-board health (Sect. 6.15) provided by anti-pandemic measures, like (Sect. 6.15.1) directed air-conditioning flows and hospital-grade virus filters. The airport screening and passenger compliance with health protection measures helped regain trust in air travel (Sect. 6.15.2). Bearing in mind that flight crew, in particular cabin attendants, fly often with many passengers, their health protection was given particular attention (Sect. 6.15.3).

Although most countries have emergency planning for natural disasters (fires, floods, earthquakes, etc.), there appears to be little planning for a pandemic, which has far greater consequences, countered in the case of COVID-19 by an assortment of improvised measures often late and not fully effective. COVID-19 may not be the last pandemic the world will see, and pandemic planning (Sect. 6.16) should cover at least the following aspects: (Sect. 6.16.1) isolation and confinement measures to limit the geographical spread, to ensure that a minimum of the affected population is supported by a maximum of the healthy population; (Sect. 6.16.2) using all the resources of the healthy population to help, recover, and cure the affected population; and (Sect. 6.16.3) making sure medical and other essential services and resources are not overwhelmed by having preplanned reserves, facilities, and ability to ramp up rapidly production of protective equipment and medicines for treatment.

As one of the worst pandemics affecting mankind, and the biggest crisis in aviation history, COVID-19 has challenged the survival and recovery prospects (Sect. 6.17) of all sectors, namely (Sect. 6.17.1) airlines with reduced passenger demand met by a fraction of the fleet, with remaining aircraft parked, or stored or likely to be disposed of, until traffic recovers sufficiently; (Sect. 6.17.2) aircraft manufacturers facing cancellations, delayed deliveries, and a lack of new orders for several years until airlines need additional or replacement aircraft; and (Sect. 6.17.3) industry in all tiers 1–4 of the supply chain with high investment to increase production at low prices before COVID-19 facing an abrupt slowdown due to the pandemic. In retrospect (Sect. 6.18) although the evolution (second wave?) and recovery (full or partial, when?) from the COVID-19 pandemic were unpredictable, its effects have shown that better planning is needed to avoid comparable disruption in the future; the aeronautical sector had plans to rapidly convert to support the fight against the pandemic and then recover quickly (Sect. 6.18.1). The government interventions during the COVID-19 pandemic have differed around the world, and support for the aeronautical sector is no exception. In the United States, the CARES bill provided large sums with temporary rules, whereas in Europe smaller support was made conditional on economic and environmental performance and innovative technology for higher competitiveness (Sect. 6.18.2).

The predictions of slow post-COVID-19 recovery on the scale of several months or years were confounded by a very sharp rebound (Sect. 6.19) to about 80% of the pre-COVID level in a few months after elimination of restrictions. For national and regional air traffic, the recovery came earlier than for international routes dependent

on the elimination of COVID restrictions in several countries (Sect. 6.19.1). The driver for sharp recovery was pent-up demand by individuals and families eager to travel after long periods of confinement or travel bans, in spite of higher ticket prices. The sharp increase in demand left operators and service providers struggling to cope (Sect. 6.19.2), with airlines and airports short of staff, following dismissals and early retirements during the COVID crisis. Staff shortages led to salary and cost increases, but revenues increased even faster due to high demand and higher ticket prices, restoring airline profitability in most cases. The increased demand for new and more environmentally friendly aircraft was met by efforts to ramp up production, which was slowed by the inability of the supply chain to ramp up rapidly after being weakened by the COVID decline, and short of the required numbers of staff and level of investment (Sect. 6.19.3). Despite the growing pains, the sharp recovery from COVID quickly restored the prospect of sustained growth of aviation (Sect. 6.20) on a broad scale with most sectors strugling to meet a boom of overdemand.

6.2 Origin

The COVID-19 pandemic started in China, and the official story (Sect. 6.3.1) is that the original carrier was a nearly extinct wild animal in a remote forest, and transmission to humans was made by bats. There is evidence to suggest that the Chinese authorities are not telling the full story, leading to various speculations, including newspaper news that the virus came out of a laboratory (Sect. 6.3.2).

6.2.1 Remote Habitat

As mankind expands into formerly virgin regions of the world, new viruses and diseases can appear that did not occur in other already inhabited regions. The case of COVID-19 is a Severe Acute Respiratoreny Syndrome (SARS) due to a virus carried by the Pangolin, an almost extinct species living in remote Chinese forests. The virus would be passed to bats, and then to humans who eat them. Bats are used as food in China and other Asian countries like Indonesia. Some questions are implied about food health standards.

COVID-19 was first reported by China in December 2019, but there is evidence it may have occurred earlier. After the disease spread to Europe, it was found that earlier cases of hospital internment in France, as far back as October 2019, were in fact instances of COVID-19, not identified at the time. Also, the analysis of waste in more than one country, including the Netherlands and Italy, showed the presence of COVID-19 at about the same time, that is, about 2 months before the Chinese announcement.

The domestic air traffic in China had a significant drop in the second half of 2019, which could be associated with some form of confinement. This suggests that COVID-19 may have appeared around the middle of 2019, and being unable to

contain or confine it, the Chinese authorities chose to become open about a pandemic that could no longer be concealed or disguised. This led to some questions not only about the calendar of the emergence of COVID-19 but also about its origins.

6.2.2 Out of a Laboratory?

The *New York Times* reported that COVID-19 had come out of a laboratory near Wuhan doing research on biological weapons of mass destruction (WMD) based on the COVID virus. In Western countries, research on biological weapons is limited to diseases for which an antidote is known, to protect those involved in such work, in case of accidental infection. It is said that in Russia, and possibly China, the research includes diseases with no known cure and that some researchers have died because of being infected during their work.

In case the laboratory in Wuhan was involved in the development of biological weapons based on COVID and without a known cure, any failure of isolation from the outside could have dire consequences. It is possible that a researcher involved in such developments could be infected without being aware, and during the incubation period of the virus, without external signs, continued a normal life outside the laboratory, unknowingly contaminating others.

President Donald Trump blamed China for the disease, without giving details; this claim may not carry much weight since he also blames the World Health Organization (WHO) that is instrumental in fighting the pandemic and decided that the United States would leave the organization. When asked about the origin of COVID-19, other Western leaders were a bit evasive, for example, (i) Emmanuel Macron replied that China was not telling the full story and (ii) Angela Merkel asked China to give a full account of the events related to COVID-19.

The prime minister of Australia was more explicit, asking for an independent international inquiry into the origins of COVID-19, drawing an angry rejection from the Chinese authorities. A British spokesman responded that these are very serious charges that require evidence. It is possible that the evidence exists in the files of Western intelligence services but is too sensitive to be revealed publicly. Some World War II secrets were made public many decades after the end of the war in 1945, and some may still be restricted, due to their sensitivity. Thus, the question of the real origin of COVID-19 may remain open to speculation for a long time while unofficial conspiracy theories abound.

China resisted for several months the proposal of WHO to send an international team to investigate the origins of COVID-19. As the team completed its visit, it was announced that the origin of COVID-19 in a laboratory could be excluded. The next day the director-general of WHO called a press conference stating that, after speaking to the team of investigators, he concluded that "all options remain on the table." Since then, the WHO has made clear on several occasions that the possibility of COVID-19 having originated in a laboratory could not be excluded, drawing angry reactions from Chinese authorities, and no overture for further investigations.

6.3 Analogies

The COVID-19 pandemic has been compared to tragic events, like other pandemics in the past (Sect. 6.3.1), the two world wars (Sect. 6.3.2), and natural disasters (Sect. 6.3.3). Its effects may be comparable to those of a biological weapon of mass destruction (Sect. 6.3.4).

6.3.1 Previous Pandemics

As of August 16, 2020, COVID-19 had caused more than 21.4 million infections in 188 countries and territories, causing 771 thousand deaths and the remaining recovered, implying a death rate of 3–6%. This can be compared with some of the worst pandemics in the history of mankind: (i) the bubonic plague of Justinian in 541–542 that killed 25–100 million or 40–50% of the population of the Roman Empire; (ii) the bubonic was also the cause of the Black Death in 1346–1353 in Medieval Europe, killing 75–200 million people or 10–60% of the population.

Large death tolls were also associated with (iii) the Cocoliztli epidemic, possibly due to *Salmonella enterica*, that killed 5–15 million people or 80% of the population in Mexico during the Spanish colonization; (iv) the influenza A virus H1N1 killed 17–200 million people worldwide in 1918 at the end of World War I, causing even more deaths than the conflict; and (v) the HIV/AIDS pandemic, still ongoing since 1981, has already killed more than 32 million people worldwide.

Although the total number of deaths and the death rate are lower, the number of people infected by COVID-19 is comparable to the worst pandemics in history, with the rate of spread around the world much faster in the age of globalization. With rising health and living standards, COVID-19 is one of the most dramatic events in the history of mankind, for which there was no warning or preparation, leading to a mixture of improvised and poorly coordinated responses to a largely unknown deadly phenomenon.

6.3.2 World Wars

The COVID-19 pandemic has been compared to the two world wars: (i) World War I killed an estimated 22 million people, including nine million combatants and 13 million civilians, over the period 1914–1918, and was fought mostly in France near Paris, though it spread over Europe and beyond to the Ottoman Empire; and (ii) World War II was the deadliest conflict in history, with 70–85 million deaths, focused in two fronts: Europe and the Mediterranean and Atlantic and Asia and the Pacific.

The COVID-19 pandemic differs in a much lower number of deaths because it is not a conflict motivated by some nations trying to annex the territory of others with armed forces trying to cause death and destruction. The number of people affected is, however, comparable for the very opposite reason of an open society free of major wars, where freedom of movement worldwide allows the fast spread of a virus facilitated by contamination even before the health effects become evident. This has led to the description of COVID-19 as a hidden enemy of mankind.

6.3.3 Natural Disasters

Most nations have emergency plans to deal with the kind of natural disasters more likely to occur in their territories, such as (i) forest fires in the Mediterranean, California, or Australia; (ii) earthquakes along the San Andreas Fault around the world; (iii) lava and ash from volcanic eruptions; (iv) sea tides from tsunamis; (v) extreme winds from tornados and tropical storms; (vi) floods from monsoons and torrential rain; and (vii) landslides and avalanches.

All these natural disasters can cause death and destruction on a large scale, but most are limited in geographical location and temporal duration; the unaffected neighboring regions of the same country and indeed other countries can come to help. COVID-19 can be considered a natural disaster caused by an unknown virus (Sect. 6.3.1), and its global spread and temporal extent affected many countries that were ill-prepared to counter it internally and less able to help others.

6.3.4 Biological Weapons

The weapons of mass destruction (WMD) are classified as ABC for Atomic, Biological, and Chemical. The atomic weapons are difficult to disguise because their development requires large facilities, and their long-term effects of radioactivity beyond a large explosion were found tragically after their use in Hiroshima and Nagasaki to end World War II; the number of victims was not higher in an incendiary bomb attack on Tokyo that had predominantly wooden housing, but it was the sole use of atomic weapons that registered most in war history.

After the infamous use of mustard gas over the trenches of northern France in World War I, the Geneva Convention banned both chemical and biological weapons. The production of chemical weapons can be disguised due to their similarity to fertilizers and pesticides, and they have been used in local wars by the regime of Saddam Hussein against the Kurds in Iraq and the regime of Bashar al Assad against the opposition in Syria. Biological weapons are the easiest to conceal in a laboratory.

It is said that attempted use of chemical weapons by Iraq during the American invasion to liberate Kuwait had unexpected consequences when the wind direction changed against the unprotected users. In the subsequent invasion of Iraq, the

Americans were prepared for chemical weapons that this time were not used. Although available to major belligerents during World War II, chemical and biological weapons were never used. It was realized that biological weapons are largelly uncontrollable in their geographical spread and temporal duration, as COVID-19 is tragically showing now.

6.4 Medical Aspects

The medical aspects of COVID-19 are not the focus of the present chapter, but it is necessary to mention some topics relevant to its effects on aviation, namely (Sect. 6.4.1) the means of protection to avoid the spreading of the pandemic; (Sect. 6.4.2) the means to cure the disease and the contribution that aviation can make providing means of support; and (Sect. 6.4.3) the development of vaccines and the extent to which they contributed to end the pandemic.

6.4.1 Medical Aspects

COVID-19 is transmitted through water droplets by coughing or sneezing or by contact with contaminated surfaces. The social distancing of 2 m prevents transmission by normal respiration, but not by coughing when water droplets can travel much farther. The most effective and guaranteed protection is the face mask covering the mouth and nose. It prevents water droplets from the wearer from reaching others and helps prevent water droplets from others reaching the mask wearer. In other words, the face mask prevents the wearer from infecting others but may be less effective at preventing others from infecting the wearer if COVID droplets bypass the mask with respiration.

The surgical mask with single use has proved unfailingly effective in millions of cases over many years. There are washable masks and masks with filters. The screen is by no means a substitute for the mask, although it can be used as a supplement. There is no evidence of the effectiveness of the screen as there is for the mask. The screen does not prevent the user from sending water droplets in certain directions nor does it prevent water droplets from coming in.

It must be assumed that a mask can be contaminated by others from the outside and could be contaminated from the inside by the wearer but has the main virtue is separating the two. The contact of droplets with the skin is another means of infection and may imply the use of gloves and indeed complete protective clothing. Putting on and removing the mask, gloves, and clothing also need to consider that they may be infected after use or by contact with the hands and must be disposed of safely.

6.4.2 Healing

COVID-19 has an incubation period of several days before the main symptoms (fever above 38 °C, respiratory difficulties, and loss of strength) become apparent. A simple temperature check with a thermometer may serve for detection after the incubation phase, but during the asymptomatic incubation phase, medical tests are needed. Tests to detect the presence of antibodies are most useful to show that the patient has been infected and is cured and therefore immune to COVID-19 in the future.

The treatment of infected cases needs quarantine and in extreme cases intensive care units with ventilators to help respiration. If the disease spreads too widely, the intensive care units may be overwhelmed. The improvisation of additional medical facilities depends on the fast supply of a sufficient quantity of means of treatment. Aviation may have contributed to the spread of COVID-19 as the worldwide means of fast transport but has also allowed the provision of urgent medical supplies across nations and continents faster than any other means of transport could do.

Some words of caution are needed about the effectiveness and availability of screening methods. Temperature measurement is the simplest but can give both (i) false negatives since the temperature is normal during the asymptomatic incubation phase; and (ii) false positives if body temperature is higher than normal for reasons other than COVID-19, such as hot climate, inside a car in the sun, and personal variations. In this case, (ii) it may be necessary to resort to medical tests that take longer and may provide a certificate valid for a stated period.

6.4.3 Vaccine

Since COVID-19 was an unknown disease when it appeared, there were no known vaccines despite dubious claims that this or other products are effective, for example, some compounds that are known to be toxic. Since the death rate from COVID-19 is, fortunately, relatively low at 3–6%, it is possible to do worse with an aggressive medication not adequately proven. This is one more reason for adherence to the WHO standards for developments of vaccines requiring thousands of tests.

The heads of foreign affairs of the United States, the United Kingdom, and Canada have held a press conference accusing hackers from Russian state-backed organizations of trying to disrupt the development of COVID-19 vaccines in their countries. The motivations for this may have been made clear barely a month later when President Vladimir Putin came on television announcing that Russia would be the first to make available a COVID-19 vaccine on January 1, 2021. He claimed that the vaccine had been used on his daughter, although the number of tests, in the tens, fell quite short of the WHO standards of thousands.

There was news of vaccine details being stolen from the European health agency before Russia announced a new vaccine Sputnik-V. The name was obviously a propaganda exercise since the vaccine had nothing to do with the spacecraft; the implication was that the vaccine was as innovative as the first spacecraft. In contrast, it was speculated that V was the fifth attempt after four failed attempts; the reply was that "V" was for victory. That was hardly the case since Russia was one of the last countries to contain the COVID pandemic due to a lack of vaccines and/or lack of confidence or reluctance of the population to take them.

Shortly after Putin's announcement, a German research institute claimed it should have a COVID-19 vaccine ready by October 2020, although earlier the German health minister had predicted that a COVID-19 vaccine would not be available before the mid of 2021. There were major well-funded efforts in several countries in the race to develop the first COVID-19 vaccine. Governments did sign sizeable contracts to produce millions of vaccines with the few pharmaceutical companies in the world able to produce such quantities.

Aside from the questions of widespread availability and cost, the three main issues about a new vaccine were (i) effectiveness: it is effective in preventing contamination in what fraction of cases? 70%, 90%, 99%? (ii) duration: the protection lasts for how long: weeks, months, years, lifetime? (iii) side effects: are there some undesirable side effects from using the vaccine and how long do they take to appear? For an old vaccine used for decades in millions of cases, the answers to these questions are well known, but for a new vaccine, the short of time of usage is a limitation.

A vaccine against COVID-19 was in any case and undoubtedly a major progress and the best if not only hope of eradicating the pandemic. On the other hand, the percentage of the population infected is small (typically 1 per 60) and the death rate after infection is also modest (3–6%), and this must be weighed against the risks of a new vaccine with a limited test of time. Other issues were whether the vaccine should be mandatory or voluntary, that is available to everyone who wishes to take it and how long it takes to vaccinate the whole willing population.

The availability of a vaccine against COVID-10 was a great success in eradicating the disease. As the percentage of immunized population exceeded about 70%, the infection rate dropped below the healing rate. The development of totally new vaccines in less than 2 years versus the past standard of one decade was due to an unprecedented level of cooperation among medical researchers worldwide and unrestricted resources made available for testing and production. With a wide variety of vaccines and high production rates, COVID-19 was eradicated first in wealthier countries able to cover the costs and soon was also eliminated worldwide with international aid contributions.

The COVID vaccines differed from the traditional vaccines that caused the body to generate antibodies by injecting a passivated version of the disease or virus. Instead, one version of the vaccine modified the DNA to lead to the production of COVID antibodies; another version did not modify the DNA but gave an instruction to the same effect. The former vaccine was much cheaper because it could be stored

in normal refrigerators, whereas the latter required much lower temperatures and more expensive means of transportation and storage. The COVID pandemic led to unprecedented large profits for pharmaceutical companies. The majority of patients did not have strong and persistent reactions to the COVID vaccines; the cases of strong and lasting negative side effects were not too numerous, and certainly a lesser evil than the spread of the pandemic, justifying a strong support for vaccination worldwide.

6.5 Confinement

Prior to the availability of the vaccine, the main protective measures against COVID-19 were confinement, at least in Europe, with other regions of the world taking similar or more relaxed measures. The confinement has implied working at home when possible (Sect. 6.5.1) and replacing face-to-face meetings with teleconferencing (Sect. 6.5.2); confinement is not possible for some essential services (Sect. 6.5.3).

6.5.1 Medical Aspects

The alternative of working at home instead of spending hours commuting to and from the job had been discussed for a long time, adopted in a few professions or instances, but never widely put into practice. The imperative need to stop the spreading of COVID-19 led at least in European countries to mandatory confinement at home. The rules were generally respected in most countries, with justified exceptions, and moderate police enforcement and the occasional fine.

Confinement has more serious consequences for industry and factories that are forced to close. In some countries like the United States, infected and noninfected workers continued to work alongside in factories with "protective" measures that are unlikely to compare with confinement. This may be one reason why the effects of COVID-19 are more widespread and have a greater scale in some countries, with the United States having the largest number of cases.

The number of official cases in China now ranks low in the world list as the 33rd country with 84,827 infections by August 16, 2020. This contrasts with 21,480,111 worldwide, 5,437,012 in the United States, 3,275,895 in Brazil, 2,589,682 in India, and 922,852 in Russia. Bearing in mind that China has the largest population in the world and was both the origin of COVID-19 and place of a second wave, the official of a number of cases smaller than Israel do not help the credibility of Chinese accounts of COVID-19.

6.5.2 *Teleconferencing*

The face-to-face meetings are easily replaced by teleconferences for small groups. Large symposia can still be organized, and some are simply canceled. Replacing the face-to-face meeting of an international committee by a teleconference saves 1 or 2 days of travel time. On the other hand, the number of teleconferences tends to increase since they are easier to attend than live meetings requiring travel.

Other live activities like shows and tourism were severely affected, waiting for deconfinement and easing of travel restrictions. Live entertainment was replaced by television shows at home. It remains to be seen whether people will be eager to return to live shows when they become possible again. Tourism was discouraged by travel restrictions and government advice that it is safer to make vacations in the home country traveling by car.

6.5.3 *Essential Services*

The essential services that must be retained include access to food and medicine, implying the that supermarkets and pharmacies must remain open. Social distancing limits capacity and can lead to long queues. All nonessential services tend to be postponed or closed unless clever workarounds are devised. Restaurants resort to supplying takeaway food, and services delivering food, goods, and other products at home proliferated.

6.6 Social Effects

The social effects of COVID-19 included the quarantine of those infected or suspected to be infected and the confinement at home or lockdown of the remaining population (Sect. 6.6.1). The resulting reduction in economic activity would lead to high unemployment and many insolvencies (Sect. 6.6.2) motivating significant government intervention that increased budget deficit and public debt (Sect. 6.6.3). Among the generally depressing decline of most sectors of activity, telecommunications contrast with the high demand and use due to confinement at home and teleconferencing.

6.6.1 *Quarantine*

The cases of infection by COVID-19 require quarantine in an intensive medical care unit of a hospital, in the more vulnerable cases of elderly people or those with respiratory problems. In order not to overwhelm hospitals, quarantine at home was

preferred for suspected infections and less serious cases. Homes for elderly people contributed to more than half of the deaths due to COVID -19 in several countries due to the proximity of a vulnerable population.

Deconfinement lead to a higher risk of a second wave of COVID-19 due to a part of the population not wearing face masks or respecting social distancing. There are also situations like public transport where social distancing is not feasible, the more so for longer travels. Some of the worst cases of COVID-19 infections have occurred in cruise ships and warships where many people occupy a limited space for a long time, leading to high infection rates.

6.6.2 Unemployment

The lockdown lead to a decline of economic activity in which many business activities cannot cope with costs or pay salaries and require government assistance to survive and avoid redundancies. Despite these measures, unemployment rose and the number of bankruptcies increased during covid-19. The number of requests for unemployment benefits rose in most countries like the United States. Despite this, in the United States, some healthy workers agreed to work alongside infected workers for fear of losing their jobs.

6.6.3 Economic Decline

The reduction in economic activity could not be fully compensated by government support, leading to economic decline. The decrease in Gross Domestic Product (GDP) for the worst second quarter of 2020 when confinement measures came into effect varied from 10% for countries like Germany, which took early measures, to 20% for the United Kingdom, which had a slow response. The figures on GDP reduction for the whole year of 2020 compared to 2019 depended very much on recovery from COVID-19 and could be worse due to the effects of a second wave.

Perhaps more important than the recession in 1 year was the recovery in the following years. If the decline in 2020 is 6–10% and 4–8% recovered in 2021, then the difference from 2019 is reduced to 2–4%. This smaller reduction over a period of 2 years was still a very negative result compared with what should have been 2 years of growth in the absence of COVID-19. A further issue was how many more years will be needed to come back to 2019 levels, with expectations of perhaps 2 or 3 with more pessimistic predictions of up to 4 or 5.

6.6.4 *Dependence on Telecommunications*

The COVID-19 crisis has affected negatively almost every sector of activity with few exceptions like telecommunications equipment and services. Work at home and teleconferencing rely almost entirely on telecommunications. The demand for more modern and efficient equipment and the much higher utilization rates translated to a boom in the sector. Confinement at home implies that both work and entertainment depend on telecommunications and live contacts on the job are replaced by teleconferencing.

6.7 Effect on Air Services

The COVID-19 pandemics caused the largest crisis in the history of aviation, with massive negative effects in almost all sectors: (Sect. 6.7.1) passenger traffic declined sharply to levels as low as an almost complete suspension; (Sect. 6.7.1) air cargo provided a much needed fast supply of medical equipment and supplies but was otherwise weak; (Sect. 6.7.3) aircraft leasing basically stopped, with airlines having large parts of their own fleets grounded; (Sect. 6.7.4) airports like airlines faced significant fixed costs while lacking revenues to match due to low passenger demand; and (Sect. 6.7.5) air navigation service providers (ANSPs) likewise had to cope with high fixed costs and no revenue from flights. The only growing sector was business aviation (Sect. 6.7.6) due to the need for safe business travel to replace unavailable or limited airline services.

6.7.1 *Airline Passengers*

The statistics on declining passenger traffic after a widespread lockdown in March 2020 due to COVID-19 were overwhelming both in the size and steepness of the change. Figure 6.1 shows the percentage of global passenger traffic by region overlaid on the number of confirmed COVID-19 cases per country on March 15, 2020, showing that the pandemic affected regions with high traffic. Figure 6.2 shows a decline in most regions of the world over 2 weeks from March 11 to 25, 2020. Figure 6.3 focuses on available seats in Europe and North America and again shows a sharp dip in March 2020.

Figure 6.4 shows stable passenger traffic in Southeast Asia for the whole of the 3 years 2017, 2018, and 2019, followed by a sharp decline to about half in the first quarter of 2020. Figure 6.5 shows that Chinese airline operations were stable in the period February 17 to April 5, 2019, and in the same period of 2020 started at one-third of the level and rose to about two-thirds with government incentives. The large

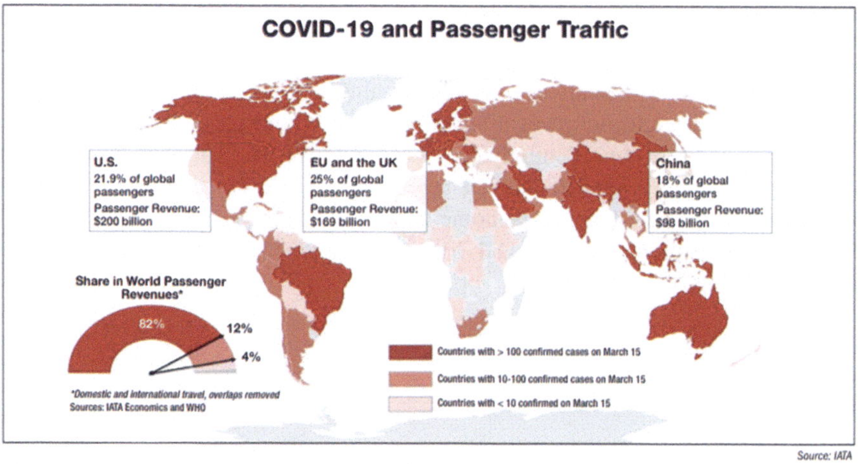

Fig. 6.1 World distribution of air traffic and confirmed COVID-19 infections on March 15, 2020

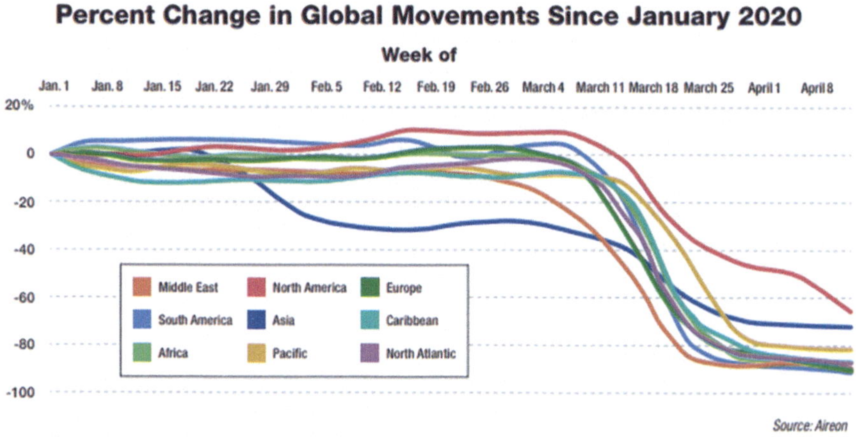

Fig. 6.2 Percent change in air traffic in different regions of the world with a sharp drop on March 11–25, 2020, due to confinement and travel restrictions

gap at the start of the period on February 17 shows a massive decline due to COVID-19 in 2020 compared to the normal level in 2019.

Figure 6.6 shows that the almost simultaneous adoption of confinement in Europe led to a sharp drop in traffic visible on a day-to-day basis between March 22 and 29, 2020. Figure 6.7 shows that the activity of the four mainline airlines in the United States dropped to less than half during March 2020 and less than a third by the end of April. With high fixed costs and a sharp reduction in revenue, the airlines could survive only a few months while consuming their existing liquidity.

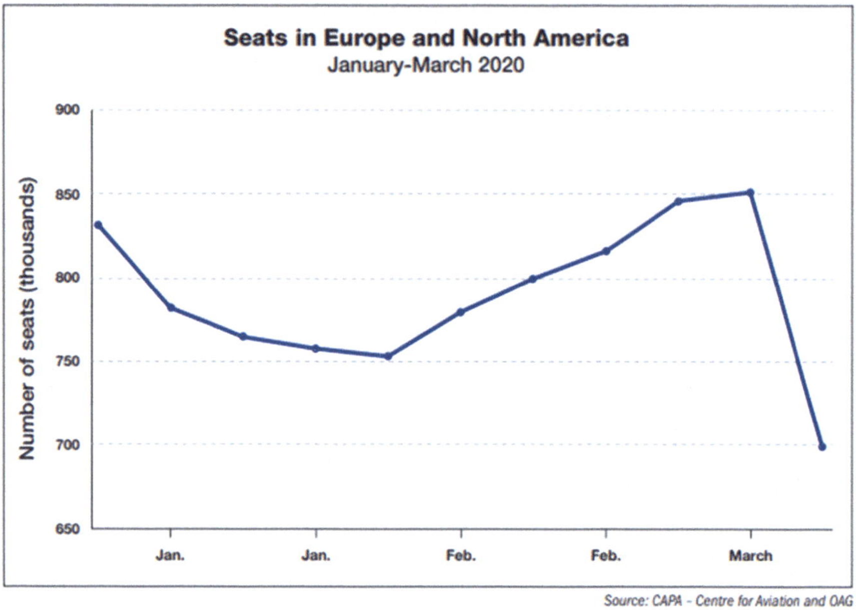

Fig. 6.3 Available seats in Europe and North America in the period January–March 2020 showing a sharp drop in March

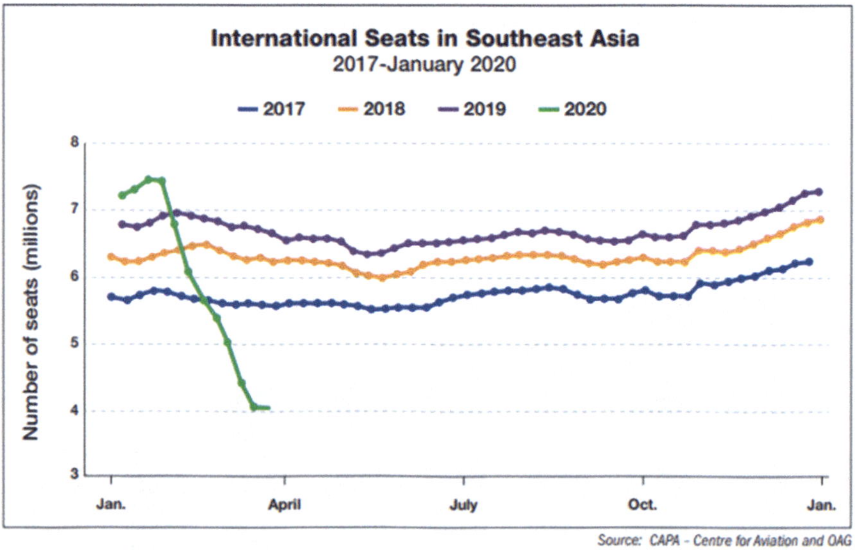

Fig. 6.4 International seats in Southeast Asia and 2017–2020 showing a sharp drop in February–April 2020

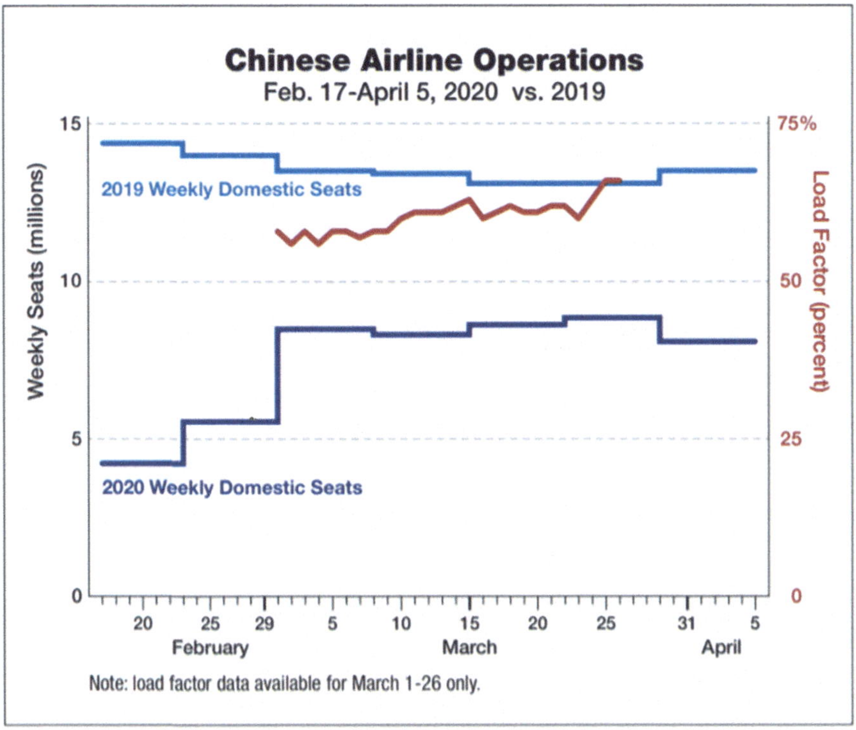

Sources: CAPA – Centre for Aviation and OAG; and internal industry statistics

Fig. 6.5 Comparison of weekly domestic seats in China in 2019 and 2020 showing the effects of the COVID-19 crisis and government-mandated recovery

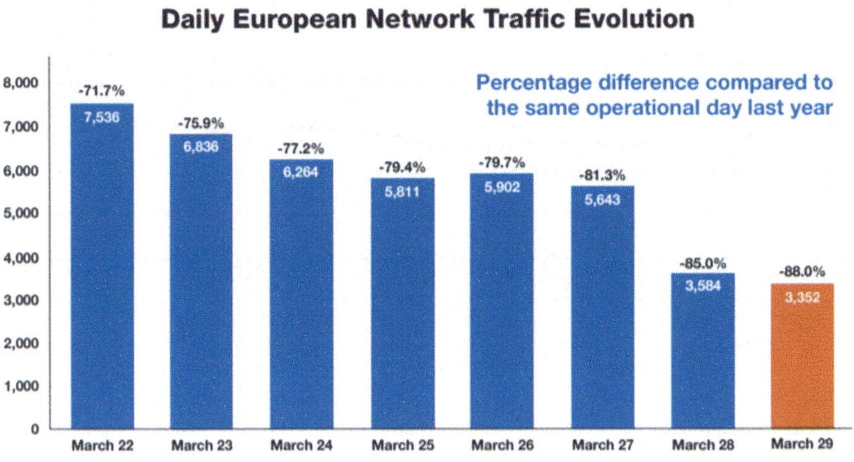

Source: Eurocontrol

Fig. 6.6 Comparison of weekly domestic seats in China in 2019 and 2020 showing the effects of the COVID-19 crisis and government-mandated recovery

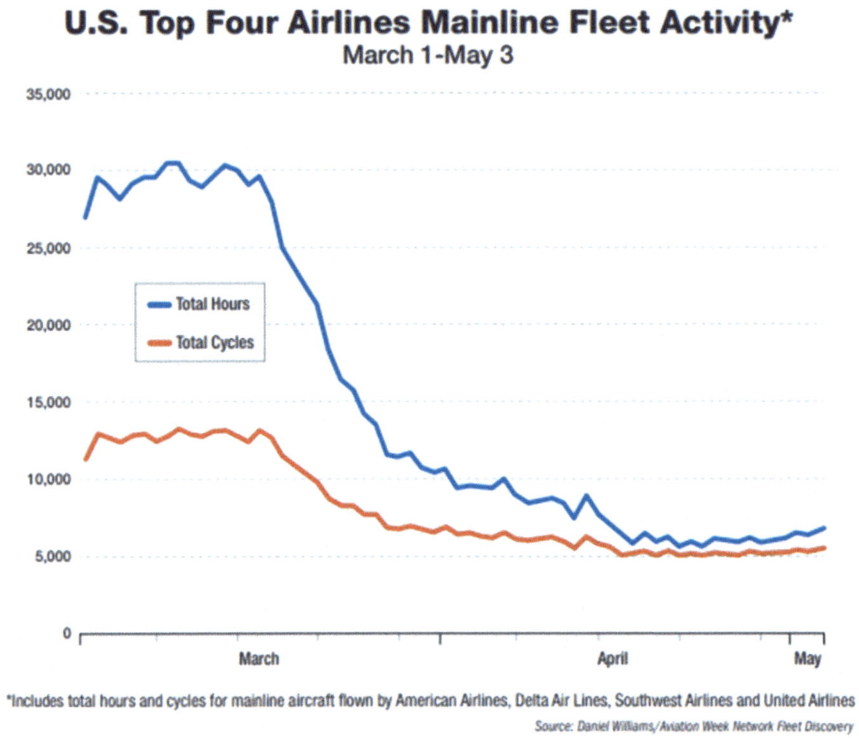

Fig. 6.7 US top four mainline traffic activity in total hours and total flight cycles in the period March 1 to May 3 of the spread of the COVID pandemic

6.7.2 Air Cargo

Airlines may have contributed to the spread of COVID-19 by providing the fastest and most convenient means of long-range transport unaware that some of their passengers were infected by an unknown disease. Air Cargo, both civil and military, was essential in efforts to stop the spread of the disease and cure those affected by providing rapidly of large quantities of medical supplies and equipment for countries with overwhelmed medical services. The normal cargo services declined due to COVID-19 much like passenger flights.

Special passenger flights were also organized, with civil and military aircraft, to bring back home passengers stranded at their destinations, either on business or tourism, by the unpredictable imposition of travel restrictions. In some countries like France, aircraft were used to transport patients from regions with overwhelmed medical services to others with spare capacity. All these flights included thorough health precautions against COVID-19, showing that this could be implemented in the context of aviation.

6.7.3 *Aircraft Leasing*

Aircraft leasing rose from about 2% of the world fleet 20 years before to about 35% at the start of the COVID pandemic. Figure 6.8 shows the rise in lessor fleets in 2009, 2014, and 2019 reaching about one-third of the world fleet. The growth in importance of lessors was due to several factors. Airlines may prefer to size their fleets to the average traffic and use leased aircraft to cover extra demand over peak periods. Lessors can use the aircraft all year-round providing aircraft to airlines around the world with different seasonal needs.

For purely financial reasons, airlines may prefer to lease new aircraft rather than a direct purchase even if they intend to fully own the aircraft later. Lessors can place large orders of hundreds of aircraft over several years providing much-appreciated production stability for aircraft manufacturers. In return, they can obtain favorable discounts on aircraft prices compared with smaller airlines that place only small orders occasionally.

Despite high demand, the leasing market was facing excessive competition due to the entry of Chinese lessors. Western lessors maximize profit, whereas Chinese lessors are subsidized by their government and maximize market share. Thus, Chinese lessors offer airlines leases at a price that Western lessors cannot possibly match. Figure 6.9 shows the in-service fleets and aircraft on order by the 10 largest lessors.

Before the COVID-19 crisis, demand for leased aircraft was increased by the Boeing 737 Max crisis. With 385 B737Max grounded and 470 undelivered airlines, there was a need to replace temporarily all these aircraft until the B737Max were ungrounded and delivered and returned to service. The COVID-19 crisis came in this period of excessive competition among lessors and high demand from airlines. The steep decline of passengers flying after the COVID-19 crisis left airlines with excessive fleets, parking or storing or disposing of part of their fleets.

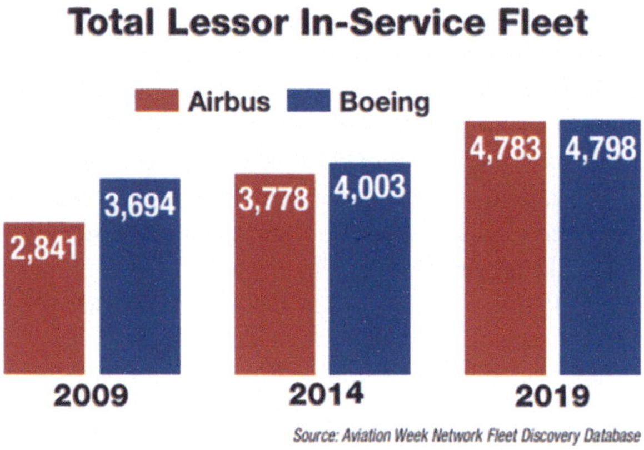

Fig. 6.8 Increase in the number of Airbus and Boeing aircraft in service with leasing companies comparing the years 2019, 2014, and 2019

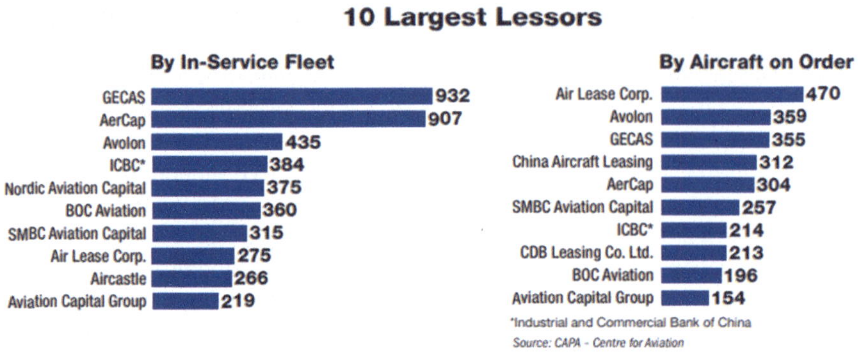

Fig. 6.9 The 10 largest lessors by the number of aircraft for leasing in service and on order

The marginal economics and environmental legislation imply that airlines use preferentially the most modern efficient and less polluting aircraft available, regardless of whether they own or lease the aircraft. Lessors are keen to keep their customer airlines, agreeing to payment delays up to 3 months or more with repayment 1 year later. At the start of the pandemic, most of the airlines (50–80%) asked for and obtained deferred payments: most did their repayments on time, generating a positive cash flow for lessors, with a smaller number of new requests for payment delays. Thus, the lessors were savvy enough to survive, and creative and open to various alternatives, like swapping delivery slots or buying stock.

6.7.4 Airports

The effects of the COVID-19 pandemic on airports were similar to that on airlines: with high fixed costs and much reduced passenger revenues to compensate, the situation became simply untenable for more than a few months, so that survival depended on government aid. Some airports became mostly parking spaces for grounded aircraft, sometimes lined along unused runways, with very little traffic in surrounding roads, in contrast with intense activity usually associated with airports.

6.7.5 Air Navigation Service Providers

The ANSPs were struggling to cope with air traffic growth before the COVID-19 crisis. In the region of densest traffic in the center of Europe, flights were diverted to longer routes to alleviate saturation. Some ANSPs were criticized for not predicting traffic growth and not having trained enough Air Traffic Controllers (ATCs) years in advance. Like airlines and airports, during the COVID pandemic, the ANSPs faced a decreased revenue due to the lack of flights and decided not to

charge for flight services at all, to avoid burdening even more cash-strapped airlines.

ANSPs have high fixed costs with salaries of personnel, including air traffic controllers who have pay comparable to airline pilots. ANSPs also have the costs of navigation infrastructure and support costs. The decision to waive charges on airline flights did not change the fact that ANSPs would need support from national governments. Also, EUROCONTROL, which coordinates air traffic in Europe and normally distributes air service charges to ANSPs, taking a fraction for itself, was left without revenue and was also funded by agreement among its 29 member states. Thus, the national governments funded both the ANSPs and EUROCONTROL.

The near-total suspension of flights during the peak of the COVID-19 crisis had implied that aviation professionals stay idle and cannot practice their skills in a real environment. When flight operations resumed, this lead to the need to relearn past practice until the same levels of efficiency were reached. Although this was a general problem across several aviation sectors, and by no means specific to ANSPs, they provide a clear example. ATCs have a demanding task practiced regularly in the safe management of traffic. An interruption of several months followed by a rapid partial ramp-up created the need to regain skills quickly enough or reassign tasks to maintain the same level of safety.

6.7.6 Business Aviation

The only sector of aviation that did not face a steep decline from COVID-19 and may have grown was business aviation, which is understandably quiet about its exceptional situation. Large corporations own business jets to transport their top executives to meetings worldwide, without dependence on airline schedules, using airports closer to headquarters and business location, and allowing work or rest on board.

COVID-19 did not end the need for business travel, and the reduced airline schedules and risk of last-minute flight cancellations were often unacceptable. In addition, corporate jets flying with a few known individuals may place less risks than an airliner with a larger number of unknown passengers. Thus corporate jets flew more hours during the COVID pandemic with executives lower down the ranks than usual. The same reasons of schedule choice and lower health risk have led to increased use of fractional jet ownership and fly-by-the-hour tickets.

6.8 Effects on Industry

The sharp decline in passenger traffic left airlines with excessive airline fleets, more conducive to cancellations and delayed deliveries than to new orders, leading to a decline of about 30% in the sizeable pre-COVID-19 order books of both Airbus and

Boeing (Sect. 6.8.1). The first-tier suppliers like engine manufacturers were affected both by production decline and reduced maintenance due to less flying, leading to larger declines of up to 50% (Sect. 6.8.2). The picture was not better in tiers 2, 3, and 4 down the supply chain depending on how exposed smaller companies were to the civil market and whether what their supply was essential for post-COVID-19 recovery (Sect. 6.8.3). With reduced flying and cash-strapped airlines deferring major services the maintenance, repair, and overhaul (MRO), companies found less work and more competition from the original equipment manufacturers (OEMs) with less orders and more overcapacity (Sect. 6.8.4). With airlines disposing of surplus older less efficient and more polluting aircraft, the used market values reached historic lows (Sect. 6.8.5). The prospects for survival (Sect. 6.8.6) were better for companies supplying unique essential parts that the OEMs cannot do without than for providers of services that overstaffed OEMs can carry out themselves.

6.8.1 Aircraft Manufacturers

COVID-19 has caused not only a sharp decline in passenger traffic but also an uncertain or unpredictable recovery that might or might not reach pre-COVID-19 levels, although the former turned out to be the case in a few years. The airlines with large parts of their fleets grounded considered to simply retire older, larger, less efficient more polluting aircraft that would be increasingly difficult to sustain in the future. With a likely cash-strapped future and reduced traffic, the smaller more modern efficient and environmentally friendly aircraft were the best choice for survival; airlines even exceptionally ordered a few of those profiting from discounts from manufacturers short of orders. The sharp post-covid recovery of traffic demand, and the delays in the availability of new aircraft, lead airlines to retain in service longer, the older less efficient aircraft that were due for retirement.

The prospects for Airbus and Boeing were similar: more cancellations, delayed deliveries, and postponed orders (Fig. 6.10), reducing the backlog by about 30%. With production rate cuts of about 30%, the backlog would not change much in the number of years. What changed radically is the transition: (i) from the pre-COVID-19 status of more demand than offer with Airbus and Boeing struggling to increase production rates from an overstretched supply chain; and (ii) to the COVID-19 period market of excess capacity over demand all the way down the supply chain; (iii) ultimately to a sharp post-Covid airline recovery, with orders far in excess of what the industry, downsized during covid-19, could provide, again struggling to increase production rates, with a supply chain understaffed and underfinanced from a marginal survival during the covid crisis.

Figure 6.11 shows the possible future scenarios for aircraft production with a large variation depending on how long the COVID-19 crisis lasts and how the recovery will look. The Airbus and Boeing deliveries are compared in Fig. 6.12 for the Asia Pacific region in 2020 and in Fig. 6.13 for deliveries worldwide in the first quarter of 2020. The deliveries of the 737Max depend on a return to service and only a few older 737NGs can be delivered.

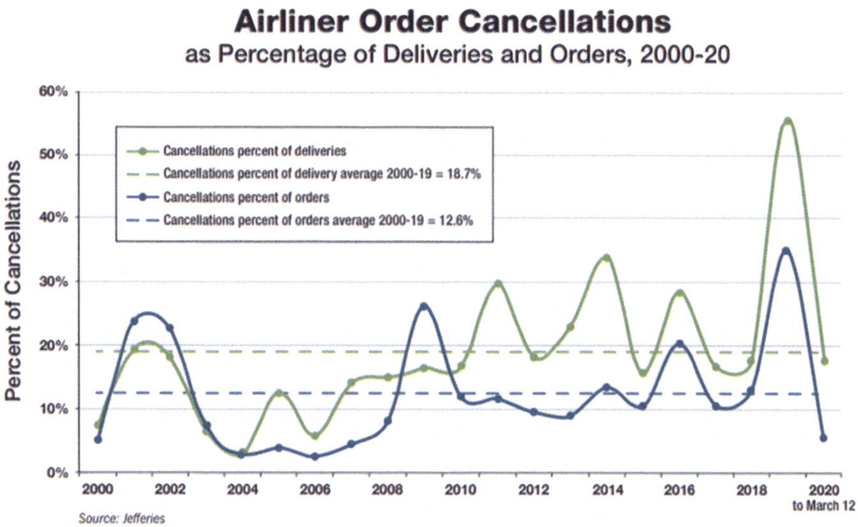

Fig. 6.10 Airliner order cancellations as a percentage of deliveries and orders 2000–2020

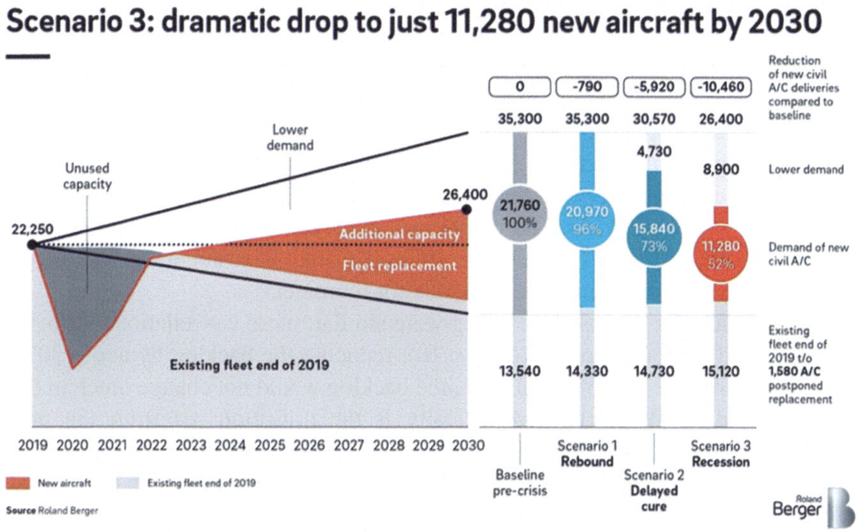

Fig. 6.11 Possible scenarios for future aircraft production

Figure 6.14 shows the current and future Boeing production rates, with significant reductions in wide bodies and uncertainties about the ungrounding of the B737Max. Figure 6.15 shows the key metrics for Airbus demonstrating the negative effects of COVID-19 in 2020 compared to 2019. Although military aircraft production is sheltered from the collapse of the civil market, Fig. 6.16 shows how the ramp-up of production of the Lockheed Martin F-35 Lightning has been affected by parts shortages and industrial issues caused by COVID-19.

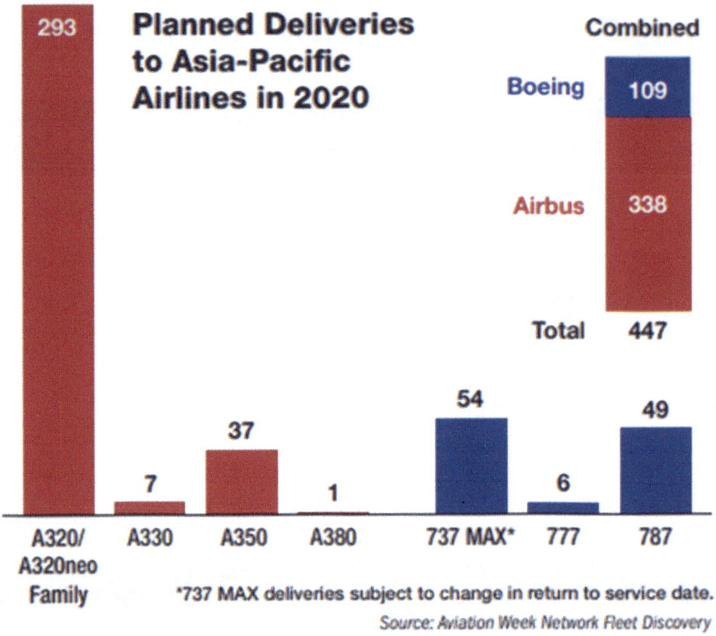

Fig. 6.12 Planned deliveries of Airbus and Boeing airliners to Asia Pacific Airlines in 2020

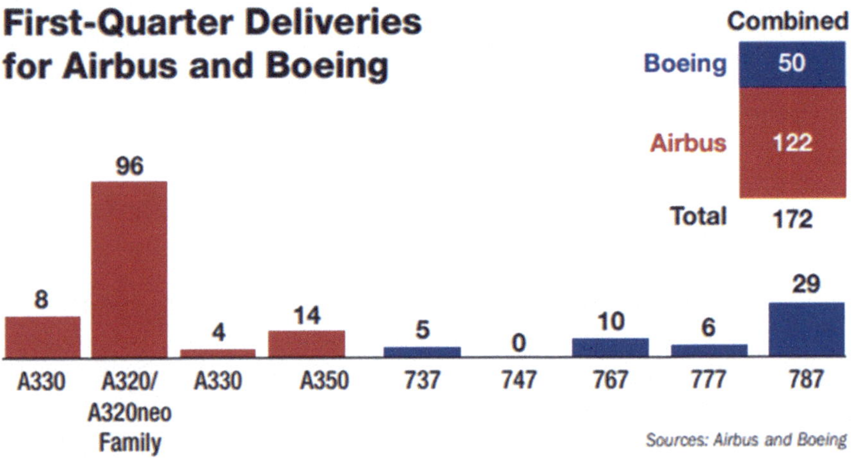

Fig. 6.13 Airbus and Boeing airline deliveries in the first quarter of 2020

6.8.2 First-Tier Suppliers

Among the first-tier suppliers, aero-engine manufacturers have a significant share of aircraft purchase and maintenance costs and can be the most critical supplier to the point of Airbus having some "gliders," that is, finished aircraft waiting for engines.

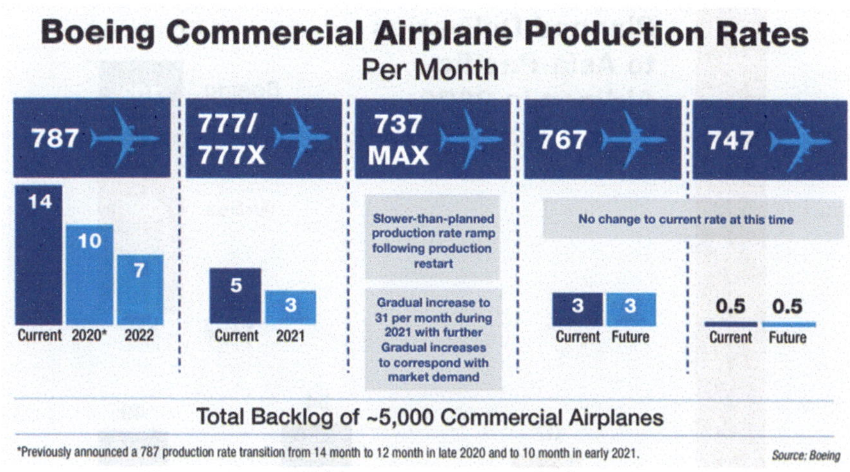

Fig. 6.14 Current and predicted future Boeing commercial airplane production rates

Airbus Key Metrics

	First Quarter 2020	First Quarter 2019 Restated*	Change
ORDER INTAKE (NET)	290	(58)	N/A
BACKLOG (UNITS)	7,650	7,357	4.0%
DELIVERIES (UNITS)	122	162	-24.7%
REVENUES**	€7,569	€9,697	-21.9%
EBIT**	€57	€319	-82.1%

*2019 financial figures restated to reflect the adoption of a new segment reporting structure for "Transversal" activities
**In € million (€1 = $1.08)
Source: Airbus

Fig. 6.15 Change of key metrics of Airbus activity between the first quarters of 2019 and 2020

The engine suppliers were doubly affected by COVID-19: (i) they supplied a smaller number of new engines due to lower aircraft production rates; and (ii) engine maintenance can be double of engine cost over the lifetime of an engine and is a significant revenue decreased by reduced flying. Thus, a revenue loss of 30% for an aircraft manufacturer can look more like 50% for an engine supplier. Post-covid recovery implied that the engine suppliers became the most hard pressed to meet demand, with simultaneous challenges: (i) the new generation of more eficient aero engines had the expected reliability issues increasing the demand for spare parts; (ii) the higher demand for new aircraft meant that engine supply was often a limiting factor; (iii) engine suppliers were in the difficult position of choosing (i) or (ii); (iv) airlines were doubly dissatisfied with grounded aircraft waiting maintenance and new aircraft deliveries delayed; (v) airlines were forced to operate longer older less efficient more reliable aircraft with costs not worse than the more modern and more efficient less reliable replacements. Engine time-on-wing became a major availability metric.

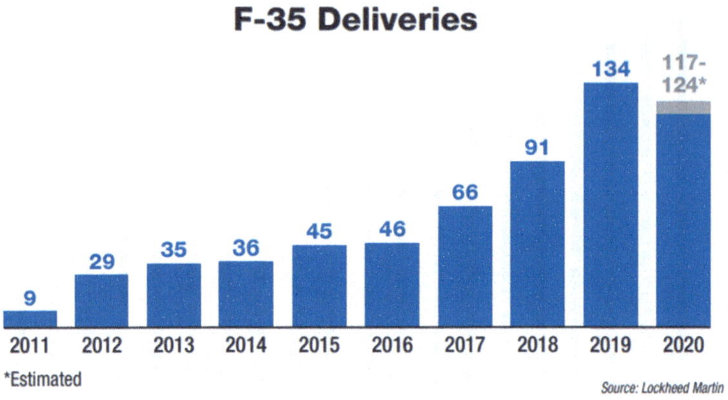

Fig. 6.16 Slowing down of production ramp-up of a military aircraft (Lockheed Martin F-35 Lightning II) due to COVID-19 restrictions (not lack of orders)

Commercial Jet Transport Engines Stored/Parked by Aircraft Class*

Engine Manufacturer	Regional Jet	Narrowbody	Widebody	Totals
CFM International		17,166	272	17,438
Engine Alliance			480	480
General Electric	3,524		3,802	7,326
International Aero Engines		4,430		4,430
Pratt & Whitney	108	1,713	925	2,746
Rolls-Royce	1,132	674	3,488	5,294
Honeywell	284			284
Totals	5,048	23,983	8,967	37,998

*As of April 18, 2020. Includes 6,077 engines in reduced-activity "parked/reserve" status that flew 1-2 of previous 7 days.

Source: Aviation Week Intelligence Network Fleet Discovery

Fig. 6.17 Commercial jet transport engines stored/parked for regional, single-aisle and wide-body aircraft

Figure 6.17 shows the number of jet transport engines stored or parked distinguishing regional and narrow- and wide-body aircraft. Figure 6.18 shows that over 21 years the value of an airliner (the Boeing B737-800 in this case) may decline to about one-quarter of its price when new, whereas the engine (a CFM-56-7B in this case) hardly devalues at all. Thus, the engine rises in value from 35% of the new aircraft to 80% for a 20-year-old aircraft. In fact, old aircraft are often bought for the spare parts of the engine.

6.8.3 Down the Supply Chain

The decline in the civil aviation market passed from airlines to aircraft manufacturers and then filtered down from tier 1 to tiers 2, 3, and 4 of the supply chain. The sharp post-covid recovery was the most challenging for lower tier suppliers that had

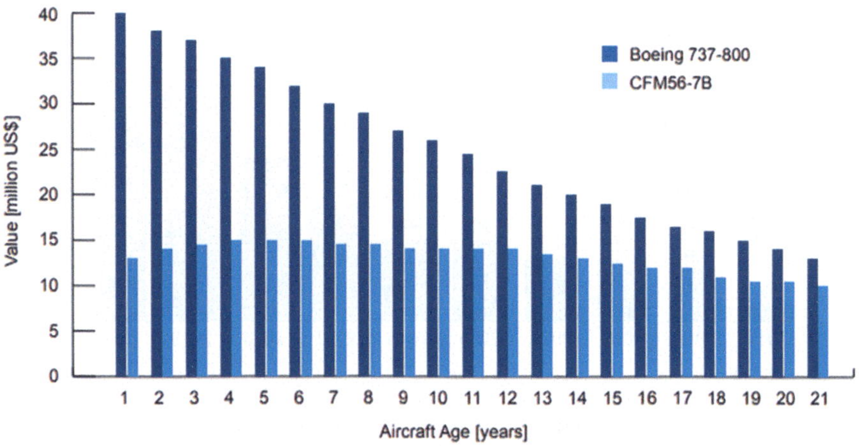

Fig. 6.18 Price evolution with age showing a steep decline for aircraft (Boeing 737–800) and a steady value for engines (CFM56-7B). (Source: International Air Transport Association (IATA))

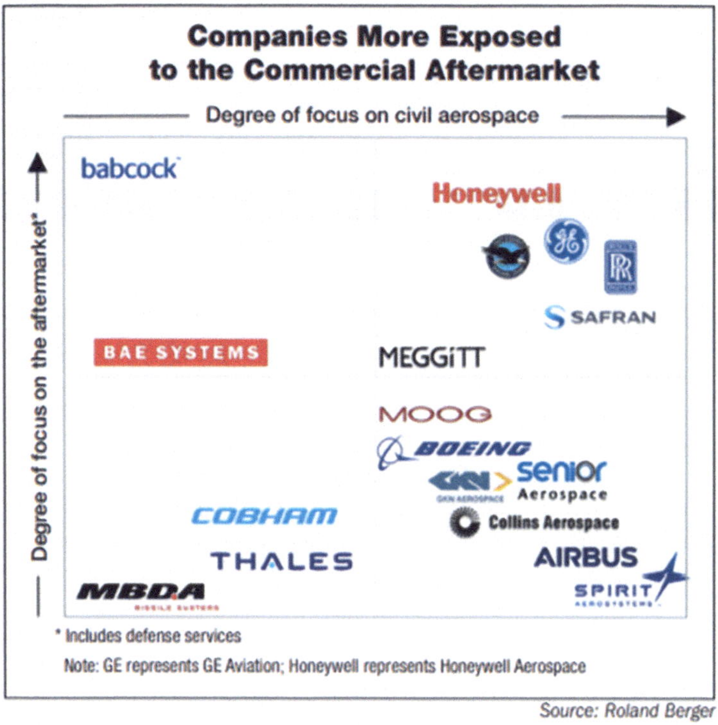

Fig. 6.19 The 10 largest lessors by the number of aircraft for leasing in service and on order

Supplier Exposure by Aircraft Platform

Percentage of 2020 Estimated Revenue

BOEING							
737		747		777		787	
Spirit AeroSystems	48.2%	Triumph Group	5.2%	Boeing	17.3%	Boeing	19.2%
Boeing	22.3	Spirit AeroSystems	1.8	Spirit AeroSystems	16.9	Spirit AeroSystems	16.1
Hexcel	10.5	Boeing	1.7	Hexcel	7.7	Hexcel	10.1
Triumph Group	8.4	Hexcel	0.3	United Technologies*	3.6	United Technologies*	5.3
Woodward	5.9	Woodward	0.3	Triumph Group	2.9	Triumph Group	3.0
TransDigm Group	3.3	United Technologies*	0.2	TransDigm Group	1.0	Woodward	2.0
United Technologies*	2.9	TransDigm Group	0.0	Woodward	0.7	Honeywell International	1.2
Honeywell International	1.6			Honeywell International	0.3	TransDigm Group	1.2

AIRBUS							
A320		A330		A350		A380	
Hexcel	8.2%	Rolls-Royce	3.0%	Hexcel	12.4%	Hexcel	3.6%
Triumph Group	7.6	Triumph Group	2.5	Spirit AeroSystems	3.8	TransDigm Group	1.1
Spirit AeroSystems	7.4	Hexcel	1.1	United Technologies*	2.7	United Technologies*	1.1
United Technologies*	4.8	Spirit AeroSystems	0.8	Honeywell International	1.6	Woodward	0.4
TransDigm Group	3.6	Woodward	0.5	TransDigm Group	1.4	Spirit AeroSystems	0.4
Woodward	3.4	United Technologies Aero*	0.4	Triumph Group	0.7	Honeywell International	0.2
Honeywell International	0.6	Honeywell International	0.1				
		TransDigm Group	0.1				

Fig. 6.20 Supplier exposure to aircraft platform for several suppliers in terms of percentage of 2020 revenues. (Source: Aviation Week Network)

lost experienced staff, and had to hire and train new employees. Figure 6.19 shows that the exposure of the supply chain depended mainly on two factors: (i) was there a strong dependence on the civil aircraft market, or was this compensated by activity in military aviation or non-aviation sectors? (ii) was the dependence only as a part supplier or also in maintenance and support?

Figure 6.20 indicates the dependence of several suppliers on the aircraft platform. For example, Spirit AeroSystems has a strong dependence on the Boeing 737 because it manufactures the fuselage in Wichita, Kansas; its dependence on the Airbus A320 is smaller because it is mainly a substructure supplier. Figure 6.21 shows the dependence of suppliers per market. Overall, there is a wide variation of the dependence of the supply chain on the platform (Fig. 6.22).

6.8.4 Maintenance, Repair, and Overhaul (MRO)

The MRO sector was doubly hit by the COVID-19 crisis: (i) less flying means less maintenance and cash-strapped airlines fighting for survival may postpone nonessential major servicing; and (ii) the original equipment suppliers (OEMs) faced with less orders and excess industrial capacity and staff made a stronger competition for maintenance to improve revenues. Engine MRO, which is a large share of the work on aircraft, faced stronger competition from OEMs for what was left of a reduced

A&D Revenue Exposure by Market

Company	Percent of 2019 Revenues by Market			
	Aerospace Original Equipment	Aerospace Aftermarket	Defense/ Space	Other Government
AAR	2%	66%	33%	0%
Aerojet Rocketdyne	0	0	100	0
AeroVironment	0	0	79	0
Astronics	43	36	18	2
CPI Aero	20	5	75	0
Crane Co.	11	5	9	0
Cubic Corp.	0	0	43	57
DLH Corp.	0	0	45	55
Ducommun	42	6	45	0
HEICO	4	49	34	0
Hexcel	68	0	20	0
Kratos Defense & Security	0	0	100	0
Mercury Systems	0	0	100	0
RADA Electronic Industries	0	0	100	0
Spirit AeroSystems	88	5	7	0
The Boeing Co.	52	11	37	0
TransDigm Group	26	32	37	0
Triumph Group	60	18	20	0

Notes: Aerospace original equipment and aerospace aftermarket include all commercial transport and business jet exposure. Defense and space includes both original equipment and aftermarket sales.

Sources: Company Reports, Canaccord Genuity estimates

Fig. 6.21 A&D revenue exposure by market. (Source: Aviation Week Network)

market. In the sharp post-covid recovery MRO organizations were overwhelmed with work requests and unable to comply timely due to shortages of spares and delays in obtaining them.

6.8.5 Used Aircraft Market

Airlines left with pre-COVID-19 fleets in excess of post-COVID-19 needs planned to dispose of surplus aircraft, starting with the (i) least efficient that may be unsustainable in a cash-strapped situation; (ii) the more polluting that are more likely to be hit by stronger environmental protection measures; and (iii) the largest that would have lower load factors in the future in routes with less passenger demand. It was thought unlikely that many B747, B767, A380, older B737, A300/310, and A330/340 would remain in airline service and used prices were expected to reach

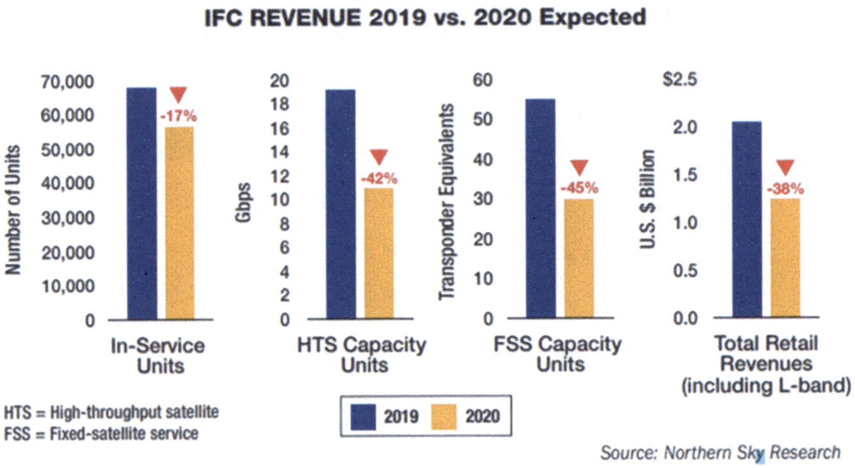

Fig. 6.22 Decline in revenue for in-flight communications (IFC) comparing 2019 records with expectations for 2020

historic lows. The sharp post-covid recovery, with an increase in aircraft orders that could not be met by industry and the supply chain, led to older aircraft being kept in service longer than planned and high demand for leased aircraft at high costs, still unable to match the growth in traffic, in spite of higher ticket costs, and bringing back airline profitability after massive losses during the covid crisis.

6.8.6 Prospects for Survival

The decline in the aviation market affects all suppliers, and made the sharp increase in demand pot-covid more challenging in several respects: (i) shortage of staff laid off during the covid crisis and not returning after; (ii) loss of experienced workers and need to train their new replacements, with lower productivity in the meantime; (iii) difficulties in recruiting the numbers of new staff needed; (iv)= finance for expansion, with higher interest rates; (v) a remaining sense of insecurity about sustainable market trends. Fig. 6.2 illustrates the effects on the in-flight communications market comparing 2019 and 2020 concerning several types of passenger and entertainment services. Figure 6.23 illustrates two possible evolutions of the consolidation of the OEMs and the four tiers of the supply chain, including vertical or horizontal integration and mixed cases depending on the value of the acquisition and financial capacity of the buyer.

For companies in the supply chain, the chances of survival during the covid crisis lied between two extremes: (i) a supplier of a unique essential part or equipment that is indispensable for recovery after the COVID-19 crisis would either survive by itself or be saved if necessary by the OEM by contracts to ensure continuity or direct purchase; and (ii) a supplier of occasional services when the OEM has excess work was unlikely to be needed when the OEM was short of work, had overcapacity, and

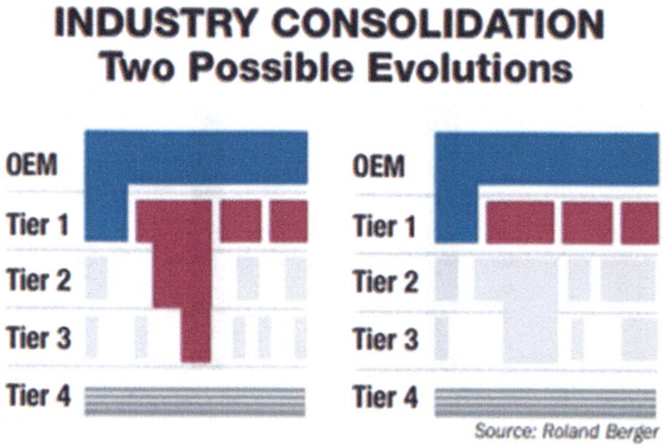

Fig. 6.23 Two possible scenarios for the consolidation of the supply chain of aviation industry

could do the work itself. The supply chain weakened during the covid crisis became unable to respond to the ramp-up demanded by the sharp pot-covid traffic recovery.

6.9 Bailout in the US CARES

Saddled with high fixed costs and reduced revenue, the aerospace sector could not survive the COVID-19 crisis without state help; the way it was provided varied around the world was as much or even more disparate as the response to other aspects of the crisis. The US Congress approved the Coronavirus Aid Relief and Economic Stability (CARES) bill. The CARES act initially covered the 6-month period ending on September 30, 2020, and imposed a number of restrictions: (Sect. 6.9.1) workers could be furloughed and pay rates could be changed; (Sect. 6.9.2) all domestic flight connections had be flown; and (Sect. 6.9.3) airports had to remain operational as before. As soon as the general rules for the distribution of funds were announced, there was no shortage of imagination on how to circumvent them, leading to widespread controversy about inequalities in their application. There were also sizeable bailouts for industry, including $60 billion for Boeing (Sect. 6.9.4). There were questions of what would happen after October 1, 2020, when the rules no longer apply, and the billions would already have been spent (Sect. 6.9.5).

6.9.1 Keeping Employment

The CARES rules forbid dismissing employees or cutting pay rates. Some major US airlines were quick to cut the working hours, thereby reducing salaries in the same proportion, with unchanged pay rates. This led to angry protests from worker unions, with workers filling the corridors of Congress to make sure they were heard.

The management of airlines eventually backed down partially as little as possible and rather grudgingly, leaving some doubts on what could happen after CARES ended. In fact the sharp post-covid recovery led to labour shortages in all categories, from pilots to cabin staff and maintenance personnel and significant pay rises.

6.9.2 Maintaining Internal Flights

The rule to maintain all internal flights hit hardest small regional airlines and tourist operators forced to keep flying nearly empty aircraft to have CARES funding. The large airlines fared much better because they needed not fly international routes and could reduce the number of internal flights by flying through their hub and reducing flight frequencies. This put the large airlines that were financially stronger at an advantage relative to small regional airlines that risked being absorbed or not surviving on their own.

As this was not enough, major airlines came up with codeshare proposals that were flatly and repeatedly rejected by the Transportation Department. The idea was that if airlines X, Y, and Z flew from A to B, they would codeshare and only X would fly the route. On a route from C to D, they would again codeshare with only airline Y flying. In this way, airlines X, Y, and Z would all cut the number of flights by three while receiving the same CARES benefit.

6.9.3 Supporting Airports

The rules for supporting airports were based on the ratio of airport liquidity to debt, leading to widely contrasting situations, for example, in the state of Alaska: (i) the largest airport received no funding, whereas a small airport would receive funding for several years of operation; and (ii) of two similarly sized airports, one would receive very little compared with the other. Thus, the $80 billion provided by the CARES bill for airlines and airports gave rise to widespread controversy and protest of unequal treatment.

6.9.4 Bailouts for Industry

The bailouts for the industry were less controversial and adjusted to the needs of each company. The size of the $60 billion bailout for Boeing did raise some eyebrows in Congress, and Boeing cleverly argued that it was not all for itself and would cover the whole supply chain as well. The $60 billion rescue package had already been mentioned in the context of the B737Max crisis before the COVID-19

pandemic, which happened to come at the right time to provide a suitable cover under the CARES bill.

6.9.5 Six-Month Duration

The CARES bill allocated more than $170 billion for aviation over a 6-month period without setting any long-term goals or objectives for industry or services. The rules for airlines and airports were generic and applicable to all, making it difficult to deal in a balanced way with very different situations, and thus, leading to controversies about the distribution of funds. The main remaining question is what happened after CARES expired on October 1, 2029 was: could employees be dismissed, salaries be cut, routes discontinued, or another sizeable bailout bill be necessary? There were estimates of up to a quarter million job losses but most of the aeronautical sector in the United States survived until the strong recovery from the COVID pandemic reversed the situation into staff shortages and higher salaries. Thus the CARES bill turned out to be a saviour out of the covid crisis.

6.10 Opposing Alternatives

The high fixed costs and small profit margins in the aviation sector mean that the loss of revenue over an extended period due to the COVID-19 crisis was the biggest crisis ever and simply not survivable without state aid (Sect. 6.10.1), although in some cases it could not or was not provided (Sect. 6.10.2). For those who survived the post-covid traffic demand turned historic large losses into historic large profits.

6.10.1 Supporting Survival

Figure 6.24 shows the minimum, maximum, and median liquidity of airlines in several regions of the world in terms of months of operation. Due to high fixed and operating costs, with small profit margins, the liquidity available to most airlines would cover only a few months of operations. An extended crisis like COVID-19 with grounded aircraft and not nearly enough revenue to cover fixed personnel and maintenance was simply not survivable without some kind of special measures.

Figure 6.25 shows the expected industry cash burn in the second quarter of 2020 of $81 billion. Figure 6.26 shows that the widespread of the COVID-19 pandemic around the world affected nearly all airlines. Figure 6.27 details the expected airline performance per region showing declines in passenger demand and capacity of the order of 50% corresponding to $85 billion losses. Figure 6.28 shows the airline revenue per region in 2019 and the percentage of state aid, with more details on the distribution of funds in the United States in Fig. 6.29.

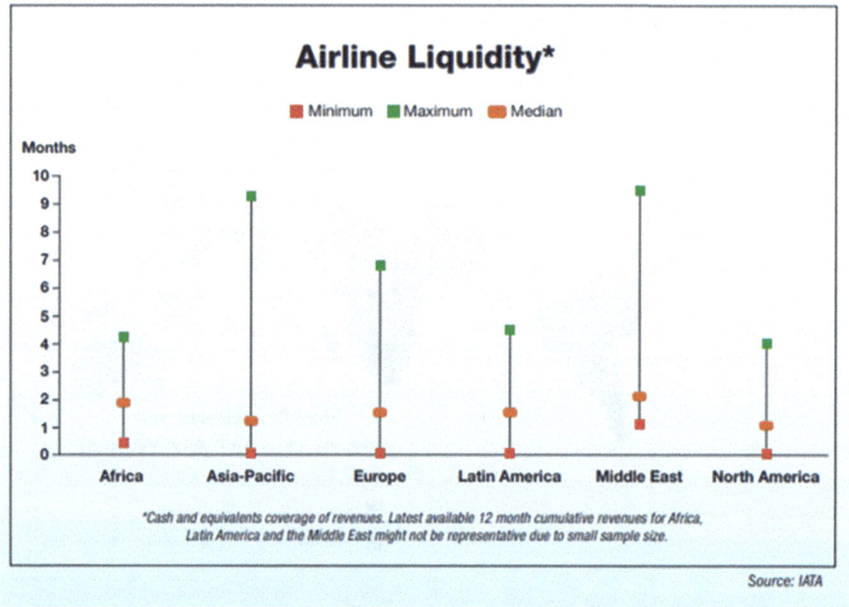

Fig. 6.24 Minimum, maximum, and median airline liquidity in terms of the monthly cost of operations for different regions

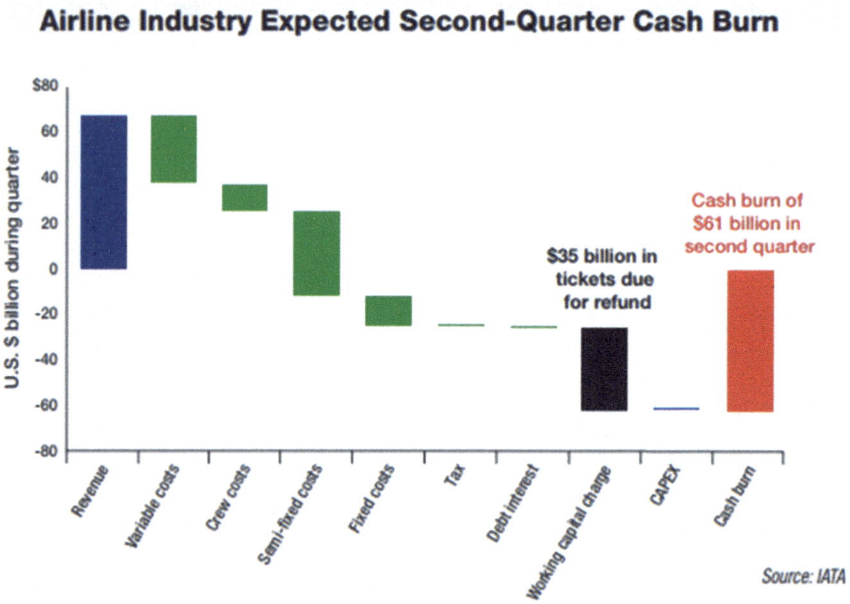

Fig. 6.25 Airline positive and negative cash flows during the second quarter of 2020 when the COVID crisis hit

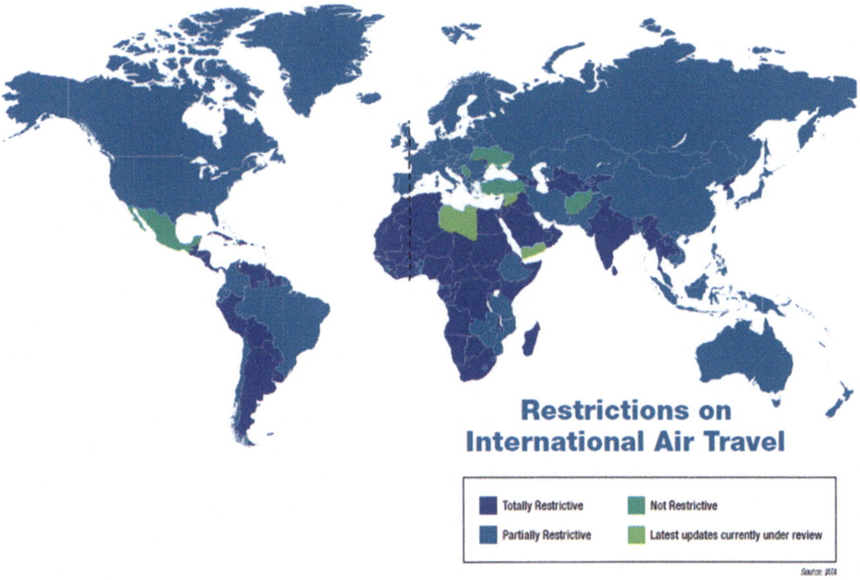

Fig. 6.26 Restrictions on air travel due to the COVID-19 pandemic in different regions of the world

Projected 2020 Airline Performance By Region			
Region	Passenger Demand (revenue passenger kilometers)	Passenger Capacity (available seat kilometers)	Net Profit (U.S. $ billion)
AFRICA	-58.5%	-50.4%	-$2.0
ASIA-PACIFIC	-53.8	-39.2	-29.0
EUROPE	-56.4	-42.9	-21.5
LATIN AMERICA	-57.4	-43.3	-4.0
MIDDLE EAST	-56.1	-46.1	-4.8
NORTH AMERICA	-52.6	-35.2	-23.1
GLOBAL	-54.7	-40.4	-$84.3

Source: IATA

Fig. 6.27 Decline in airline performance per region in terms of passenger demand and capacity, in 2020 relative to 2019, and net losses due to the COVID-19 pandemic

6.10.2 Allowing Bankruptcies

The government reactions to the unsustainable airline situation were very different even in the same region. In Asia, Singapore Airlines and Cathay Pacific were in a particularly difficult situation, with no domestic market, flying out of a single airport, with only international flights from Singapore and Hong Kong, respectively, which would take the longest to recover from the uncoordinated worldwide

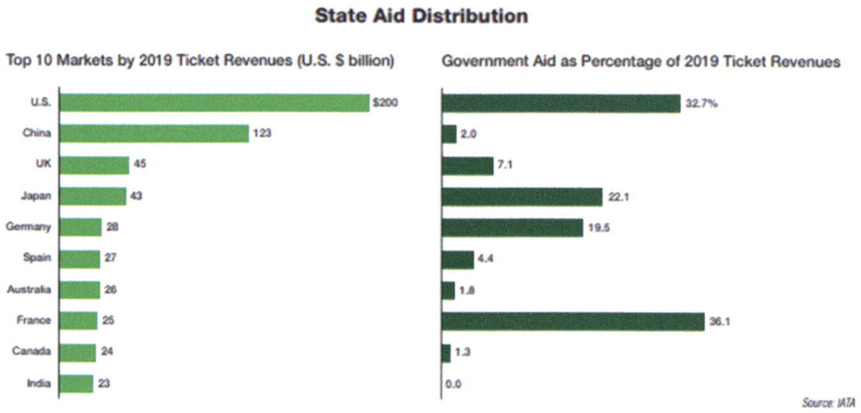

Fig. 6.28 Airline ticket revenues in 2019 in the top 10 countries and percentage of support given in 2020 to help survive the COVID-19 crisis

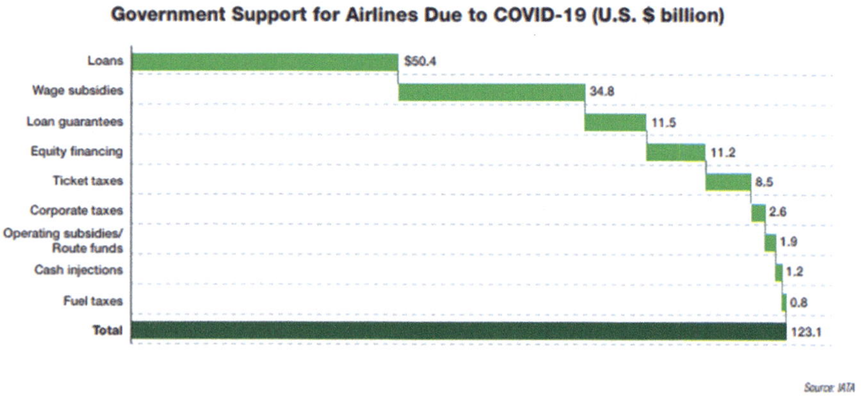

Fig. 6.29 US government support to airlines under the CARE act in the 6 months from April 1 to September 30

deconfinement. Both airlines had a long record of profitability and high reputation for service, and their wealthy parent states were in a good position to provide sizeable financial support to ensure long-term survival.

Not so far away Thai Airlines with a long record of poor financial performance was let into receivership by a government lacking either the confidence or the means for an effective bailout. The situation was similar with some airlines in Latin America, including JENAM despite partial ownership by American Airlines. The local governments with weak debt-ridden pre-COVID-19 economies, further strained by the pandemic, did not find the funds or priorities for the required sizeable bailout of airlines, even flag carriers.

6.11 Government Policies

In Europe, most but not all governments took steps to rescue the national flag carrier, as shown by the cases of Lufthansa in Germany (Sect. 6.11.1), Air France/KLM in France/the Netherlands (Sect. 6.11.2), and Alitalia in Italy (Sect. 6.11.3), with Britain/Spain less keen on British Airways/Iberia (Sect. 6.11.4) and private airlines like low-cost carriers (LCCs) left to their own devices (Sect. 6.11.5).

6.11.1 Germany: Lufthansa

Germany provided a €9 billion rescue package that Lufthansa eventually accepted with reluctance for lack of a better alternative swallowing a bitter pill of conditions, including a reduction of fleet size to better match future predicted demand and loss of precious slots at key airports like Frankfurt and München. The government share of the company might evolve over time depending on future performance ranging from nationalization to full reprivatization.

6.11.2 France/the Netherlands: Air France/KLM

The $7 billion rescue of Air France/KLM were shared 80% by France and 20% by the Netherlands based on the valuation of the contributions of the two airlines to their joint alliance. The conditions were less severe than for Lufthansa but included a reduction in domestic flying in favor of high-speed trains. Although Lufthansa had a better profitability record than Air France/KLM, it got a more demanding deal from the German government.

6.11.3 Italy: Alitalia

The rescue of Alitalia perennially in trouble with labor union issues remained as uncertain as ever, with successive take-over schemes collapsing. The COVID-19 crisis did give the Italian government the opportunity to intervene without the objections and controversies surrounding previous subsidies. The plans were unclear, ranging from a small 100 aircraft fleet of almost irrelevant size to grand rebuilding plans looking rather dubious when compared with earlier similar attempts. The fact remained that Italy is a major tourist and travel destination providing a stable market that should support a flag carrier, but has instead been increasingly dominated by foreign low cost carriers while the Alitalia sales prospects took longer to materialize, until the recent deal with Lufthansa to buy the Alitalia successor airline ITA:

6.11.4 Britain/Spain: British Airways/Iberia

The relations between British Airways and the British government and Iberia and the Spanish government looked more strained, with the airlines benefiting from generic COVID-19 support measures but no rescue plan, and rather complaining about the negative impact of government regulations. In the case of British Airways, the imposition of quarantine on passengers from a blacklist of countries led to threats of court action against the government.

6.11.5 Low-Cost Carriers

The LCCs have been as vocal as British Airways about quarantine and travel restrictions in Britain and elsewhere and have in some cases rejected social distancing in aircraft cabins as leading to impossibly low load factors that are uneconomic at low ticket prices. The LCCs do not have the benefits of government support for flag carriers, and neither do they have the restrictions, being free to cut jobs, dismiss, and rehire with lower wages; this is a model that British Airways may have comtemplated in the absence of government support.

6.12 Comprehensive Support

The case of France is a good example of a comprehensive support package for the aerospace sector including (Sect. 6.12.1) military contracts, (Sect. 6.12.2) industry support, and (Sect. 6.12.3) research for competitiveness.

6.12.1 Military Contracts

The military contracts are usually large investments for a batch of aircraft spread over several years and give much-needed stability to aerospace companies active both in the defense and civil sectors, providing a buffer against fluctuations in the civil market, which amounted to an almost collapse in the case of COVID-19. The French plan reasserted military spending and provided basic support to dual-role aerospace industry, assuming future defense budgets resist the pressures of post-COVID-19 economic vulnerabilities.

6.12.2 Industry Support

Unlike Boeing, which claimed a rescue package of $60 billion to compensate a $41 billion debt and avoid the risk of bankruptcy due to the B737Max crisis before COVID-19, Airbus was in a healthy liquidity situation before COVID-19 and more concerned about support for its airline customers and essential elements of the supply chain. Airbus never had a reduction of production in its history until the COVID-19 crisis, whereas in the same period since the foundation of Airbus, Boeing has coped with three periods of reduced airliner production (Fig. 6.30). Post-covid both Airbus and Boeing aimed at sharp increases in production rates, limited by the supply chain for both, with Boeing hampered to a much larger extent by its own in-house quality issues; Airbus did not have major quality issues by itself, but its subcontracts to Spirit AeroSystems were affected by its reaquisition by Boeing, following quality problems with both.

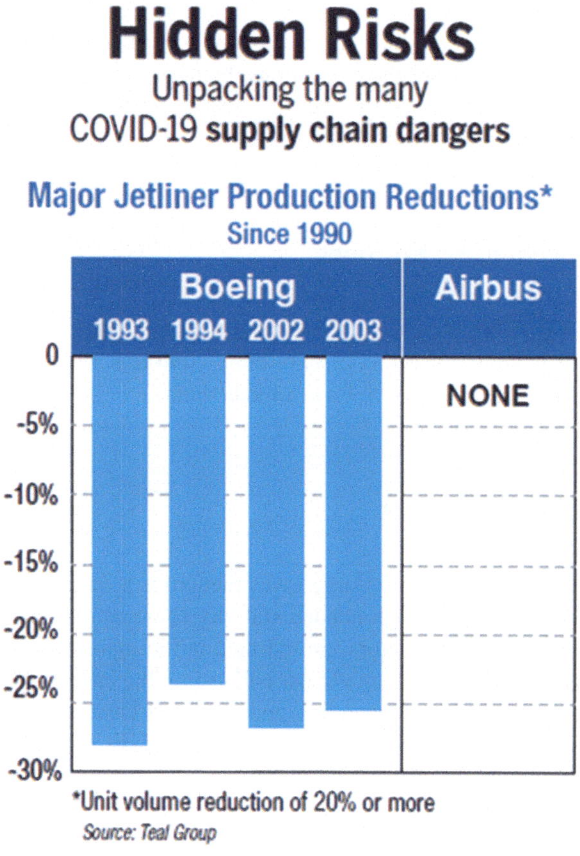

Fig. 6.30 Major jetliner production reductions. (Source: Aviation Week Network)

6.12.3 *Research for Competitiveness*

The €15 billion package for support of the French aerospace sector included €7 billion for Air France, €4.5 billion for industry supply chain, and €1.5 billion for research of development, preparing also for the development of a 20–30% more efficient successor to the A320 family in 2030–2035, with two alternatives: (i) a conventional aircraft with a up to 25:1 bypass ratio engine (up from 12 in the current CFM Leap engine) and slender high aspect ratio wings; and (ii) a novel configuration like a flying wing with large internal volume for hydrogen as fuel for gas turbines or fuel cells. While this was preparatory research rather in advance of an eventual aircraft program, it did provide a focus for the research and development community in future competitiveness. The next generation of Airbus e-zero aircraft could comprise could be three families of aircraft: (i) a regional airliner replacement for the ATR with turboelectric propulsion, (ii) a hydrogen-powered short-haul aircraft to replace the A220 and lower end of the A320 family, and (iii) a long-range single-aisle aircraft using synthetic sustainable fuels as a replacement for the A321neo/LR/XLR at the upper end of the A320 family.

6.13 Support from the European Union

The comprehensive support plan for the aerospace sector of France, like a comparable plan from Germany, was conceived to dovetail with the initiatives of the European Union to bring the benefits of wider cooperation and coordination. The support of the European Union, in general, was based on the fund for economic and financial stability post-COVID-19 (Sect. 6.13.1) and specifically included research in Horizon Europe for a more efficient and environmentally friendly aviation (Sect. 6.13.2) and coordination measures to deconfine and open-air travel in a coordinated way across Europe (Sect. 6.13.3).

6.13.1 *Financial Stability*

The European Union fund for recovery from the COVID-19 crisis provided the background of economic and financial stability to enable member states to move toward future efficiency and environmental protection goals, including the hydrogen economy, energy conservation, more recycling, reduced consumption of finite resources, and greater emphasis on renewable and sustainable approaches and practices.

6.13.2 Research and Environment

The convergence toward the hydrogen society included the hydrogen-powered aircraft benefiting from the experience with cryogenic propellants in the Ariane 5/6 launchers with considerable additional developments in aircraft systems and turbine propulsion. For shorter ranges, electric and hybrid propulsion provided an alternative path dependent on progress with fuel cells and batteries. Both approaches should bring unquestionable environmental benefits and pose integration challenges.

6.13.3 Coordinated Measures

The prospects for recovery of long-haul air travel were strongly compromised due to the uncoordinated deconfinement measures around the world, leaving regional and short-haul transport as the nearest term prospect for return regular operations. The measures of the European Union to coordinate deconfinement across Europe were an essential contribution for the re-emergence of the airline market in the near term. In the sequel the strong post-covid market recovery brought aircraft industry and airlines into the opposite situation of a strong boom struggling to meet demand.

6.14 Recovery Prospects

The prospects for air transport recovery from the COVID-19 crisis were better in the short term for regional and short-haul transport leading to a V-shape (Sect. 6.14.1), whereas long-haul flight continued to be hampered by uncoordinated deconfinement worldwide causing a delayed rise in a U-shape (Sect. 6.14.2). The fundamental uncertainty concerning the long-term level relative to pre-COVID-19 traffic (Sect. 6.14.3) from full recovery (100%) to partial (70–90%) and the number of years to reach that level was dispelled by a strong post-covid recovery in a couple of years to near pre-covid levels of traffic and much higher revenues turning unprecedented losses into unprecedented profits due to higher ticket prices rising faster than costs.

6.14.1 Short Haul: V-Shape

Figure 6.31 shows the status of the worldwide jet transport fleet on April 30, 2020, with only about one-third of the aircraft in service. Figure 6.32 shows the value of airline deliveries in billion dollars since 2003, with the steady growth interrupted by the decline during the COVID-19 crisis in 2019–2020; the decline is larger in the for twin-aisle aircraft than for single-aisle, and the predictions for the 2020–2029

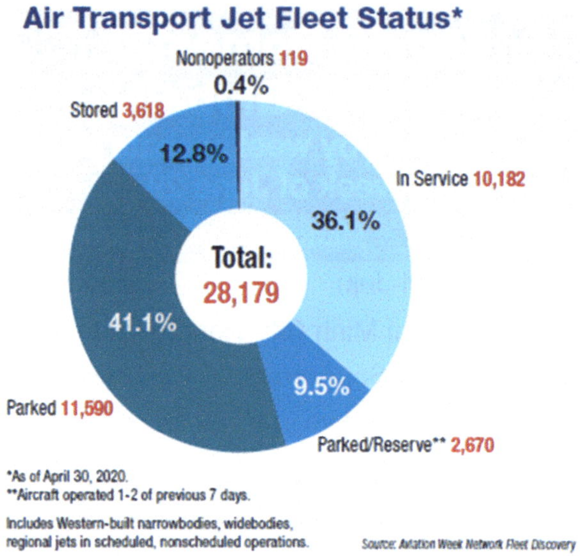

Fig. 6.31 Status of air transport fleet on April 30, 2020, distinguishing aircraft in service, stored, parked, and in reserve

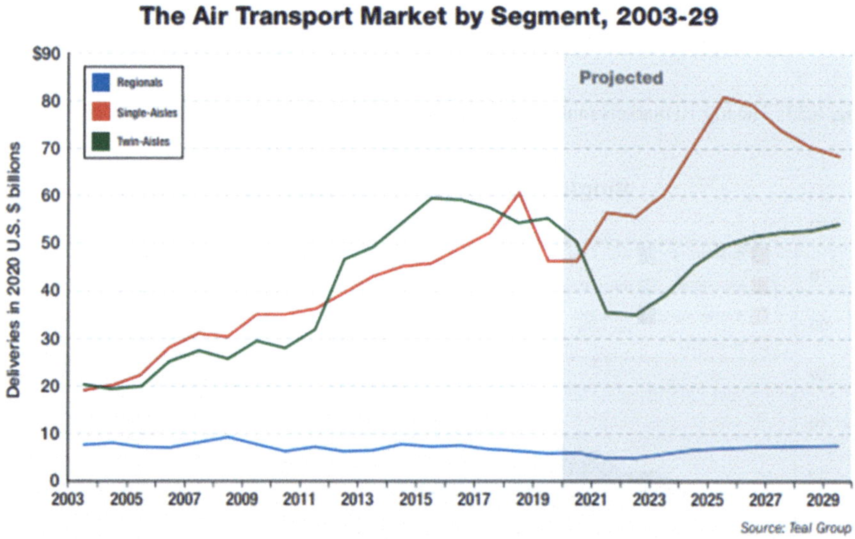

Fig. 6.32 Number of regional single-aisle and twin-aisle aircraft in service over the period 2003–2020 and projections for 2020–2029

period show a slower recovery as well, with the reduced market favoring the economics of smaller planes.

Figure 6.33 shows that the 10 largest domestic markets ranked by weekly seats are all in the Far East where faster recovery can be expected. Figure 6.34 shows the

Top 10 Domestic Markets Globally
Ranked by weekly seats for week of June 29

Rank	Market	Country
1	Seoul-Jeju	S. Korea
2	Ho Chi Minh City-Hanoi	Vietnam
3	Tokyo-Sapporo	Japan
4	Osaka-Tokyo	Japan
5	Fukuoka-Tokyo	Japan
6	Shanghai-Guangzhou	China
7	Shanghai-Shenzhen	China
8	Tokyo-Okinawa	Japan
9	Hanoi-Da Nang	Vietnam
10	Shanghai-Chengdu	China

Source: CAPA and OAG

Fig. 6.33 The top 10 markets ranked by weekly seats on June 29 are all in the Asian Far East

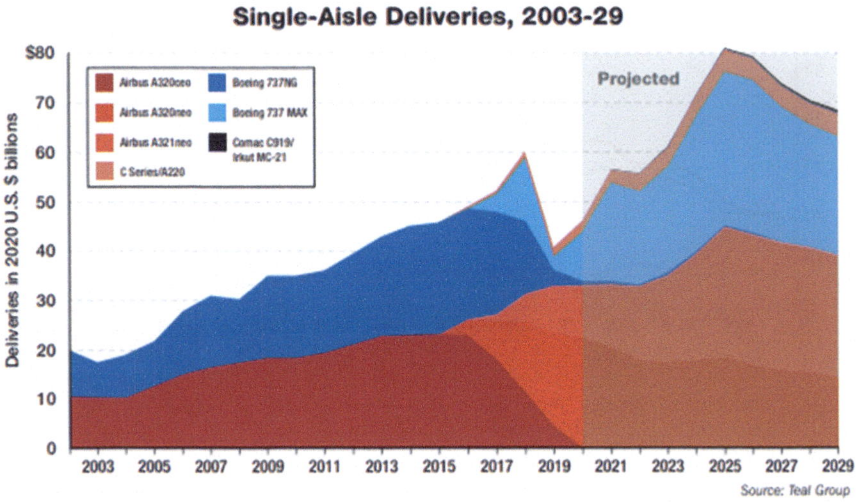

Fig. 6.34 Deliveries of single-aisle aircraft from Airbus, Boeing, and COMAC over the period 2002–2020 and predictions for 2020—2029

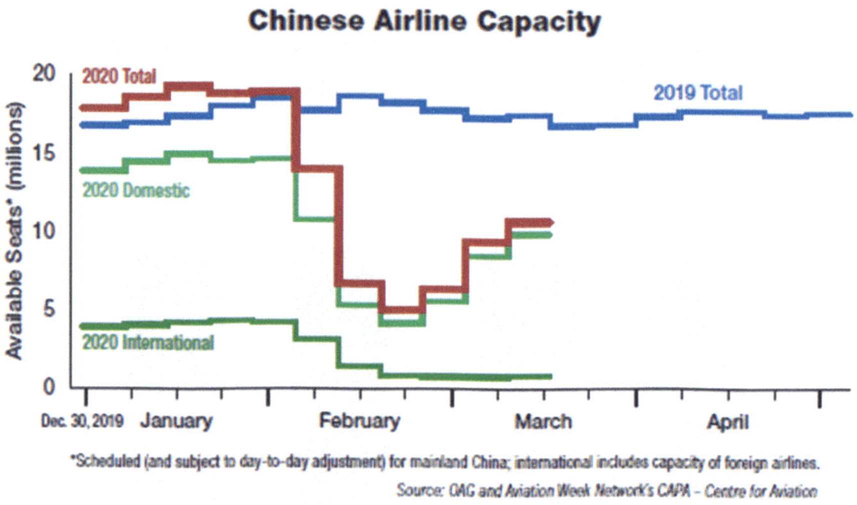

Fig. 6.35 Chinese airline domestic, international, and total capacity comparing 2019 to 2020 and showing the decline due to COVID-19 and the prospects for domestic recovery and lag of international recovery

single-aisle deliveries in the period 2003–2020 with predictions up to 2029 for Airbus, Boeing, and Comac models. Figure 6.35 shows the evolution of the Chinese domestic and international markets in the first 4 months of 2020, with the sharp decline due to COVID-19, followed by a quick partial recovery in the domestic market and a further decline in the international market.

Figure 6.36 compares the share of total passengers associated with domestic and international traffic, demonstrating a clear shift toward domestic and away from international from 2019 to 2020 for two Japanese airlines: All Nippon Airlines (ANA) and Japan Air Lines (JAL). Figure 6.37 shows global revenue passenger kilometers by quarter in the period 2016–2021, with the sharp COVID-19 decline of about 80% in 2020. The apparent better recovery in the international than domestic market is due to the use of passenger kilometers rather than the number of seats or flights.

6.14.2 *Long Haul: U-Shape*

The coordinated deconfinement allowed a more rapid resumption of regional and short-haul flights leading to a V-shaped recovery curve. The lack of worldwide coordination left long-haul flights in a stagnant position corresponding to the horizontal branch of a U-curve. It was only when deconfinement was generalized worldwide allowing long-haul flights between almost any city pair that the sharp rise became possible in the vertical branch of the U-shape.

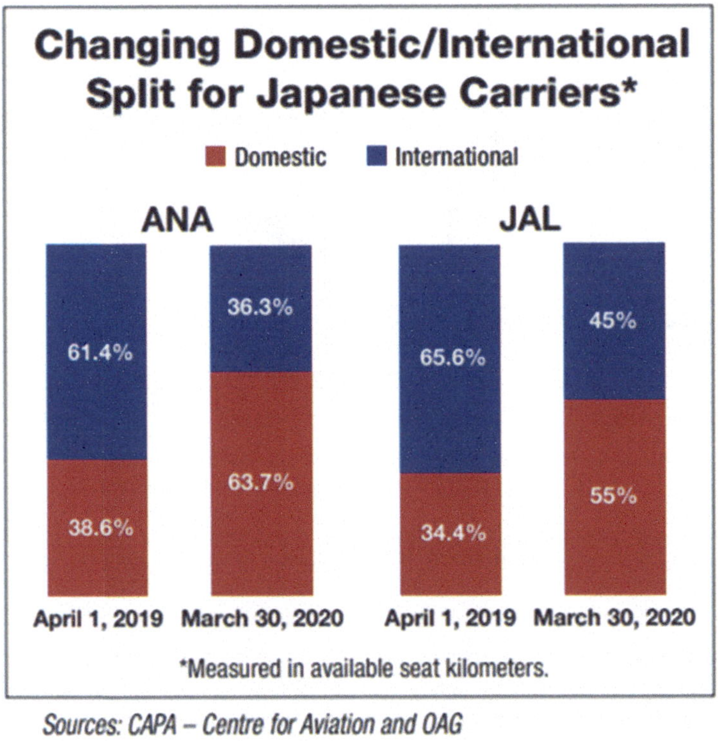

Fig. 6.36 Decline of international relative to domestic flights for two Japanese airlines, comparing April 1, 2019, to March 30, 2020

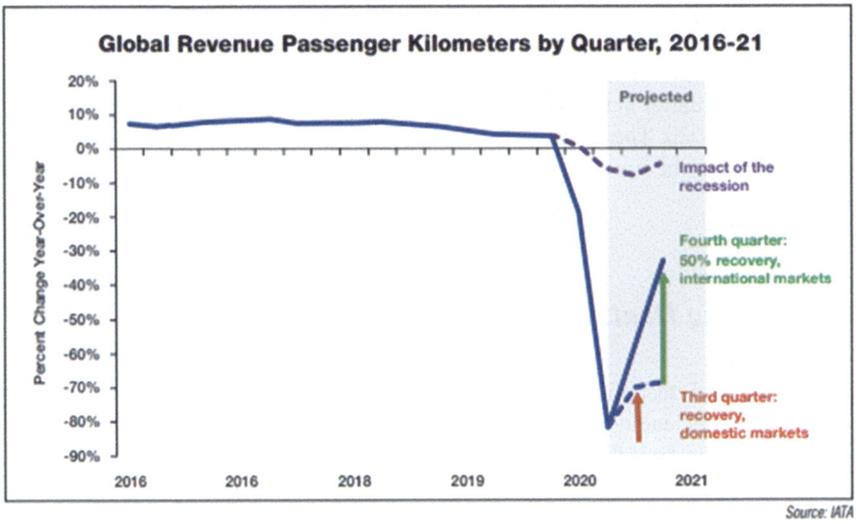

Fig. 6.37 Global revenue passenger kilometers by quarter 2016–2021, showing a sharp drop in 2020 due to COVID-19 and recovery in domestic and international markets and the impact of recession. (Source: Aviation Week Network)

6.14.3 Long-Term Level

Airlines started recovery from the covid crisis as soon as possible, offering incentives to passengers such as low fares despite lower revenues. In order to capture as much traffic as possible, airlines flew up to 80% of the route network with only 30% of seat capacity, using the smallest possible aircraft to help load factors, and hope that traffic growth gradually improves profitability. The main open questions were whether recovery to 100% of pre-COVID-19 levels was possible or traffic would stabilize at a lower level in the 70–100% range. Also, how many years would elapse until that ultimate level would be reached? In fact the sharp post covid airline travel demand and higher ticket prices meant a rapid rise to near pre-covid levels with higher profits.

Figure 6.38 shows the projected US domestic traffic recovery by types of airlines with the drop to 40% of pre-COVID-19 level recovered to 85% in 3 years (2023) and full recovery of 100% expected in 7 years (2027), and likely to be achieved earlier given the current strength of demand. Figure 6.39 shows global passenger kilometer traffic in 2020, with a drop to 10% due to COVID-19 in March and a predicted recovery in December to 64% in the baseline case and 47% in the case of higher pandemic risk.

Figure 6.40 shows that revenue passengers' kilometers increased at the same rate for domestic and international flights starting in 2016 at 80% of the pre-COVID-19 level. The COVID-19 crisis in 2020 caused a large decline, relatively smaller at 60% for domestic rather than 40% for international traffic. The predicted recovery art the

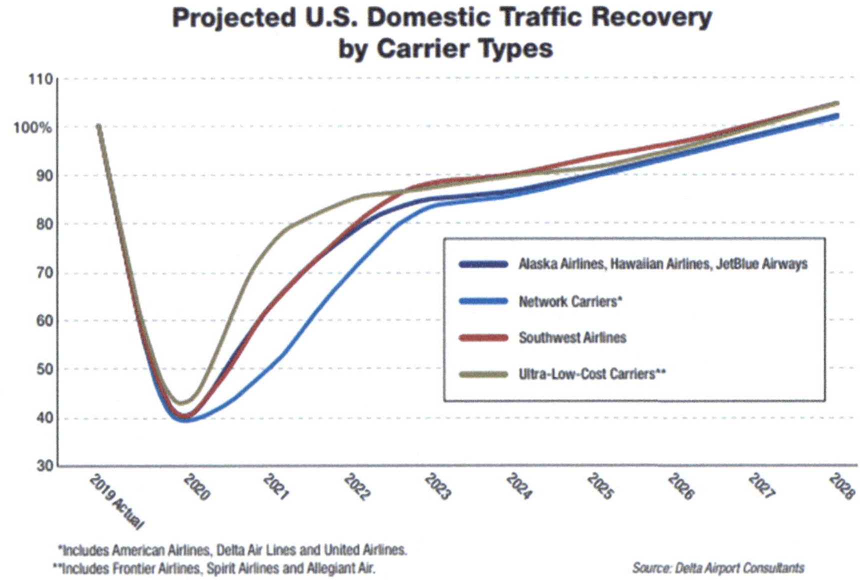

Fig. 6.38 Recovery from the COVID-19 crisis of the US domestic air traffic by airline or airline groups

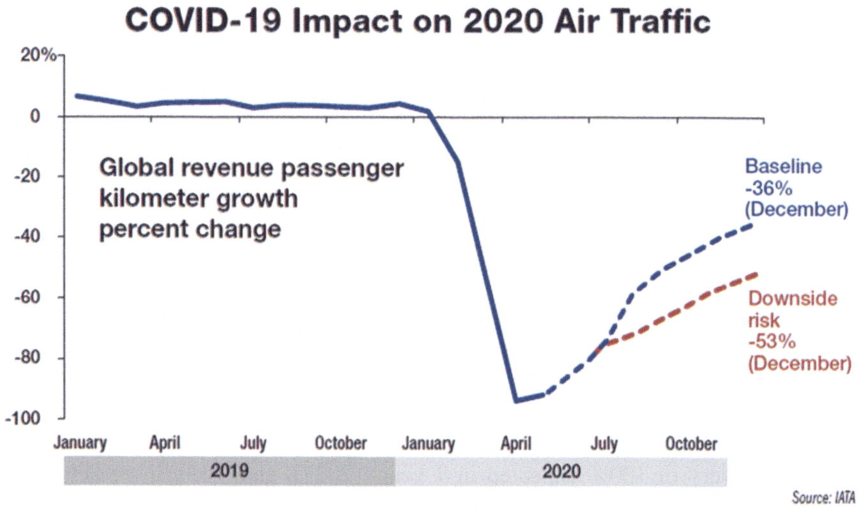

Fig. 6.39 Impact of COVID-19 on 2020 air traffic, with a sharp decline to 10% in March 2020 and recovery to 47–64% by December with the higher figure as a baseline and lower figure associated with slow recovery from the pandemic or second wave

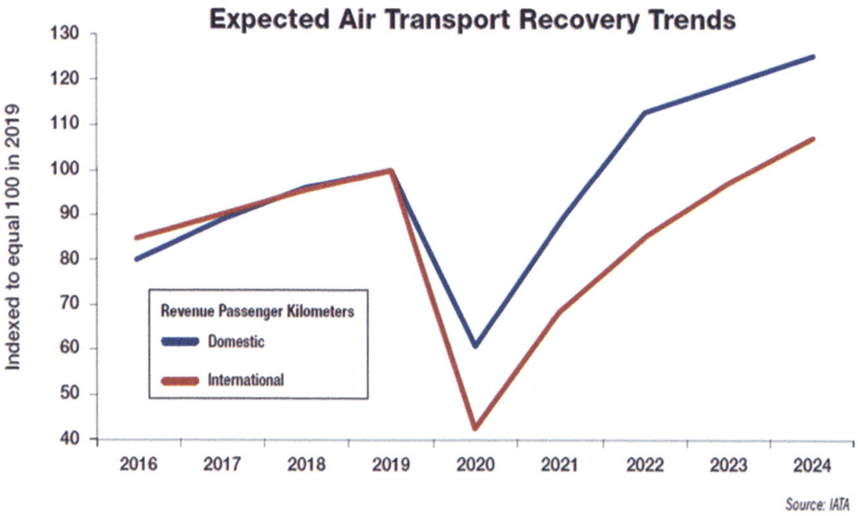

Fig. 6.40 Expected air traffic recovery trends following a sharp decline in March 2020 due to COVID-19 and quicker ramp-up of domestic flying due to consistent deconfinement and slower ramp-up of international travel due to uncoordinated deconfinement

time was that domestic flights reach 2019 levels in 1–2 years in 2022 and international flights recover the 2019 levels in 2–3 years in 2023. Beyond these dates, air traffic was expected to grow beyond pre-COVID-19 levels. In fact air traffic took a bit longer to reach near pre-Covid levels and higher ticket prices lead to strong profits despite higher costs like salaries.

6.15 Onboard Health

Airlines were keenly aware that to regain traffic besides incentives like low fares it was necessary to gain passenger trust (Sect. 6.15.2) that anti-pandemic measures (Sect. 6.15.1) provided sufficient health protection. The flight crew protection was particularly relevant for cabin attendants (Sect. 6.15.3) who make frequent flights.

6.15.1 Anti-pandemic Measures

The air-conditioning systems aboard aircraft use high-efficiency particulate air (HEPA) filters similar to hospitals. Air-conditioning uses 50% recycled and 50% external air and completely renews cabin air every 3–4 minutes, which is faster than most buildings. Airbus indicates that its air-conditioning systems send air vertically down from the roof that is extracted from the floor. This minimizes sideways flow to passengers in the same row; the effect can be increased by turning on air vents. The high seat backs also prevent air from flowing forward or backward to other seat rows.

All the preceding measures diminish the likelihood of water droplets from one passenger reaching another. Some airlines considered using transparent plastic barriers between passenger seats. Others left the middle seat in each row empty to provide greater separation between window and aisle passengers. Some airlines adopted more elaborate seating patterns to achieve greater social separation among passengers at the expense of reduced load factors.

6.15.2 Passenger Trust

To regain passenger trust that there was no health risk in flying, airlines put an increasing focus on cleaning the aircraft cabin and toilets and showing to passengers perfectly clean cabins. The use of face masks throughout the flight was mandatory in the European Union, though not in Britain, where, for example, British Airways made them optional. The permanent use of masks and the control of air-conditioning flows and seat arrangements do nothing against contamination of seats, tray tables, display screens, and toilets.

For the latter, at least three cleaning methods were available: (i) sweeping with disinfectants, (ii) filling the cabin with gaseous antiviral products, and (iii) using ultraviolet (UV) light that kills microbes. The demand for these products increased significantly during the COVID-19 pandemic, and most could only be used with no passengers in the cabin. Some of the products could also degrade cabin materials like seat fabrics. A basic cabin cleaning takes half an hour and a thorough one 3–4 h. Cleaning between flights or stops may require adjusted flight schedules.

6.15.3 Flight Crew Protection

The flight crew, the cabin attendants, make many flights with many passengers and thus health protection is more critical. Another issue is how to deal with non-cooperative or recalcitrant passengers both in the aircraft and at the airport in some situations. For example, if a passenger tests positive and cannot fly with the family, or the case of a child who tests positive, or a passenger who tests positive at an intermediate stop-over; all of these can become difficult situations.

There are also limits to health protection. Putting all carry-on luggage in the hold would prevent passenger contamination by touching other's baggage. This may be resented by passengers used to carry essentials in cabin baggage to work during the flight or avoid stray luggage undermining their work at their destination. This would risk sending electronic devices with lithium-ion batteries in the luggage hold. These batteries can easily catch fire if crushed. Such a fire can quickly be extinguished in the cabin, but in a luggage hold, it is another matter.

6.16 Pandemic Planning

The COVID-19 pandemic has caught the world and the aviation community by surprise, leading to improvised and uncoordinated measures whose effectiveness could be improved based on a tragically earned experience. COVID-19 may not be the last pandemic the world may see, and as for other disasters and emergencies, planning could cover: (Sect. 6.16.1) early detection to apply quarantine and containment measures on the small fraction of the population; (Sect. 6.16.2) this allows the use of the resources of the majority of the unaffected population to support and heal the infected population; and (Sect. 6.16.3) some backup reserves and preplanned actions could enable a rapid reaction in a suitably large scale.

6.16.1 Isolation and Containment

Early detection is fundamental in limiting the number of infected and potentially infected people. COVID-19 has been detected in waste products and a posteriori in hospital patients. Routine analysis of waste products and special attention to abnormal hospital cases could be used for early detection of new viruses and potential pandemics. The potentially infected population would be quickly isolated, and severe cases sent to the hospital.

The isolated area would be separated from the rest by a hard border that could be approached or crossed only by authorized personnel specially trained to work in contaminated conditions that would transport all supplies between the outside and inside of the border. No one else would be allowed to approach the border either

from the outside or the inside. The border would be open only when the pandemic is fully eradicated inside.

6.16.2 *Marshalling of Support*

Most of the population outside the border would have an overwhelming priority in helping to recover those inside. Thus all available resources could be allocated to provide medicine, food, and any other essentials on any scale needed. Telecommunication and psychological support would ensure those inside the border do not feel neglected and are aware that their well-being is the priority of those outside the border who would endeavor to satisfy every reasonable wish of those inside.

Preplanning would ensure that a basic set of contingency medical supplies and equipment of general use be available. Buildings would be surveyed and available for the rapid expansion of hospital capacity. Plans could be made for the rapid ramp-up of production of specialized equipment according to the needs of the pandemic. All these measures have already been used and proved, so it is mainly a matter of collecting them in a comprehensive plan with fast implementation when needed.

6.16.3 *Backup Reserves*

Every sector of activity should be included in the planning to ensure any relevant capability is used when needed. Two examples are given from aviation. Cabin attendants are trained in basic health care. Obviously, they could not replace trained hospital staff in intensive care units. But with some more training, they could take over some simpler hospital tasks such as surveillance of patients or help in simpler treatments, freeing the professional hospital staff to concentrate on more demanding tasks.

A second example concerns the industrial aerospace supply chain with a multitude of design and production skills and sophisticated programmable production facilities. These skills could be used to design medical equipment and plan the rapid ramp-up of production when needed. The combination of pre-stored general supplies and an array of preplanned production ramp-ups would form the emergency plan to deal with pandemics.

6.17 Survival in a Crisis Period

As the biggest crisis in the history of aviation, the COVID-19 pandemic has put in a survival and recovery state the main sectors: (Sect. 6.17.1) starting with airlines, with the decline in passenger demand met by a fraction of the active fleet, and the

remaining aircraft parked or stored (or to be disposed of) pending a traffic recovery that might take several years until new aircraft are needed; (Sect. 6.17.2) continuing with aircraft manufacturers, facing canceled orders, delayed deliveries, and a lack of new orders until traffic recovers and older less efficient and/or more polluting aircraft have to be replaced; and (Sect. 6.17.3) filtering down the four tiers of the supply chain that had been under pressure to lower prices and invest in increased capacity before COVID-19, finding afterward a sharp decline in orders and revenue.

6.17.1 Airline Fleets

Before the COVID-19 pandemic, the long-haul traffic was softening with continuing strong demand in the short-haul, leading airlines to increase their fleets; the sharp decline after the pandemic forced airlines to rethink their fleet plans. The sharp decline in passenger demand required only a fraction of the fleet to remain active. Some older aircraft, with higher fuel consumption and/or emissions, were grounded with the expectation of disposal since by the time they could be flown again the economics would be unattractive and environmental restrictions unsustainable.

The total fleet minus active aircraft and those awaiting disposal was grounded in one of two states: (i) parked, which allows quick return to service and has low initial costs, but higher maintenance costs over a longer time; and (ii) storage, which has lower maintenance costs, but a higher initial cost and longer time to return to service. It is not easy for airlines to decide what part of the fleet to park or store, with low traffic levels and unpredictable ramp-up of recovery. No new aircraft were expected to be required until traffic demand recovered, or existing aircraft would be replaced by new more efficient and less polluting types.

6.17.2 Aircraft Manufacturers

Both Airbus and Boeing admited to being in discussions with almost every customer or airline on revising the order book and delivery schedule, and not in a desirable way: (i) in some cases, order cancellations could be avoided due to low traffic volume and expectations of slow recovery to levels below the pre-pandemic status; (ii) delivery delays were preferable to adjust to variable and hopefully improving traffic demand; and (iii) new orders were expected to wait 2–3 years for single-aisle and 3–5 years for twin-aisle, as long-haul traffic recovers more slowly than short-haul.

Boeing's and Airbus's initial reactions to COVID-19 have been to lower production rates by about 30% corresponding to the reduction in backlog and keeping the same period 5–8 years to fulfill all orders. In fact, twin-aisle production rates were already softening before COVID-19 and were cut to less than half afterward. The main difference was in single-aisle where Airbus plans to ramp up above 62 up to 75 and Boeing plans to ramp up to 52 (after ungrounding of the 737Max) have were

revised to about 30 per month during covid with production rates may be adjusted temporarily to the lowest possible economic level awaiting recovery. The sharp post-covid traffic recovery left Airbus and Boeing in the opposite situation of struggling to ramp up production.

6.17.3 Down the Supply Chain

Before the unpredictable arrival of the Covid crisis, both Boeing and Airbus aggressively pursued plans to cut costs, putting pressure on suppliers to reduce prices to single-percent-digit profit margins. At the same time, suppliers were asked to make significant investments in increasing production capacity to meet the increasing demand for airliners. Thus, much of the supply chain was high on debt for expansion and low on liquidity from reduced prices when the COVID-19 pandemic suddenly hit with a sharp reduction in orders—very much an anticlimax.

The situation of thousands of companies in the supply chain depended on size, products they supplied, and exposure to civil aviation. At one end, tier 1 large engine manufacturers could not be allowed to fail as they are essential for recovery, and some of them have military contracts and non-aviation activities as a buffer. Down tiers 2–4, smaller companies may fail, leave the market, or be absorbed. Some with low exposure to civil aviation may survive by other means. Those that supply essential parts needed for recovery have to be rescued, if necessary, by primes. Airbus and Boeing were thoroughly examining their supply chains in a "reorganization" aimed at ensuring the survival of essential suppliers by sustaining orders, making capital available, or taking over control. All this downsizing left the whole aviation production sector struggling to meet airline orders in the sharp post-covid traffic recovery.

6.18 Rebound to Near Pre-COVID Levels

The deep recession in the aeronautical sector due to low demand during the COVID pandemic (Sect. 6.18.1) has been followed by a sharp recovery, struggling to keep up with renewed demand (Sect. 6.18.2).

6.18.1 Effects of COVID-Induced Depression

The COVID-19 pandemic has had tragic consequences due to its unexpected rapid spread and lack of preparation to deal with the situation. These hard-earned lessons should be incorporated into an emergency plan to detect pandemics as early as possible and eradicate them quickly and effectively. Another set of preventive measures

would be to adopt procedures that prevent contamination, for example, electronic or biometric screening that avoids the handling of documents at airports. The COVID-19 pandemic has affected severely aviation prompting government intervention, sometimes in very different forms.

In the United States, the CARES bill provided more than $190 billion for aviation in bailout funds without long-term objectives and only short-term requirements that were not always fully met and generated controversies: (i) airlines were required not to furlough employees or reduce pay rates, but some tried to reduce working hours, thus decreasing monthly salaries and provoking protests in the corridors of congress; (ii) airlines and airports were required to keep domestic flight networks, but the rules affected differently small regional and large international airlines and there were funding discrepancies between large and small airports; and (iii) industry received significant rescue funds, such as $60 billion for Boeing and $24 billion for General Electric without strategic technological objectives. The CARES bill covered only 6 months and helped the US aeronautical sector survive until the post-COVID recovery.

In Europe, national interventions did not exceed €15 billion, supplemented at the level of the European Union, and were applied based on case-by-case studies with significant economic and market constraints and long-term objectives of greater efficiency and reduced environmental impact: (i) airlines such as flag carriers received support packages up to $10 billion requiring better economic and environmental performance together with other measures such as fleet reductions, loss of airport slots, and closing of domestic routes; and (ii) industry support strengthened the military contracts and covered all levels of the supply chain to enable the maturation of new technologies, like electrification and hydrogen power, to enable a future generation of up to 30% more efficient aircraft with environmental impact reduced by 25–75%.

The world military aircraft market (Fig. 6.41) is about one-third to one-fourth of the civil aircraft market and thus only a partial buffer. On shorter routes up to 1000 km in Europe, high-speed rail (Fig. 6.42) is a strong competitor; for continental distances above 1000 km and transoceanic routes up to 20000 km, there is no credible alternative to aircraft in terms of speed and travel time, and indeed safety and efficienc, and also flexibility in joining different population centers. The state and EU support allowed the European aeronautical sector to survive the COVID depression, and as in the United States, enter a phase of strong recovery with sharp renewal of demand.

6.18.2 Reversal into Strong Demand

The recovery from the COVID depression was much quicker than most expectations due to the strong effect of "pent-up" demand. During the COVID pandemic, families, relatives, friends, and associates were prevented from traveling, creating an urge to come together as soon as the restrictions were lifted. While there were job

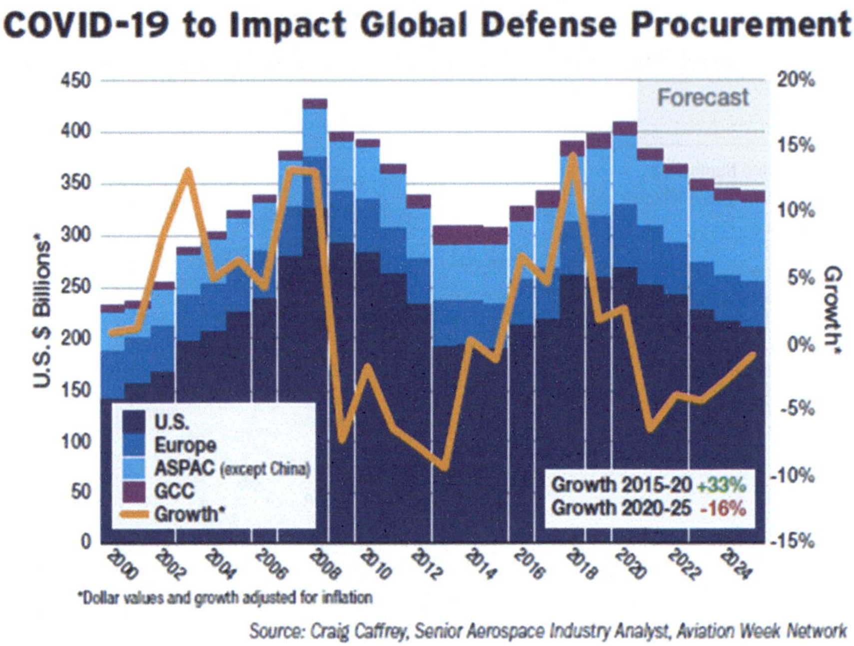

Fig. 6.41 COVID-19 to impact global defense procurement. (Source: Aviation Week Network)

losses in private service sectors, many workers kept their jobs and revenues and spent rather less in travel and leisure activities limited by COVID restrictions. The long periods of home confinement and mobility restrictions led to a strong urge to travel, which was not hindered by higher prices, as demand exceeded the supply from an aeronautical sector struggling to recover from depression.

The low ticket prices to encourage travel during the COVID pandemic gave way post-COVID to higher ticket prices and airlines and airports struggling to fully satisfy demand. During COVID, pilots, cabin crew, and airport staff took early retirement, were dismissed, or moved to other sectors not to return to what they came to see as "unstable employment." The rehiring and new recruitment efforts provided crews and staff to operate the main more profitable routes; service to local or regional routes, for which grounded aircraft were available, were postponed until more staff could be hired. Airports faced bottlenecks and delays, and as airlines needed to hire more staff with higher wages, which could be mostly compensated by higher revenues from higher ticket prices. Some airlines exceeded pre-COVID revenues with about 80% of pre-COVID traffic due to higher ticket prices, while wage increases, and other costs meant the return to profits was not as high or widespread.

The recovery of airline traffic post-COVID led to increased orders first for single-aisle aircraft on domestic and transcontinental routes in regions with less COVID restrictions and later international flights as the COVID pandemic was contained worldwide. The order books of Airbus and Boeing swelled beyond pre-COVID

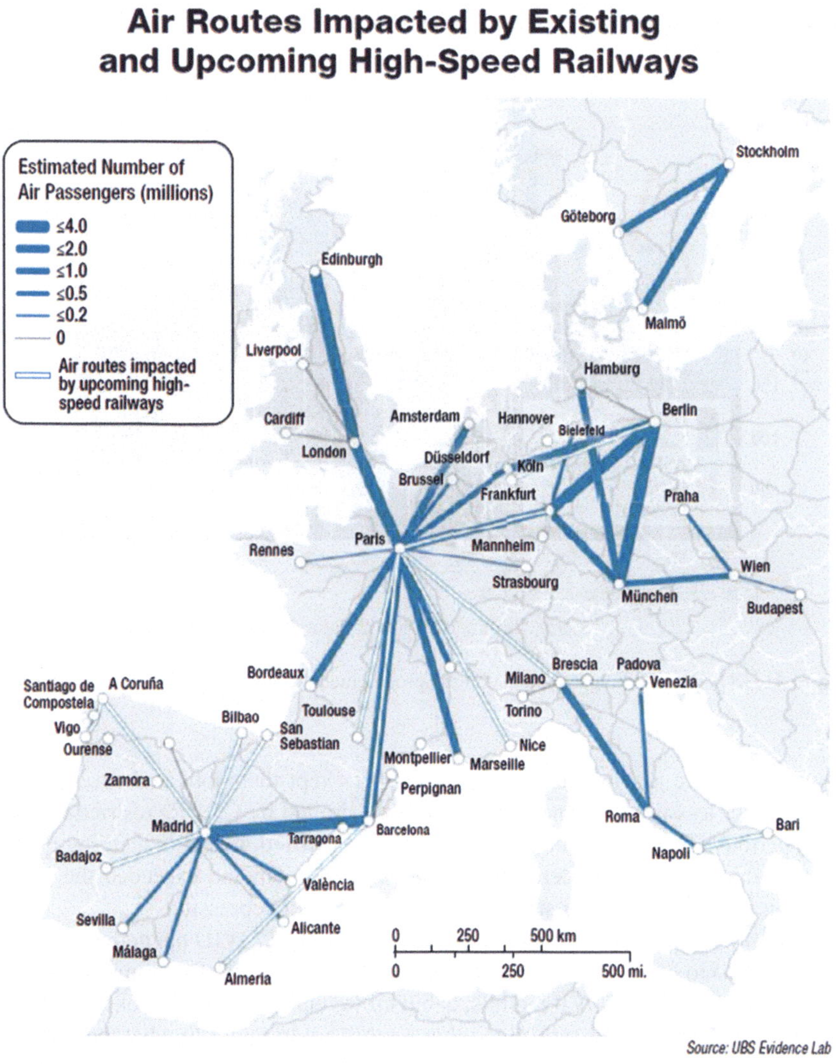

Fig. 6.42 Air routes impacted by existing and upcoming high-speed railways. (Source: Aviation Week Network)

levels, overwhelming cancellations during the COVID period. The efforts to increase production rates, for example, to 75 from 50 for the Airbus A320 family, and from 30 to 52 for the Boeing B737, pre-crisis levels struggled with supply chain issues both for Airbus and Boeing, plus production quality issues at Boeing. Most noticeable for the new generation of more efficient engines are the CFM Leap and Pratt & Whitney PW1200, with initial in-service issues compounding the difficulties in increasing production. At all tiers of the supply chain, companies that had cut off production and laid off workers to survive in a precarious financial situation

were now asked to quickly ramp up production, requiring larger investments, more abundant raw materials, and an enlarged less experienced workforce, which take time to train to former productivity levels. Thus, the transition from a lack of demand and supply cut-back during the COVID pandemic to a sharp recovery of demand and a need for sustained supply may require a ramp-up lasting several years.

6.19 Sharp Post-COVID Recovery

The sharp post-COVID recovery (Sect. 6.19.1) is driven by pent-up and enduring demand (Sect. 6.19.2) requiring efforts toward meeting a sustained growth (Sect. 6.19.3).

6.19.1 Regional and International Air Traffic

Figure 6.43 shows for the "Big Four" airlines of the United States the sharp decline due to COVID around March 2020 and the steady recovery to near pre-COVID levels by the end of 2022. Figure 6.44 shows flights in Europe per market segment with a steady recovery for airline traffic from mid-2021. The cargo traffic did not decline during the COVID pandemic due to the need to carry urgent medical supplies and the increase of e-commerce and associated package transport during COVID confinement. Business aviation was the quickest to recover and was less affected than airline traffic. The main cloud on the Horizon for civil aviation is the need to cut emissions (Fig. 6.45), currently at 3% of the world total, but that could rise to 6% by 2050 due to an annual traffic growth of 5%, making urgent the drive toward net zero.

6.19.2 Pent-Up and Enduring Demand

The pent-up demand following the lifting of COVID restrictions has led to a demand for air travel that has not relented and may remain unless a global financial crisis occurs. Capacity may initially limit traffic levels to about 80% of 2019 pre-COVID levels but ticket prices and revenues are higher, and demand may remain ahead of supply for another 1–2 years and be sustained afterward. Figure 6.46 shows the post-COVID rebound of passenger movements at airports without reaching yet 2019 pre-COVID levels. The high demand and limited capacity have led to larger waiting times before and after flights (Fig. 6.47). Full recovery to 2019 pre-COVID levels can sustain expansion and profitability for several years to come if environmental challenges are met.

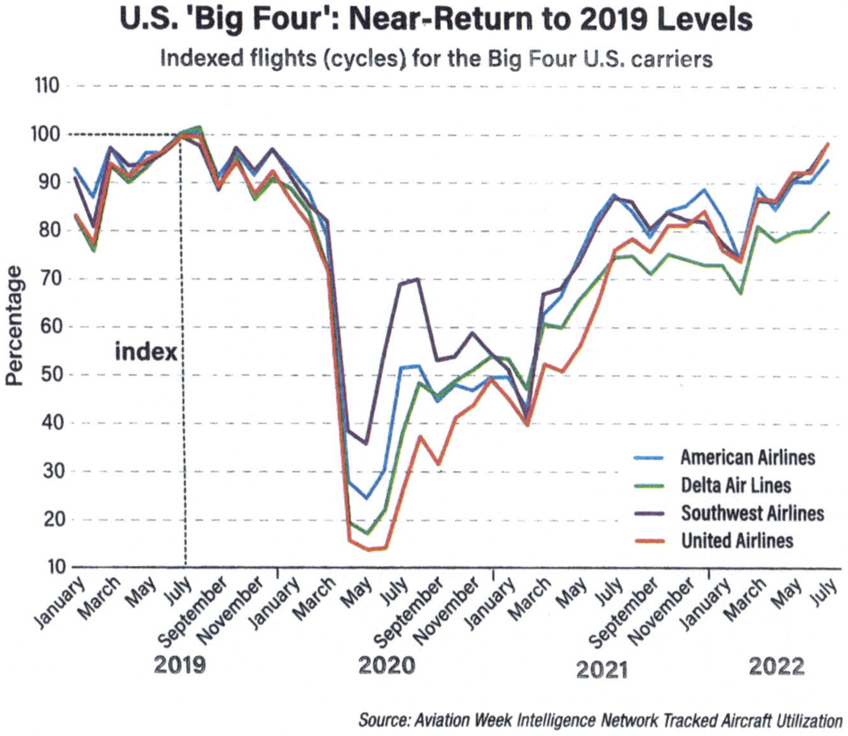

Fig. 6.43 US “Big Four”: near-return to 2019 levels

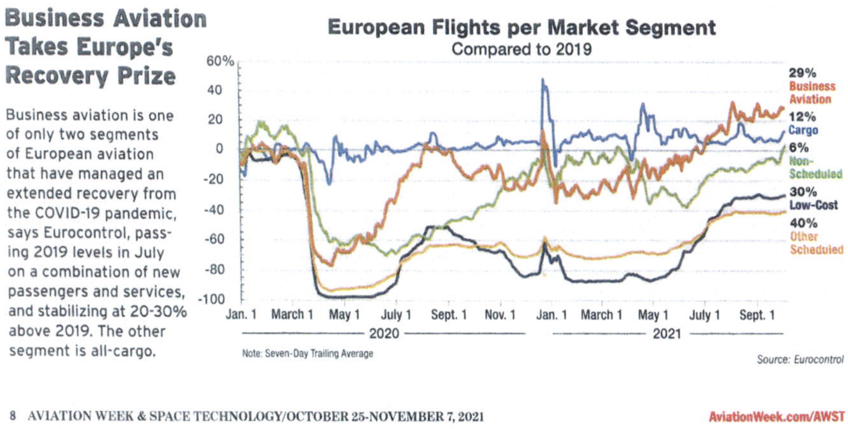

Fig. 6.44 European flights per market segment

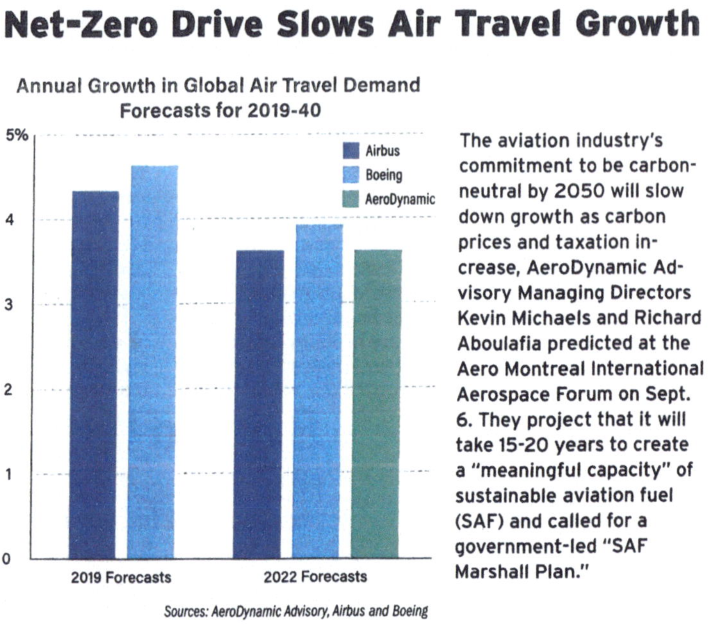

Fig. 6.45 Net-zero drive slows air travel growth

World's Busiest Airports in 2021

Airport	Passengers (Millions)	Versus 2020	Versus 2019
Atlanta (ATL)	75.7	▲ 76%	▼ 32%
Dallas/Fort Worth (DFW)	62.5	▲ 59%	▼ 17%
Denver (DEN)	58.8	▲ 74%	▼ 15%
Chicago (ORD)	54	▲ 75%	▼ 36%
Los Angeles (LAX)	48	▲ 67%	▼ 46%
Charlotte (CLT)	43.3	▲ 59%	▼ 13%
Orlando (MCO)	40.3	▲ 87%	▼ 20%
Guangzhou, China (CAN)	40.3	▼ 8%	▼ 45%
Chengdu, China (CTU)	40.1	▼ 2%	▼ 28%
Las Vegas (LAS)	39.7	▲ 79%	▼ 23%

Source: Airports Council International

Fig. 6.46 World's busiest airports in 2021

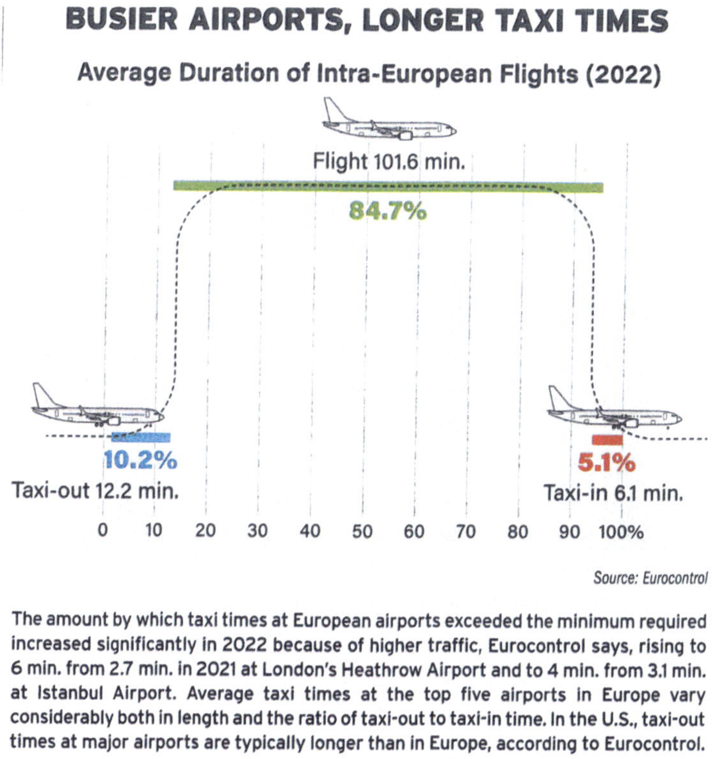

Fig. 6.47 Busier airports, longer taxi times

6.19.3 Efforts for Sustained Expansion

The efforts for sustained expansion can be seen in all sectors of aviation starting with the increasing backlogs of Airbus and Boeing single-aisle aircraft, which are recovering from the relatively modest dip during the COVID pandemic (Fig. 6.48) leading to increased deliveries (Fig. 6.49). The twin-aisle deliveries are also on the rise (Fig. 6.50). The growth in the aviation sector applies across all market sectors (Fig. 6.51). The most notable case is business aviation (Fig. 6.52). It was much less affected than airlines during the COVID pandemic due to the demand for travel without schedule limitations and sharing the flight with familiar passengers. The post-COVID recovery has not changed private flying preferences and has led to an unprecedented excess of demand over supply for business jets, with waiting periods of up to 2 years for new aircraft and used aircraft sold in a few weeks. The recovery of the supply chain is the main factor delaying growth, and the persistent Boeing production quality issues have lead to fluctuating production and uncertainties at suppliers. The yearly deliveries of airliners have not increased steadily, with the occasional fall back to lower production levels.

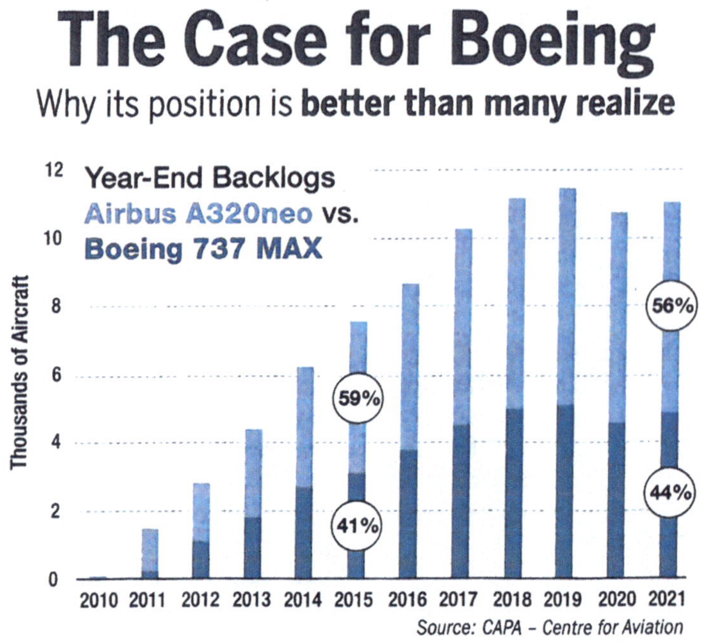

Fig. 6.48 The case for Boeing

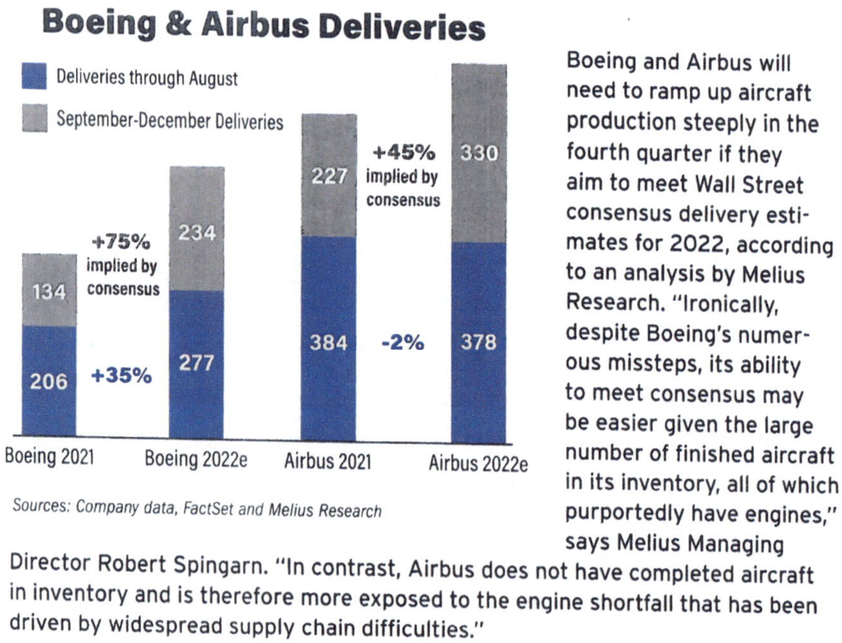

Fig. 6.49 Boeing and Airbus deliveries

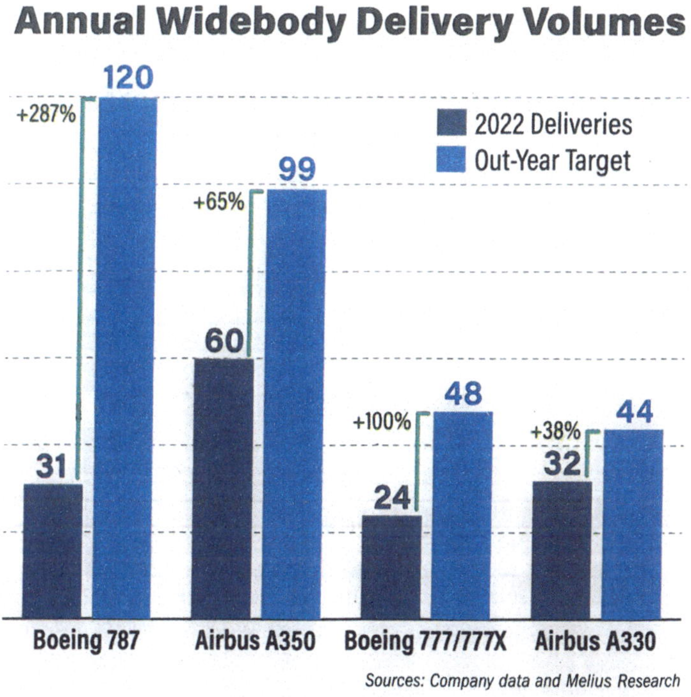

Fig. 6.50 Annual wide-body delivery volumes

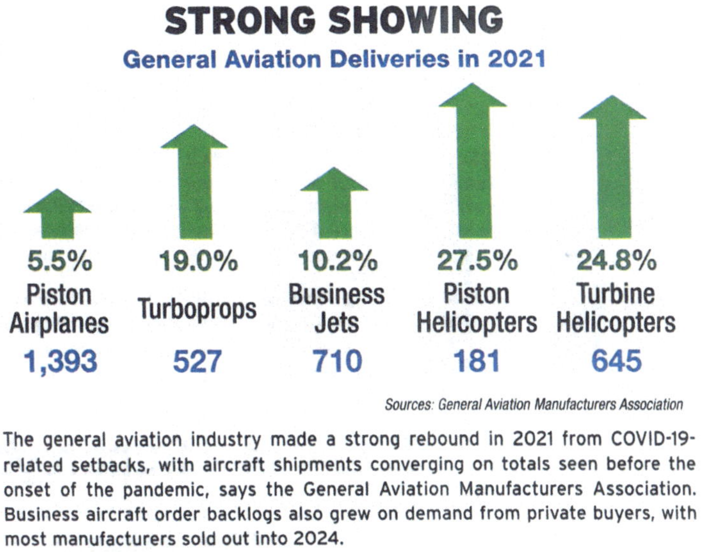

Fig. 6.51 Strong showing

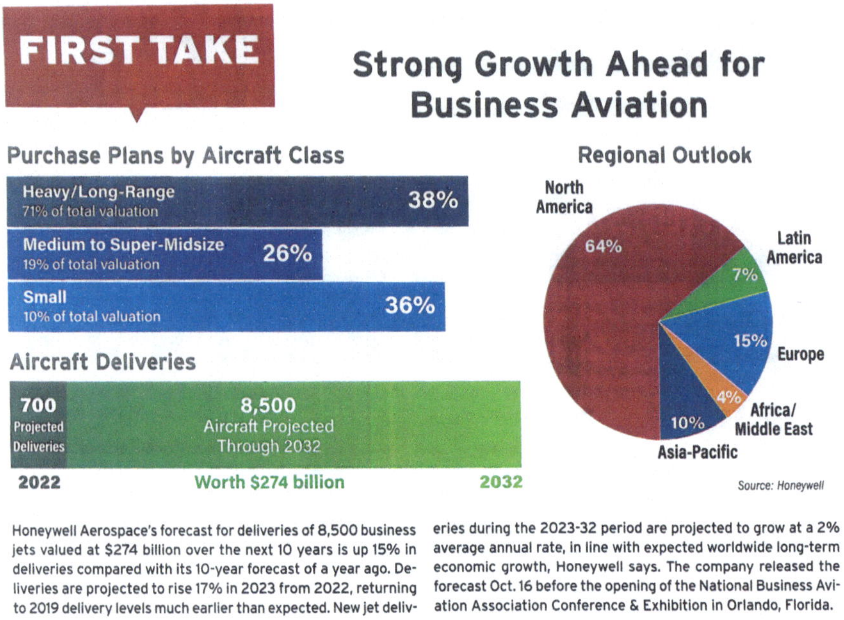

Fig. 6.52 Strong growth ahead for business aviation

6.20 Conclusion

The COVID pandemic, the biggest crisis in the history of aviation, has shown the resilience of the aeronautical sector to survive large fluctuations of demand that can be seen as a major recession. The government support helped key airlines like flag carriers to overcome a difficult COVID period until traffic resumed healthy growth post-COVID; the positive prospects for civil aviation in the coming years remain, provided that environmental challenges can be met. The military sector is more sheltered from the fluctuations in demand than civil aviation, and defense contacts stable over years of development and production helped the aeronautical industry survive through the COVID period. The deterioration of the security situation in Europe following the Russian annexation of Crimea and the invasion of Ukraine has led to a strong increase in defense spending to be sustained over the coming years (Fig. 6.53). The aeronautical sector may have sustained growth in the coming years, with a stable background of strong military spending, and a healthy civil aviation market, provided that environmental challenges are met.

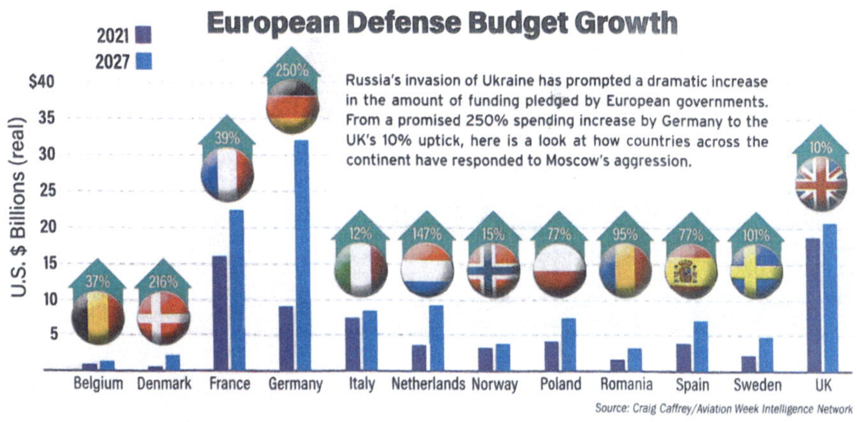

Fig. 6.53 European defense budget growth

References[1]

1. World Sailing. Coronavirus (COVID-19) World Sailing Guidance for Event Organizers for the Protection of the Health of Athletes Version 6,0 – 10 February 2022 (here).
2. World Health Organisation – WHO Official COVID-19 info. https://www.who.int/emergencies/diseases/novel-coronavirus-2019.
3. Nico D'Auterive. The pandemic accord explained: What countries are going to protect against future global health emergencies. United Nations Foundation, May, 19, 2023 (here).
4. Mobilizing Social Science: SSRC Responds to Covid-19. Social Science Research Council (here).
5. WHO Official COVID-19 info. World Health Organization.
6. Coronavirus (COVID-19) Information from FAA. Federal Aviation Administration (here).
7. COVID-19 Aviation Health Safety Protocol: Operational guidelines for the management of air passengers and aviation personnel in relation to the COVID-19 pandemic. European Centre for Disease Preventiopn and Control. 11 May 2022 (here).
8. Review of Aviation Safety Issues Arising from the COVID-19 Pandemic. EASA. 30 Apr 2021 (here).
9. VIDEO: Flying and Covid-19: what you need to know. Amanda Schwengel. MSU Denver RED. June 16, 2020 (here).
10. Taking stock of the pandemic's impact on global aviation. McKinsey & Company article, March 31, 2022 (here).
11. Impact of the COVID-19 pandemic on commercial air transport. Wikipedia (here).
12. Economic Impacts of COVID-19 on Civil Aviation. International Civil Aviation Organisation (here).
13. The Impact of COVID-19 on the Airline Industry. Aero Sweep. May 9, 2020.
14. The Impact of COVID-19 on aviation. Cirium Jan 12, 2021.
15. COVID-19's impact on the aviation and travel industry. The Gap Partnership.
16. The impact of COVID-19 pandemic on air transport. Taylor & Francis Online.
17. The impact of the coronavirus pandemic on the aviation. UK Parliament. May 13, 2021.

[1]The effects of the COVID-19 pandemic, in general, and in particular, on aviation, are addressed in the following 30 references.

18. Coronavirus: impact on the aviation industry worldwide – statistics & facts. Statista (here).
19. Aviation industry suffers 'worst year in history' as COVID-19 grounds international travel. World Economic Forum Feb 18, 2021 (here).
20. The 6 challenges of the Aviation Industry: The COVID-19 Era. Global Market Insights.
21. The impact of COVID-19 on the airport business – and the path to recovery. Airports Council International. Advisory Bulletins. Feb 24, 2022 (here).
22. Impact of Covid-19 on the aviation insurance industry International Bar Association. (here).
23. COVID-19 and the aviation industry: Impact and policy responses. OECD Policy Responses to Coronavirus (COVID-19). OECD. 15 October 2020.
24. X. Sun et al. COVID-19 pandemic and air transportation: successfully navigating the paper hurricane. J. Air Transport. Manag. (2021).
25. Abbot. R. 5 Ways COVID-19 has changed the aviation and airline industry. VHR Global Technical Recruitment. Jul 14, 2021 (here).
26. COVID-19 Support to the Aviation Sector. Australian National Audit Sector.
27. Gole, I. Catalin Dobrea, R. Cristina Combos, C. Aviation industry – challenges and uncertainties after the Covid-19 pandemic. SHS Web of Conferences. Volume 92, 2021.
28. Sun et al., 2023 A data-driven analysis of the aviation recovery from the COVID-19 pandemic. Air Transp. Manag., 109 (2023), Article 102401.
29. Aviation trends post Covid-19: Nive issues to watch as the industry prepares for take-off. Allianz Report. July 2021 (here).
30. COVID-19 Pandemic: Observations on the Ongoing Recovery of the Aviation Industry. Government Accountability Office. Oct 21, 2021 (here).

Chapter 7
Strategies and Economics for Decarbonizing Aviation by 2050

Rosa María Arnaldo Valdés and Victor Fernando Gómez Comendador

Contents

7.1 Introduction

According to the International Civil Aviation Organization (ICAO), aviation is responsible for 2% of global CO_2 emissions. The international aviation community aspires to 2% annual fuel efficiency improvement and carbon-neutral growth by 2020. However, aviation is expected to occupy an increasingly large share if it continues to grow as foreseen (5% annually), as other sectors are seeking to reduce their emissions in line with their carbon budgets. The achievement of this goal is still at risk, and even if achieved, aviation will still produce by 2050 a significant amount of CO_2, which might endanger the achievement of the goals of the Paris Agreement.

R. M. A. Valdés (✉) · V. F. G. Comendador
Universidad Politécnica de Madrid, Madrid, Spain
e-mail: rosamaria.arnaldo@upm.es; fernando.gcomendador@upm.es

L. M. Braga da Costa Campos (ed.), *Accelerating Sustainable Aviation Initiatives*, Sustainable Aviation, https://doi.org/10.1007/978-3-031-73993-4_7

The ICAO 2019 report [28] quantified emissions of the aviation sector over the period 2016–2050 could be between 56 gigatonnes of carbon dioxide (GtCO2) in a business-as-usual scenario and 12 GtCO2 in an optimistic, but unlikely to met, scenario with technological improvements and 100% biofuel use. As claimed by Carbon Brief, these figures would imply that aviation emissions from 2016 to 2050 could consume between 27% and 12% of the remaining carbon budget to keep global temperature rise below 1.5C above pre-industrial levels.

Additionally, carbon dioxide is not the only way aviation affects the climate. Aircraft emit other gases and aerosols that change the composition of the atmosphere, such as NOx and water vapour. They also produce "contrails", which impact the cloudiness of the sky and how much solar radiation reaches the surface of the Earth. The extent to which these extra factors amplify the CO_2 effect is still poorly understood and not captured in ICAO's estimates of aviation's impact on the climate.

Understanding and minimizing the environmental impacts of aviation and protecting the environment is a long-term commitment for aviation in Europe. The European Flightpath 2050 goals considered environmental protection a key challenge for the future of the aviation industry [20]. Flightpath 2050 vision aimed to continuously reduce the environmental impact in the face of continuing expansion in demand for aviation. It was envisaged that towards 2050, the forecast growth in the aviation industry will drive the need to deliver revolutionary technology solutions at an increasing rate and secure the path to sustainable energy supplies that can displace today's fossil fuels to fully mitigate the potential impact on the atmosphere. This vision was translated into five ambitus goals. The ambition settled in the Flightpath 2050 document for CO_2 reduction was that in 2050, technologies and procedures available will allow a 75% reduction in CO_2 emissions per passenger kilometre, and some phases of the aircraft operation, such as taxiing, will be emission-free.

However, experts and studies are calling the aviation community's attention towards the need to establish even more ambitus goals for aviation CO_2 emissions in the path to decarbonize the aviation sector by mid-century. Aviation is being challenged to immediately start to reduce its in-sector emissions, then sharply reduce its CO_2 emissions and fully decarbonize towards the second half of this century. Governments, aviation industry, international and research institutions, private sector, and civil society must do more to harness viable technological and policy solutions to face this challenge.

The EU has already started to modernize and transform the economy with the aim of climate neutrality. Between 1990 and 2018, it reduced greenhouse gas emissions by 23%, while the economy grew by 61%. However, current policies will only reduce greenhouse gas emissions by 60% by 2050. Much remains to be done, starting with more ambitious climate action in the coming decade.

It is a crucial moment in the global response to climate and biodiversity emergencies, and this is the last generation that can still act on time. This decade is a decisive moment to fulfil commitments under the Paris Agreement, in the interest of the health, wellbeing, and prosperity of all.

Answering to this need, very recently the EU has taken two initiatives to foster decarbonization and set more ambitious targets to reduce emissions: the "Green Deal Communication" [9] and the "Fit for 55 Declaration" [16].

Up to now, the aviation industry approach had focused on four pillars of climate action: reducing fuel use (and CO_2 emissions) through new technology and alternative fuels; better operations of existing aircraft; and improvements in infrastructure. For all emissions that cannot be reduced through these pillars, global market-based measures could be used to offset the remaining emissions in order to meet the targets set by the industry.

However, both recent legislative initiatives, Green Deal and Fit or 55, have inverted the focus of attention, and made pricing and market-based instruments one key element in the renovated EU environmental policy. The experience of the last 16 years has shown that emissions trading is a highly effective mechanism to bring down emissions in a cost-effective way, while the revenues it generates can be used to support the transition to cleaner production and to stimulate innovation. Reinforcing and applying emissions trading to new sectors has become a necessary complement to other policies to avoid much stronger regulatory measures in all sectors, notably in fuel standards, renewable energy, and energy efficiency, as well as taxation.

While other chapters of this book have dealt with technological or operational alternatives, following the new European attention to reinforced market-based measures, this chapter focusses on economic strategies for aviation decarbonization.

7.2 "The European Green Deal" Communication

This EU Communication sets out a European Green Deal for the European Union (EU) and its citizens. It resets the Commission's commitment to tackling climate and environmental-related issues. The Green Deal is a new growth strategy that aims to transform the EU into a fair and prosperous society with a modern, resource-efficient, and competitive economy where there are no net emissions of greenhouse gases in 2050 and where economic growth is decoupled from resource use.

This communication presents an initial roadmap of the key policies and measures needed to achieve the European Green Deal. It will be updated as needs evolve and the policy responses are formulated. Figure 7.1 illustrates the various elements of the Green Deal.

By summer 2020, the Commission had presented an impact assessment plan to increase the EU's greenhouse gas emission reduction target for 2030 to at least 50% and towards 55% compared with 1990 levels in a responsible way. To deliver these additional greenhouse gas emissions reductions, the Commission, by June 2021, reviewed all relevant climate-related policy instruments. This comprised, among other measures, the reinforcement of the Emissions Trading System (ETS) [13], including a possible extension to new sectors; Member State targets to reduce

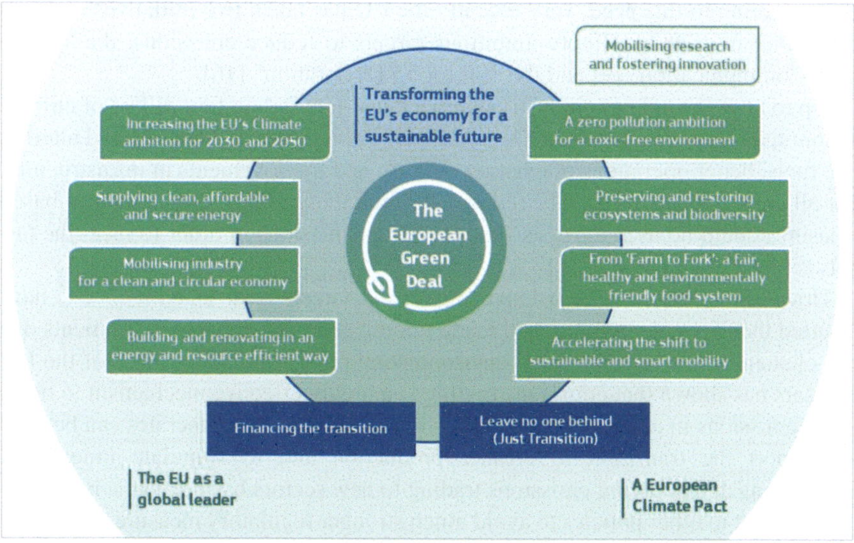

Fig. 7.1 The European Green Deal. (Source: European Commission)

emissions in sectors outside the Emissions Trading System [14]; and the regulation on land use, land use change, and forestry [14]. These policy reforms help to ensure effective carbon pricing throughout the economy. It is expected that these reforms will encourage changes in consumer and business behaviour and facilitate an increase in sustainable public and private investment.

This initiative includes different pricing instruments intended to complement each other and jointly provide a coherent policy framework. Ensuring that taxation is aligned with climate objectives is also essential. The Commission proposed to revise the Energy Taxation Directive [13], focusing on environmental issues and proposing to use the provisions in the Treaties that allow the European Parliament and the Council to adopt proposals in this area through the ordinary legislative procedure by qualified majority voting rather than by unanimity. To avoid the risk of carbon leakage ambition, the Commission proposed a carbon border adjustment mechanism for selected sectors to reduce the risk of carbon leakage.

Transport accounts for a quarter of the EU's greenhouse gas emissions and is still growing. To achieve climate neutrality, a 90% reduction in transport emissions is needed by 2050. The price of transport must reflect the impact it has on the environment and on health. Fossil-fuel subsidies should end, and, in the context of the revision of the Energy Taxation Directive, the Commission will look closely at the current tax exemptions, including for aviation and maritime fuels, and at how best to close any loopholes.

Similarly, the Commission proposed to extend European emissions trading to the maritime sector and to reduce the EU Emissions Trading System allowances

allocated for free to airlines. The EU should, in parallel, ramp up the production and deployment of sustainable alternative transport fuels.

To achieve the ambition set by the European Green Deal, there are significant investment needs. The Commission has estimated that achieving the current 2030 climate and energy targets will require €260 billion of additional annual investment [15], about 1.5% of 2018 gross domestic product (GDP). This flow of investment will need to be sustained over time. The Commission will present a Sustainable Europe Investment Plan to help meet the additional funding requirements. At least 30% of the InvestEU Fund will contribute to fighting climate change. As part of the Sustainable Europe Investment Plan, the Commission proposes a Just Transition Mechanism, including a Just Transition Fund. The need for a socially just transition must also be reflected in policies at the EU and national level. Coherence of climate and environment policies and a holistic approach are often a precondition for ensuring they are perceived as fair, as illustrated by the debate on taxation of various modes of transport.

7.3 Declaration "Fit for 55: Delivering the EU's 2030 Climate Target on the Way to Climate Neutrality"

In the declaration "Fit for 55: delivering the EU's 2030 Climate Target on the way to climate neutrality", the EU has deployed the ambitious target to reduce net emissions by at least 55% by 2030 compared to 1990 and for being the first climate-neutral continent by 2050. These targets are no longer aspirations or ambitions but obligations established in the first European climate law that create new opportunities for innovation, investment, and employment.

This package of measures is the most comprehensive set of proposals the Commission has ever tabled on climate and energy. Among the most innovative measures, compared to previous proposals, it includes a greater emphasis on economic and market-based measures. For example, carbon is priced in more sectors, which is expected to bring additional revenues to facilitate a just transition and to make clean solutions cheaper. It focuses on taxing energy sources in line with climate and environmental goals. The package translates into practice the principle that the polluter pays.

The package of measures also supports greater use of renewable energies and greater energy savings. It facilitates increased sales of new clean vehicles and cleaner fuels for transportation. It ensures that the industry can lead the transition and gives the security needed to drive investment and innovation. Its goal is to reverse nature's growing inability to remove carbon from the atmosphere. And it helps to ensure that climate goals are not undermined by the threat of carbon leakage, furthering the global fight against climate change.

Figure 7.2 presents the Fit for 55 Package at a glance. Overall, the package strengthens eight existing pieces of legislation and presents five new initiatives

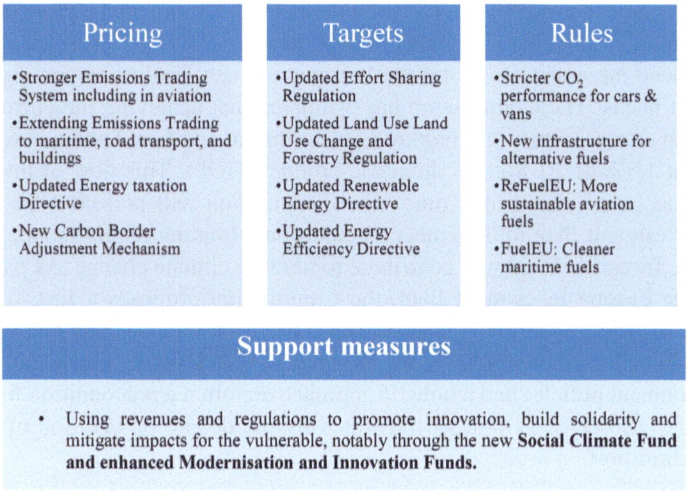

Fig. 7.2 Fit for 55 Package at a glance. (Source: European Commission)

across a range of policy areas and economic sectors: climate, energy and fuels, transport, buildings, land use, and forestry.

The Fit for 55 package consists of a set of interconnected proposals that constitute a careful balance between pricing, targets, standards, and support measures. The EU recognizes that over-reliance on strengthened regulatory policies would lead to unnecessarily high economic burdens, while carbon pricing alone would not overcome persistent market failures and non-market barriers.

As can be seen, pricing and market-based instruments are one key element in the renovated EU environmental policy. New policies advocate for:

- Stronger Emissions Trading System, including in aviation
- Extending Emissions Trading to maritime, road transport, and buildings
- Updated Energy Taxation Directive
- New Carbon Border Adjustment Mechanism

A cornerstone of the package is to build on the important achievements of the EU Emissions Trading System by strengthening it and applying it to new sectors where so far emission reductions have been lacking. To strengthen the role of carbon pricing in the transport sector, the Commission proposes to gradually extend the current EU ETS to the maritime sector over the period 2023 to 2025.

Greater efforts will also be required from aviation operators to reduce their emissions, which is why the Commission is proposing to phase out the free emissions allowances this sector currently receives. To also address aviation emissions at the global level, the Carbon Offsetting and Reduction Scheme for International Aviation (CORSIA) will be implemented through the EU Emissions Trading System Directive.

Additionally, Commission proposes to start applying emissions trading from 2026 for road transport and buildings, and similarly, apply emissions trading to

fuels in the buildings sector to help bring cleaner heating fuels to the market and shorten payback periods for investments in renovation and accelerate fuel-switching in heating and cooling in existing buildings.

The carbon price, however, does not guarantee a quick transition to zero emission mobility, for which complementary policies, including on charging infrastructure, are needed. Next to carbon pricing, other measures are needed. Emissions remain higher than in 1990 and a 90% reduction in overall transport emissions by 2050 will be required to reach climate neutrality.

The Commission is also proposing to promote the uptake of sustainable fuels in the aviation and maritime sectors, complementing the ETS for the aviation and maritime sectors, which makes polluting fuels more expensive for suppliers. The ReFuelEU Aviation to promote sustainable aviation fuel will oblige fuel suppliers to blend an increasingly high level of sustainable aviation fuels into existing jet fuel uploaded at EU airports, as well as incentivize the uptake of synthetic fuels, known as e-fuels. The upcoming zero emission aviation Alliance will complement this work to ensure market readiness for disruptive aircraft configurations (e.g., hydrogen, electric). The Commission also calls on the Council and the European Parliament to agree quickly on the updated Single European Sky regulatory framework, which is estimated to help cut aviation emissions by up to 10%.

The Fit for 55 proposals contains their own dedicated financial instruments to support a fair transition, based on the revenues generated by the extension and strengthening of emissions trading.

The Commission has also put forward a proposal for a Carbon Border Adjustment Mechanism (CBAM), which addresses the risk of carbon leakage for a targeted number of sectors (iron and steel, cement, fertilizer, aluminium, and electricity generation), by pricing the embedded emissions in certain goods imported to the EU (less any carbon price paid at source). The intention is to ensure that the EU's climate ambition is not undermined by less ambitious climate policies in other parts of the world.

7.4 Market-Based Measures

Because of the new European attention to reinforced market-based measures, the following sections of this chapter will analyze in more detail the most important market-based instruments applied in aviation and the achievements that these measures can bring.

Market-based measures are instruments designed to address the climate impact of aviation beyond what operational and technological measures or sustainable aviation fuels can achieve [10]. They are part of the comprehensive approach needed to reduce aviation's emissions, as technological and operational measures alone are currently not sufficient to tackle the growing impact of the aviation sector on climate change. Market-based measures, comprising both cap and trading as well as offsetting schemes, are designed to mitigate climate change through in-sector emission reductions or through incentivizing efforts outside the aviation sector.

Market-based instruments are indirect regulatory instruments that influence actors' behaviour by changing their economic incentive structure. Costs from environmental externalities, such as greenhouse gas (GHG) emissions, are usually not reflected in consumption or investment decisions but are nonetheless imposed on third parties. Putting a price on greenhouse gas emissions is important to harness market forces and achieve cost-effective emission reductions. Therefore, these types of policies work by reflecting the environmental impact of a certain action by attaching a cost to it to signal and provide an incentive to the polluter to reduce this impact.

Market-based solutions are the main concern of this chapter. Figure 7.3 shows the breadth of market-based instruments used in the EU. The most important instrument types here are:

- Emissions trading schemes
- Climate change levy schemes

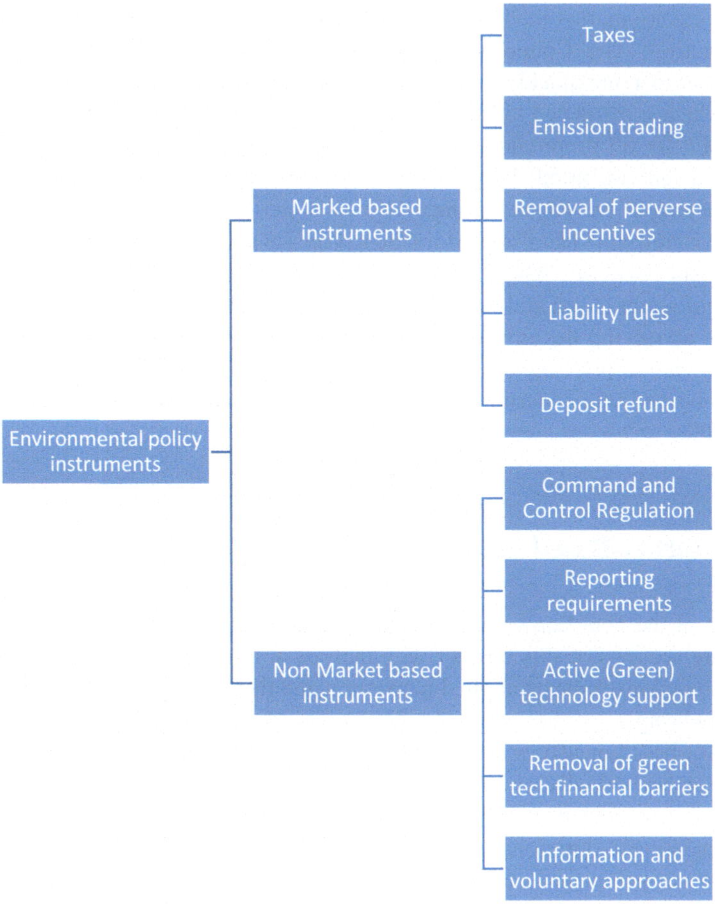

Fig. 7.3 Taxonomy of climate policy instruments. (Source: CECILIA 2050)

Both tax and emissions trading schemes therefore offer efficiency advantages over simple regulation and technological solutions. With taxation, policymakers effectively set the pollution price but leave the resulting emissions level uncertain, whereas with trading, the level is fixed and the resulting price uncertain, determined by the market trading patterns. Both of them are discussed below in more detail.

Regarding the removal of perverse incentives, this instrument refers to the removal of subsidies for environmentally harmful activities and products, such as subsidies for the extraction or production of fossil fuels, exemptions from energy taxes, or fossil fuel subsidies to keep the price level low. Furthermore, this also includes the removal of further incentives rewarding harmful activities, such as the abolition of liability exemptions [22]. Regarding the extent of these so-called perverse incentives, the environmental impact of their abolishment can be potentially large, as the following example shows: in 2012, total support provided by EU Member States to coal power generation (hard coal and lignite) summed up to some EUR 10 billion (not including historic support), improving the profitability of the investment, mainly in the form of investment grants, exemptions from fuel taxes, and support for decommissioning and waste disposal. Meanwhile, the total external costs of coal power generation are estimated at EUR 86 billion [2].

Also of relevance are liability instruments that "impel concerned parties to internalize external costs through the threat of consequential costs" [22]. Deposit refund systems charge consumers an upfront payment for improper waste disposal. The refund is a reward for returning the waste to the right collection point.

7.5 Emissions Trading Systems

Cap and trade systems generate economic incentives to change the behaviour of societal actors and reduce pollution. The starting point in a cap-and-trade system is a limit on the physical emission quantity (cap) of a harmful substance, introduced by a regulatory authority, which issues emission allowances according to the maximum limit it has defined. This is either done by grandfathering (free allocation according to certain criteria, such as past pollution) or auctioning, which generates revenue for the governments. The emission allowances can be traded on a market, and polluters can buy and sell them according to their needs, with the price of the allowances determined by supply and demand. Polluters carry out mitigation actions until it is cheaper to buy an allowance on the market than to mitigate a further emission unit. As a result, those polluters with the cheapest mitigation options will reduce the most.

Therefore, the environmental outcome of an aviation emissions trading system is determined by the emissions cap. Aircraft operators are able to use allowances from outside the aviation sector to cover their emissions. The absolute level of CO_2 emissions from the aviation sector itself can exceed the number of allowances allocated to it, as the increase is offset by CO_2 emissions reductions in other sectors of the economy. In emissions trading schemes, the total level of pollution is guaranteed by

the number of permits allocated to the companies. Then the permits are traded such that more efficient abaters can sell excess allowances to less efficient abaters, giving a certain outcome at the lowest cost.

Cap and trade systems theoretically guarantee the most cost-effective mitigation pathway for society [40] and are more dependable than taxes since emissions are capped. However, in real applications, market design can become very complex. Furthermore, the demand for allowances also depends on the economic situation. In a recession, economic activity will decrease and thereby emissions, which will drive down allowance prices. The low allowance price will not serve as an incentive to invest in low-carbon technologies, which can create a lock-in effect as no further action is required to achieve the emission cap. This can make future emission reductions more expensive, threatening the theoretical cost-effectiveness of an emissions trading system (ETS).

Cap and trade systems have been most developed worldwide through exemplary programs such as CORSIA at the international level or ETS at the European level. This alternative, however, may not deliver sufficient CO_2 emissions reductions to address the problem of aviation decarbonization.

7.5.1 CORSIA: ICAO Carbon Offsetting and Reduction Scheme for International Aviation

The 37th Session of the ICAO Assembly (2010) adopted two aspirational goals:

- To improve energy efficiency by 2% per year until 2050
- To achieve carbon-neutral growth from 2020 onwards.

Measures include technological innovations, operational improvements, sustainable aviation fuels, and market-based measures [25, 28]. Figure 7.4 illustrates the expected contribution of measures for reducing international aviation net CO_2 emissions.

The 38th Session of the ICAO Assembly (2013) decided to develop a global market-based measure for international aviation; further discussions on its design features and implementation mechanisms were undertaken, including possible means to address special circumstances and respective capabilities of states [26, 28].

The 39th Session of the ICAO Assembly (2016). States finally adopted a global market-based measure scheme for international aviation, in the form of the Carbon Offsetting and Reduction Scheme for International Aviation (CORSIA), to address the increase in total CO_2 emissions from international aviation above the 2020 levels (Assembly Resolution A39-3) (ICAO, 2019) [27, 28].

The 40th Session of the ICAO Assembly (2019) acknowledged the progress achieved on all elements of the basket of measures available to address CO_2 emissions from international aviation, including aircraft technologies, operational improvements, sustainable aviation fuels, and CORSIA. The 40th Session affirms

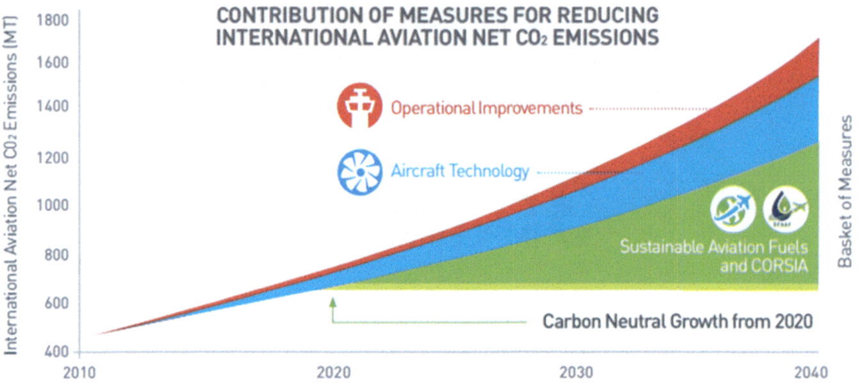

Fig. 7.4 Contribution of measures for reducing international aviation net CO_2 emissions. (Source: Carbon Offsetting and Reduction Scheme for International Aviation (CORSIA) Implementation Plan. [28])

the preference for the use of aircraft technologies, operational improvements, and sustainable aviation fuels that provide environmental benefits within the aviation sector. Marked that the environmental benefits from aircraft technologies, operational improvements, and sustainable aviation fuels may not deliver sufficient CO_2 emissions reductions to address the growth of international air traffic. Recalls its decision at the 39th Session to implement a Global Market Based Measure (GMBM) scheme in the form of the Carbon Offsetting and Reduction Scheme for International Aviation (CORSIA). Determines that the CORSIA is the first and only global market-based measure applying to CO_2 emissions from international aviation so as to avoid a possible patchwork of duplicative State or regional Market Based Measures (MBMs) [28]. It represents a cooperative approach that moves away from a "patchwork" of national or regional regulatory initiatives through the implementation of a global scheme that has been developed through global consensus among governments, industry, and international organizations. It offers a harmonized way to reduce emissions from international aviation, ensuring that there is no market distortion while respecting the special circumstances and respective capabilities of ICAO Member States [28].

- Phase implementation for CORSIA

ICAO MEMBER STATES PARTICIPATING IN CORSIA need to ensure that their aeroplane operators comply with the CORSIA offsetting requirements every 3 years, in addition to annual CO_2 MRV (Fig. 7.5).

The 39th and 40th Sessions of the ICAO Assembly introduced an approach based on the use of a phased implementation (Fig. 7.6) for the CORSIA to accommodate the special circumstances and respective capabilities of states, in particular developing states, while minimizing market distortion, as follows:

- The pilot phase applies from 2021 through 2023 to states that have volunteered to participate in the scheme.

Fig. 7.5 Phases of CORSIA implementation. (Source: Carbon Offsetting and Reduction Scheme for International Aviation (CORSIA) implementation plan. [28])

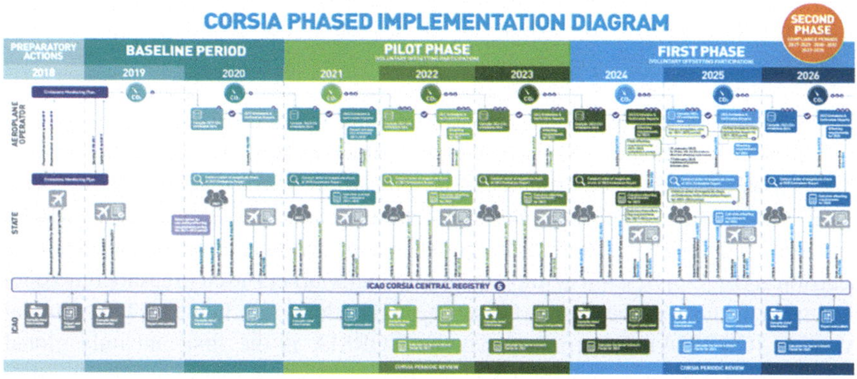

Fig. 7.6 CORSIA phased implementation diagram. (Source: CORSIA Brochure, ICAO)

- The first phase applies from 2024 through 2026 to states that voluntarily participate in the pilot phase, as well as any other states that volunteer to participate in this phase.
- The second phase applies from 2027 through 2035 to all states that have an individual share of international aviation activities in revenue tonne kilometres (RTKs) in 2018 above 0.5% of total RTKs, or those states whose cumulative share in the list of states from the highest to the lowest amount of RTKs reaches 90% of total RTKs [except Least Developed Countries (LDCs), Small Island Developing States (SIDS), and Landlocked Developing Countries (LLDCs) unless they volunteer to participate in this phase].

Starting in 2022, the Council will conduct a review of the implementation of CORSIA every 3 years, including its impact on the growth of international aviation. This review will serve as an important basis for the Council to consider whether it is necessary to adjust the next phase or compliance cycle and, as appropriate, to recommend such adjustments to the Assembly for its decision [27].

- CORSIA performance

The success of the implementation of CORSIA relies on the establishment of robust and transparent monitoring and on the eligible offset credit programs. CO_2 offsetting is calculated as the operator's annual emissions multiplied by a growth factor determined by ICAO as a combination of the individual operator's and the sector's performances.

By 2020, global international aviation emissions are projected to be around 70% higher than in 2005, and the International Civil Aviation Organization (ICAO) forecasts that by 2050 they could grow by a further 300–700% [16].

During the period 2021–2035, and based on expected participation, the scheme is estimated to offset around 80% of the emissions above 2020 levels. This is because participation in the first phases is voluntary for states, and there are exemptions for those with low aviation activity. All EU countries will join the scheme from the start.

However, despite the great international effort involved in CORSIA, according to the Air Transport Action Group (ATAG), CORSIA shouldn't have any real impact on the growth of air traffic (i.e., on passenger demand), as the global scheme was developed as a way to help ensure aviation growth continues (to drive economic development through international trade and tourism), whilst also taking care of its environmental responsibility [5].

Before the establishment of CORSIA in 2012, ICAO assessed the impacts of the implementation of three Global Market-Based Measure (GMBM) options: global mandatory offsetting, global mandatory offsetting with revenue, and global emissions trading. The following assumptions were considered: (i) impacts would be assessed from 2020 to 2036; (ii) the future per-tonne prices of CO_2 (emissions units to be purchased) would be USD $30 in 2020, USD $40 in 2030, and USD $45 in 2040; and (iii) the costs resulting from purchasing emissions units would be passed through to ticket prices (i.e., 100% cost-pass-through).

Considering a GMBM without revenue (the modelling did not distinguish between offsetting and emissions trading), which is CORSIA's case, and the assumptions above, the estimated air traffic demand (measured in revenue ton-kilometres (RTK)) would be reduced by 1.2% and the profits for the international aviation sector would be USD $0.4 billion, or 1.2% lower, in 2036 [26].

The assessment also demonstrated that the differences between GMBM impacts by world regions or groups of states were marginal. For example, the GMBM impacts on traffic demand and operating results (profit) in the six regions (Europe, Asia/Pacific, North America, Latin America, Africa, and the Middle East) were generally consistent with the global average of a 1.2% reduction for both. The Middle East would be the region with a higher reduction in traffic demand (of 1.4%), whereas Africa and Asia/Pacific would have slightly better results in profitability, lowered by 1% (Table 7.1).

A more recent assessment [5] made by ICAO's Committee on Aviation Environment Protection (CAEP) suggests that, in a scenario of high carbon prices in the future (USD $33 per ton in 2025 and USD $40 per ton in 2035) and industry CO_2 growth, the CORSIA scheme (i.e., the purchasing of carbon offsets) will cost airliners USD $12.4 billion in 2030 (approximately around 1.1% of the projected

Table 7.1 Results of ICAO's 2012 GMBM assessment

Assessment/Scenario	Without GMBM scheme (baseline) 2020	2036	With GMBM scheme 2036
Air traffic demand, billion RTK	743	1561	1543
Operating results, billion USD $ (profit)	16.2	33.7 (2.8%)	33.3 (2.7%)

Table 7.2 Result of the CE Delft 2016 GMBM assessment

	WEO—New Policies Scenario		
Year	2020	2030	2040
Carbon prices, €2015 per tonne of CO_2	18.0	33.4	45.1
Total costs to industry (in billion €2015)	–	7.9	14.0

annual industry revenue) and USD $23.9 billion in 2035 (~1.8% of projected annual industry revenue).

Another recent study [5] made by CE Delft considers similar carbon prices (compared to previous ICAO's assessments), based on the World Energy Outlook (WEO) New Policies Scenario, and estimates lower yearly costs for the aviation sector resulting from buying carbon offsets (Table 7.2).

In any case, considering that in 2015 the aviation industry spent around USD $181 billion on fuel (around a third of operating costs), the overall financial impact or burden to the industry will be relatively insignificant and a lot less than the cost of fuel for any airline (and less than a tax). This figure also reflects the insufficient contribution of CORSIA to meet the 2 °C target under the 2015 Paris Agreement [33].

However, for each airline individually, the burden will vary depending on the airline's business model as well as its share of international flights, growth rate, fleet age, and operational efficiency. Airlines that grow fast (being the majority of low-cost carriers) and have high greenhouse gas (GHG) emissions will be affected more by CORSIA. Strategies for monitoring, reporting, and verification, as well as biofuel and other efficiency measures, may impact reported performance significantly [38]. It is to be noted that, according to data from the EU's Transport and Environment group in 2019, Ryanair has become the first airline to be included in a list of Europe's top polluters, which had until now been exclusively occupied by coal plants [46].

In these circumstances, it will be necessary to wait until reliable information about the effectiveness of the programme might be available after the pilot phase, which applies from 2021 through 2023 to states that have volunteered to participate in the scheme.

- CORSIA and COVID-19

While the impact of COVID-19 is still under development and more information is needed to accurately quantify the impact of the COVID-19 pandemic on the aviation sector's CO_2 emissions in 2020 and subsequent years, this impact has resulted so far in a sharp decline in aviation activity, with an expected related decrease in the

sector's CO_2 emissions in 2020 compared to the forecasted 2020 CO_2 emissions prior to the pandemic.

CORSIA's sectoral baseline is defined as the average of total CO_2 emissions for the years 2019 and 2020 on the routes covered by CORSIA offsetting each year from 2021 onwards (Assembly Resolution A40-19, paragraph 11). Therefore, the expected reduction of 2020 CO_2 emissions from international aviation due to the COVID-19 pandemic will lead to a decrease in the CORSIA baseline compared to the non-COVID-19 scenario [29].

The impact of the COVID-19 pandemic on the CO_2 emissions from international aviation in 2020 and subsequent years could affect the following three CORSIA's design features, defined in Assembly Resolution A40-19, as follows:

CORSIA calculates annual offsetting requirements for individual aeroplane operators every year from 2022 based on an annual Sector's Growth Factor (SGF), which represents the CO_2 emissions growth of international aviation each year from 2021, compared to CORSIA's sectoral baseline (average of 2019 and 2020). The impact of COVID-19 on a given year's SGF will be reflected through the impact on CORSIA's sectoral baseline and the impact on the given year's CO_2 emissions from international aviation, which will depend on the aviation sector's recovery pattern from 2021 onwards. All things considered, a given year's SGF could be higher or lower than that year's SGF under a non-COVID-19 scenario. The compounded effect on the year's SGF will determine the magnitude and nature of the impact on the total CORSIA offsetting requirements for that year, as well as the associated costs for the industry [29].

The calculation of an aeroplane operator's annual offsetting requirements during CORSIA's pilot phase (2021–2023) requires multiplying the corresponding year's SGF with the emissions value of the operator. For the latter, states can choose, for their attributed operators, between two options: an operator's emissions in a given year (i.e., 2021, 2022, and 2023 emissions) or the operator's 2020 emissions for each of the 3 years during the pilot phase. Therefore, the impact of COVID-19 on the annual offsetting requirements during CORSIA's pilot phase is not only through the impact on a given year's SGF (as described above), but also through a state's choice of reference emissions.

CORSIA refers to a certain percentage (0.1%) of global CO_2 emissions in 2020 as a threshold for a new entrant operator to be covered by the CORSIA offsetting requirements. Therefore, any impact on 2020 emissions due to COVID-19 may create a need to review this provision [29].

Considerations on the need and timing to adjust the CORSIA baseline and other design features above, if appropriate, need to be based on proper technical data and assessment. On the matter of timing, for potential adjustments, due consideration has to be given to the timeline for CORSIA implementation, as described in Appendix 1 of Annex 16, Volume IV, and to the related milestones, as follows:

- 31 October 2022: the SGF value corresponding to year 2021 CO_2 emissions will be provided by ICAO in order for states to calculate the 2021 offsetting requirements of individual aeroplane operators attributed to them; this will be the

first instance in which information on CORSIA's sectoral baseline will be applied.
- 31 January 2025: deadline for aeroplane operators to cancel eligible emissions units to comply with their total offsetting requirements for CORSIA's pilot phase (2021–2023). Annex 16 of Volume IV does not specify any date for the purchase of units by operators.

Any adjustment to CORSIA's design features is a matter that requires careful consideration by the relevant ICAO bodies. The ICAO Council, at the 220th Session in June 2020, considered an analysis by the Committee on Aviation Environmental Protection (CAEP) on the COVID-19 impact on CORSIA's design features [29].

In addition to the impact assessment, the Council also considered the legal and reputational aspects related to the various options, as well as the importance of maintaining the originally agreed balance between the scheme's economic impacts and environmental benefits, together with its simplicity and practicality, whilst responding to this unprecedented crisis.

The Council agreed that, to safeguard against an inappropriate economic burden on aeroplane operators, 2020 emissions should not be used for the three CORSIA design features listed above. In this regard, the Council decided that during the pilot phase, 2019 emissions shall be used for 2020 emissions and published in all relevant ICAO documents referenced in Annex 16, Volume IV. There was no change to the provisions of Annex 16, Volume IV, or Assembly Resolution A40-19 [28].

In addition to the above, it is important to note that paragraph 17 of Assembly Resolution A40-19 includes a provision that the ICAO Council will conduct a periodic review of the CORSIA every 3 years starting from 2022. In this regard, the Council highlighted the importance of undertaking the 2022 periodic review, with the technical contribution of the Committee on Aviation Environmental Protection (CAEP), which will offer an opportunity to examine the impact of COVID-19 on CORSIA on various issues, including the impact on the baseline beyond the pilot phase. The Council will consider the structure, process, and methodology of the 2022 periodic review, including the work programme of CAEP, at its 222nd Session in March 2021 [29].

On 1 July 2020, the ICAO Council voted by a large majority to adopt an industry proposal to change the crucial emissions baseline rule for the CORSIA carbon offsetting scheme for international aviation. The move is aimed at protecting airlines from what the ICAO describes as the "inappropriate economic burden" airlines are likely to face from a lower baseline as a result of the collapse of international air traffic this year, leading to increased offsetting costs in future years. Instead of the baseline being calculated on the 2019–2020 average CO_2 emissions from international flights covered by CORSIA, it will now be based on 2019 emissions only. The Council also voted to remove 2020 emissions from two other design features of the scheme: the selection of the reference year for calculating offsetting requirements in the pilot phase and the emissions threshold for new CORSIA entrants. However, the Council did not remove the requirement for operators to monitor and report their verified 2020 emissions to their national authorities. The adjustment, which will

increase the baseline by around 30%, will likely delay the implementation of CORSIA carbon offsets by 3 to 5 years.

It also agreed to consider, following further analysis, amendments to A40-19 to also use only 2019 emissions for the three design features beyond the pilot phase, which would then be presented to the next Assembly in 2022 for a decision. The Council also decided to initiate the process for establishing the 2022 periodic review of the scheme called for in A40-19 and requested the ICAO Secretariat to present it with a review structure, process, and methodology for consideration at its 22nd Session in March 2021. The 2022 review would also consider whether it would be necessary to adjust the next phase or compliance cycle and, if so, submit recommendations to the Assembly. It would also examine the impact of COVID-19 on CORSIA, including consideration of the baseline beyond the pilot phase, on the different phases of CORSIA implementation, and on the growth factors.

What appears to be a technical change will, in fact, postpone the start of the scheme by at least 3 years, until 2024 or later, depending on how fast the sector recovers from the current crisis and on whether governments decide to extend this change to the subsequent phases, pointed out ICSA member Carbon Market Watch.

7.5.2 EU Emissions Trading System

The EU Emissions Trading System (EU ETS) is the cornerstone of the European Union's policy to tackle climate change and a key tool for reducing greenhouse gas emissions cost-effectively, including from the aviation sector. It is the first and so far the biggest international system for capping greenhouse gas emissions. It operates in 31 countries: the 28 EU Member States, Iceland, Liechtenstein, and Norway, and around 500 commercial and non-commercial aircraft operators that fly between airports in the European Economic Area (EEA).

The EU ETS began operation in 2005. EU countries have amended the original legislation several times as the system has evolved. As regards aviation, legislation to include aviation in the EU ETS was adopted in 2008 by the European Parliament and the Council [19]. The 2006 proposal to include aviation in the EU ETS, in line with the resolution of the 2004 ICAO Assembly deciding not to develop a global measure but to favour the inclusion of aviation in open regional systems, was accompanied by a detailed impact assessment [4]. After careful analysis of the different options, it was concluded that this was the most cost-efficient and environmentally effective option for addressing aviation emissions. In October 2013, the Assembly of the International Civil Aviation Organization (ICAO) decided to develop a global market-based mechanism for international aviation emissions. Following this agreement, the EU decided to limit the scope of the EU ETS to flights between airports located in the European Economic Area (EEA) for the period 2013–2016 (Regulation 421/2014) and to carry out a new revision in light of the outcome of the 2016 ICAO Assembly. The temporary limitation follows on from

the April 2013 'stop the clock' decision [3] adopted to promote progress on global action at the 2013 ICAO Assembly.

European Council Conclusions of October 2014 [17] expected that the EU ETS will be the main European instrument to achieve the EU's binding 2030 target of an at least 40% domestic reduction of greenhouse gases compared to 1990 [19]. The European Commission assessed the outcome of the 39th ICAO Assembly and, in that light, made a new legislative proposal on the scope of the EU ETS. Following the EU legislative process, this Regulation was adopted in December 2017 [19]. The legislation maintains the scope of the EU ETS for aviation limited to intra-EEA flights. It foresees that once there is clarity on the nature and content of the legal instruments adopted by ICAO for the implementation of CORSIA, as well as about the intentions of other states regarding its implementation, a further assessment should take place and a report be presented to the European Parliament and to the Council considering how to implement CORSIA in Union law through a revision of the EU ETS Directive. The regulation also sets out the basis for the implementation of CORSIA.

The EU ETS working principle is based on a "cap", or limit, on the total amount of certain greenhouse gases that can be emitted by factories, power plants, other installations, and aircraft operators in the system. Within this cap, companies can sell or buy emission allowances from one another. The limit on allowances available provides certainty that the environmental objective is achieved and gives allowances a market value, in line with the 2015 Paris Agreement on climate change. Aircraft operators are entitled to free allocation based on an efficiency benchmark, but this might not cover the totality of emissions. The remaining allowances need to be purchased from auctions or from the secondary market. The system allows aircraft operators to use aviation allowances or general (stationary installations) allowances to cover their emissions. By 30th April each year, companies, including aircraft operators, must surrender allowances to cover their emissions from the previous calendar year. If a company reduces its emissions, it can keep the spare allowances to cover its future needs or sell them to another company that is short of allowances. The flexibility that trading brings ensures that emissions are cut where it costs least to do so. The number of allowances reduces over time, so total emissions fall.

For the aviation sector, there were 3 distinct periods defined: first period (phase II): 2012; second period (phase III): 2013–2020; and third period (phase IV): 2021–2030.

For 2012, the aviation sector allowances' cap has been set at 97% of the historical emissions (the mean between emissions of 2004, 2005, and 2006) and was originally set at 210 MtCO2e/year. This cap was meant to reflect the initial inclusion of all flights from, to, and within the EEA in the EU ETS. However, following the "stop the clock" temporary suspension until the end of 2016, the number of aviation allowances put into circulation in 2013–2016 was significantly lower than the original cap. In 2017, the intra-EEA scope for aviation was prolonged until 2023. The adjusted approach for determining the annual aviation cap still applies.

For the second period (2013–2020), the cap has been set at 95%. Airlines continue to receive the large majority of their allowances for free in the period

2013–2020. For this phase, a linear cap reduction factor of 1.4% annually will be applied. The annual cap on aviation allowances for phase 3 of the EU ETS (2013–20) was originally 210,349,264 allowances (plus 116,524 allowances from 2014 to account for Croatia's integration) [19]. Aircraft flying to airports in the EU, Iceland, or Norway from elsewhere in the world are exempt from the EU ETS until 31 December 2023.

During the fourth phase of the EU ETS (2021–2030), the EU was aiming to cut its emissions by at least 40% by 2030, in line with the 2015 Paris Agreement on climate change. For this phase, a linear cap reduction factor of 2.2% (48.4 million allowances) annually is considered for both stationary sources and the aviation sector. The linear reduction factor does not have a sunset clause, and the cap will continue to decline beyond 2030. From 2021 on, 57% of allowances are to be auctioned. At least half of the EU countries' proceeds must be used for climate-related purposes.

The EU ETS has been effectively implemented over recent years on intra-EEA flights and has ensured a level playing field with a very high level of compliance. It will continue to be a central element of the EU policy to address aviation CO_2 emissions in the coming years. The complete, consistent, transparent, and accurate monitoring, reporting, and verification of greenhouse gas emissions remains fundamental for the effective operation of the EU ETS. Aviation operators, verifiers, and competent authorities have already gained wide experience with monitoring and reporting; detailed rules are prescribed by Regulations (EU) N°600/2012 [11] and 601/2012 [18].

With the inclusion of intra-European flights in the EU ETS, it has delivered around 100 MT of CO_2 reductions/offsets between 2012 and 2018. The total amount of annual allowances to be issued will be around 38 million, whilst verified CO_2 emissions from aviation activities carried out between aerodromes located in the EEA have fluctuated between 53.5 MT CO_2 in 2013 and 61 MT in 2016. This means that the EU ETS is now contributing more than 23 MT CO_2 of emission reductions annually [11], or around 100 MT CO_2 over 2012–2018, partly within the sector (airlines reduce their emissions to avoid paying for additional units) or in other sectors (airlines purchase units from other ETS sectors, which would have to reduce their emissions consistently). While some reductions are likely to be within the aviation sector, encouraged by the EU ETS's economic incentive for limiting emissions or the use of aviation biofuels, the majority of reductions are expected to occur in other sectors.

- EU ETS vs. CORSIA: Why ICAO and CORSIA cannot deliver on climate?

As CORSIA is an international agreement that touches on areas of EU competence (climate and aviation), it is ultimately for the EU institutions to determine whether that minimum level of environmental integrity has been reached. This has been codified in the 2017 revision to the EU ETS Directive 2003/87/EC. In line with that legislation, European member states wrote to ICAO at the end of 2018 to state that they were not yet in a position to confirm, either way, whether they will participate in CORSIA. This reflected the EU position that such a decision can only be

made once all the CORSIA rules have been decided and following an evaluation by the European Commission of the scheme's potential environmental effectiveness.

Target: The EU's 2030 economy-wide target commits the bloc to achieving an emissions reduction target of −40% against 1990 levels. That target is a combination of emission reductions under the bloc's ETS and targets for each member state for sectors of their economy not covered by the ETS under the Climate Action Regulation. In establishing the ETS target, the Council of the EU included outbound aviation emissions, i.e., emissions from all flights departing EU airports to any destination either within the EU or beyond the EU, with a target of 111 Mtons by 2030. As outbound aviation emissions are currently (2017) estimated to be 174 Mtons (UNFCCC), that target is a reduction of over 36% from current emissions. It should be noted that while the initial target includes all emissions from outbound aviation, the scope of aviation's inclusion in EU ETS has been limited to outbound flights within Europe until at least 2024.

By contrast, the CORSIA target has been set to stabilize emissions from international aviation at 2020 levels, allowing airlines to continue growing their emissions. Setting aside how each measure will achieve their respective target (airlines purchasing allowance reductions from other sectors covered by ETS and airlines purchasing offsets approved by ICAO under CORSIA), the CORSIA target is a weaker target than the EU 2030 target. Therefore, any move to replace the existing ETS target with the CORSIA target would represent a regression in Europe's climate ambition, a move that the Paris Agreement explicitly prohibits.

A further distinction between the EU 2030 targets and the CORSIA scheme is that the former is to be solely achieved through reductions within the EU and therefore does not recognize the use of international credits. In contrast, CORSIA is built explicitly around the use of such international credits.

As a result, the implementation of CORSIA into EU law, in a manner that replaces existing EU legislations, risks creating a situation where all sectors of Europe's economy, bar aviation, are legally obliged to achieve emission reductions without offsetting. As the commitment to reduce emisions from domestic flights reductions was stated in Europe's NDC, which explicitly states "No contribution from international credits" a backtracking on this for any part of its economy would, like with a weaker target, count as backsliding which the agreement prohibits.

Due to the weaker target and the use of offsets, implementing CORSIA in EU law in a manner that replaces existing legal commitments would weaken Europe's overall climate ambition. According to independent research commissioned by T&E, over the period 2021–2030, Europe's aviation emissions would increase by 683.8 Mtons CO_2 (TAKS, 2019), which is equivalent to the 2017 CO_2 emissions of Poland and France combined (Global Carbon Atlas, 2019) [13, 16, 36, 48].

- The EU ETS and the COVID-19

The EU emission allowance (EUA) price dropped from €24.07/t to €15.25/t in March 2020, at the start of the COVID-19 pandemic in Europe. This is not the first time that we have observed such a dramatic EUA price drop. The 2008 financial crisis caused an even higher price drop with a quite similar reduction in economic

activity [45]. Yet, in the aftermath of the 2008 financial crisis, the EUA price remained below €10/t for over 9 years, even when economic activity had already started recovering. In the case of COVID-19, however, the price recovered almost completely to €22.10/t already in June 2020, just 2 months after the drastic drop [6].

One possible explanation for the difference in EUA price dynamics during the two crises might be attributed to the legislative changes introduced in the EU Emissions Trading System (ETS), for instance, the establishment of the Market Stability Reserve (MSR).

In the short term, emissions being significantly reduced due to COVID-19 will lead to an additional surplus of allowances in 2020 and 2021. However, most of these surpluses (over 80%) will be cancelled by the year 2030 due to the automatic invalidation mechanism of the MSR.

If COVID-19 leads to sustained low emissions, the current MSR will not be able to prevent the TNAC ('total number of allowances in circulation) from growing fast, which may lead to price declines. In general, the effectiveness of the MSR to handle supply-demand imbalances will rely heavily on future emission trends. The larger these imbalances get, the more important it will become to revisit and recalibrate the different mechanisms for dealing with supply–demand imbalances.

According to Central Europe Pipeline System (CEPS) Energy Climate House, the first feature of the MSR that should be revisited is the withdrawal rate. A 12% withdrawal rate after 2023 would be insufficient to address the growing supply-demand imbalances following a COVID-19 economic downturn. The MSR review should consider increasing the withdrawal rate further from 24%. If there is a desire to support low-carbon investments more directly through the ETSprice signal, a carbon price floor should be considered. This price floor can be set as an auction reserve price below the current carbon price (close to €20) to act primarily as an insurance mechanism so that the carbon price is sustained at a level that incentivizes private sector investments, which trigger energy transformation [52].

- The EU ETS and the Fit for 55

The European Commission has revised the ETS aviation rules as part of the Fit for 55 legislative packages, to help ensure that the sector contributes to the more ambitious target of achieving net emission reductions of at least 55% by 2030, compared to 1990 levels. The key changes are:

- An expansion of the EU ETS to cover maritime transport
- An increase in the linear reduction factor to 4.2% (from the previous 2.2%)
- Introduction of emissions trading for buildings and road transport
- Confirmation that the normal intake rate for the MSR at 24% will apply until 2030

Regarding aviation, the key elements of this revision are:

- The Commission proposes that flights within the European Economic Area (EEA) continue to be covered by the EU ETS, as well as flights to Switzerland and the UK.

- Aside from those constituting public service obligations, flights between the EU's outermost regions (OMRs) will be included in the EU ETS.
- International flights between OMRs and EEA countries will also be included. The total number of aviation allowances in the ETS will be capped at current levels and reduced annually under the ETS linear reduction factor.
- The number of free allowances allocated to aircraft operators will also be reduced progressively to reach full auctioning by 2027.
- In parallel, the Commission proposal will appropriately implement the Carbon Offset and Reduction Scheme for International Aviation (CORSIA) under the International Civil Aviation Organization for extra-European flights.
- Under the proposal, the Commission will apply CORSIA to flights that are outside the EU ETS and depart or arrive in countries that apply CORSIA. Emissions from these flights will be offset once collective international emissions exceed 2019 levels. The eligible offset units need to originate from countries that participate in the Paris Agreement and in CORSIA. Offsets must be reliably accounted for to avoid them being counted twice.
- To avoid market distortions, the Commission will maintain the route-based approach, ensuring that airlines operating flights on the same routes are treated equally, regardless of nationality.
- In addition, the Commission proposes to implement Member States' CORSIA notification to EU-based airlines of the offsetting for their 2021 emissions.

7.6 Climate Change Levy Schemes: Carbon-Based Charges, Taxes, and Levies

In this state, much attention has been devoted to the other main market-based instrument, the climate change levy schemes, mostly at the national and local level.

In general, this category can be instantiated as charges, taxes, and levies. The term charge usually refers to a mandatory payment of an amount related to carbon emissions, whether implemented as a tax or as a levy. A tax usually means a compulsory payment that is not fully required of those paying it. A levy usually means a charge that is fully rebated to the payer, in cash or in kind. For the sake of this study, all three terms are considered equivalents and referred to indistinctively.

The primary rationale for environmental taxation is the externality argument. By leaving a tax on the pollution-generating activity, the social costs of pollution can be 'internalized' to the agent (who must pay the tax), and the socially optimal level of pollution would occur. Carbon taxes are a type of Pigouvian tax [49]. This type of tax seeks to correct a negative or positive externality, in this case, the decarbonization of aviation. The imposition of carbon taxes would increase the price of a flight, which could lead to a reduction in demand.

Carbon-based taxes generally refer to taxes on aviation jet fuel through price signals to airliners (through higher fuel costs) and/or consumers (through higher

ticket prices). There is also the air passenger tax, which in practice implies altering the price relationship between air travel and other transport modes, along with other categories of consumption. If the air passenger tax is distance-based, it incentivizes customers to opt for shorter flights.

The recent study by CE Delft for the European Commission provides a more updated and comprehensive inventory of taxes applied in aviation [8]. Although aviation in many countries is exempted from all taxes, a significant number of countries levy taxes on certain aviation activities. For example, in the European Union, Value Added Taxes (VAT) or taxes on domestic flights are the most prevalent and applied in at least 17 states [8]. Six EU states applied some kind of tax on international aviation, normally in the form of passenger's ticket taxes departing from airports in the country. Outside the EU, the study evaluated counted 13 countries (including Australia, Canada, the USA, Hong Kong, Brazil, and Japan) tax aviation activities, in most of cases in the form of ticket or departure taxes, normally a fixed amount per passenger, depending on the destination or travel class. A small number of countries apply VAT or sales taxes (a levy proportional to the value of the ticket), as in, for example, Japan, Mexico, the USA, and Canada. Additionally, commercial air transport, both for passengers and cargo operations, is subject to various charges and fees that are not seen as taxes, as these are levied to cover the costs of the provided services.

However, it is true that aviation is currently undercharged from an environmental perspective. This low-charge regime is even more important for international aviation. When it comes to the taxation of aircraft fuel, different schemes are applied. Fuel on domestic flights is sometimes taxed (e.g., in the USA or France, which charges air freight with a tax of €1.33 per tonne of freight).

In contrast, according to Article 24 of the International Civil Aviation Convention (Chicago Convention), jet fuel for international commercial flights (e.g., kerosene) is legally exempt from taxes, and therefore only with bilateral agreements between two or more countries would it be possible to introduce a tax on jet fuel for international flights [13, 34]. Unlike domestic transportation fuels, they are subject to no excise tax to reflect environmental damages in fuel prices.

The Chicago Convention provides that fuel and lubricating oils on board that are carried on an aircraft of a contracting state when they reach the territory of another contracting state and are still on board such aircraft when it leaves said state shall be exempt from the payment of customs duties, inspection charges, or other similar taxes and fees, if national or local [44]. Besides, the Chicago Convention prohibited the application of taxes to kerosene so as not to slow down the development of international traffic. However, 76 years after the Chicago Convention, the aviation sector contributes more than 2% of global CO_2 emissions and is expected to exceed 20% by 2050.

Although this measure is generally recognized as positive, there is not yet a clear agreement on what could be the most convenient type of tax, what should be its value, and what would be the expected impacts. Under different considerations of the price elasticity of aviation, certain authors have recently proposed different tax values as effective market-based instrument (MBI). The authors [33] claim that a

tax of an order of magnitude of €150/ton CO_2 could be an effective measure to reduce air transport demand and therefore aviation emissions. At the International Monetary Fund [31], a uniform, globally applied CO_2 price of USD $25 per ton is considered. This emissions price corresponds to the medium damage scenario studied by AGF [1] and is consistent with the US 2010 interagency assessment of environmental damages per ton. The authors claim that a USD $25 per ton emissions price would add about US 6 cents per litre, or about 8%, to the price of jet fuel.

According to Hosoda [24], the effects of a fuel excise duty on kerosene, the equivalent is €330/kilolitre, are estimated. The introduction of this duty EU-wide (considering 28 Member States (MS)) would result in a 10% increase in the average ticket price and an 11% decline in passenger demand. The aviation-related fiscal revenue would increase from €10 billion to €27 billion (an increase of 1.7%), while on the other hand, there is a relative reduction of 11% in the number of direct jobs and the value added by the aviation sector. The introduction of excise duty on fuel would have an impact of less than 0.1% for most MS, although some outliers have a contraction of GDP by 0.6% or an increase of 0.7%. This study also analyzed other types of taxes, such as VAT on flight tickets and the abolition of the ticket tax. Overall, the analysis shows that new or increased aviation taxes would have a negative impact on the aviation industry's economic growth (lower direct employment and direct value-added), but their impact on the overall employment within an MS, on fiscal revenue, and GDP would be close to zero.

Between 1972 and 2013, the Japanese Aviation Fuel Tax was levied on aviation fuel for domestic flights, amounting to JPY 26000/kilolitre (approximately €193/kilolitre, considering an exchange rate of 1 EUR = 135 JPY) until April 2011. Afterwards, the Japanese government implemented a 30% reduction in the tax, resulting in a fuel tax of JPY 18000/kilolitre (approximately €133/kilolitre or USD $3/ton CO_2). Hosoda [24] analysed the effect that this tax had on the national demand for aviation fuel using a Bayesian structural time series model based on monthly observations of fuel consumption between 2004 and 2013 and domestic data on aviation provided by the Japanese government. From 1994 to 2013, the real-term price of jet fuel quadrupled, approximately doubling each year, and accordingly, airlines' operational costs attributable to fuel increased from 13.6% in 2003 to 33.1% in 2013. However, the price of airline tickets remained relatively unaltered for the same period, probably due to a low-cost pass-through assumed by airlines. Consequently, between 1991 and 2013, Japan's domestic revenue (RTK) grew at an average annual growth rate of 13%, and the volume of passengers increased by 23 million. Despite the increase in fuel costs, during this period, Japan also had a positive evolution in fuel efficiency. With the tax adjustment in April 2011, the study estimated that there was an increase of 9.7% in fuel consumption until December 2013 [37, 51] (Table 7.3).

Another study [39] made by the National Technical University of Athens analyzed the impact of the introduction of a carbon pricing policy (i.e., a tax on CO_2 emissions for domestic flights) on competitive US air transport in 2012. The carbon price at the time was USD $10 per ton CO_2, and to account for the uncertainties related to carbon price (and the recommended levels of a carbon price to achieve the

Table 7.3 Effects of the carbon pricing policy in the US domestic airline network

Effect/Scenario	Low USD \$10/ ton CO_2	Medium USD \$20/ ton CO_2	High USD \$50/ ton CO_2
Average carbon cost imposed on the airline for each passenger (Δcost)	4.75	9.51	23.77
Average price increase (%Δprice)	1.07%	2.15%	5.36%
Average demand change of within-group connections (%ΔMSyx)	−0.22%	−0.45%	−1.13%
Reduction in total air travel (%Δpassengers)	−1.47%	−2.91%	−7.07%

Table 7.4 Sectoral and macroeconomic impacts (in percentage change compared to the BaU scenario) of subsidy removal and carbon tax addition in Ireland

	Sectoral Real Value—added in 2030			Consumer Price Index in 2030	
Scenario	ELC	TRP	Real GDP in 2030	Energy	Non-energy
PT_ATS	−0.26%	−9.84%	−0.34%	−0.18%	0.27%
ST_KRS	−2.28%	−3.30%	−0.23%	2.86%	−0.26%
PT_ATS_tax	−2.48%	−13.18%	−0.66%	9.60%	−0.06%
ST_KRS_tax	−4.41%	−7.03%	−0.82%	13.04%	−0.50%
ALL	−6.10%	−14.31%	−1.32%	8.73%	−0.53%
ALL_tax	−8.08%	−17.22%	−1.83%	19.94%	−0.61%

goals of the Paris Agreement), the study considered three scenarios for the carbon price: (i) low scenario, USD \$10 per ton CO_2; (ii) medium scenario, USD \$20 per ton CO_2; and (iii) high scenario, USD \$50 and USD \$100 per tonne CO_2. Regarding the carbon cost pass-through, the study assumed that ticket prices would be adjusted as a response to the carbon emissions cost based on the profit maximization behaviour of airlines. Table 7.4 presents the simulation results obtained. The findings of the study suggested that, on average, US airliners would pass on the full carbon cost to passengers. On average, ticket prices increase by 1.07–10.73% and passenger demand decreases by 1.47–13.50%, depending on the carbon price set, respectively. Within-group market shares, on average, an itinerary may lose 0.22–2.23% of its market due to ticket price changes among the competitors. These relatively low percentage changes in the conditional market share could be additionally explained by the lack of competitive transport options in the USA (i.e., no high-speed rail).

The study also assessed the reduction in air carbon emissions, which accounted for 17.05% of the higher carbon price scenario. Considering the aviation industry's ambitious goal to reduce net aviation CO_2 emissions by 50% until 2050 (relative to 2005 levels), the study concluded that the carbon price should be as high as USD \$295/ton of CO_2. With this price, the simulation estimated an intense increase in ticket prices of about 32% and a reduction in total passenger demand of about 39.6%. On an airline-level analysis, the results of the study indicate that the effects differ depending on the airline type and that low-cost airlines, except Southwest

Airlines, face the largest increase. Considering the medium scenario, low-cost airlines would face a price increase between 1.9% and 3.4%, an average price increase of 2.15% considering legacy airlines. Nevertheless, the study concluded that the considered carbon policy wasn't expected to influence the competition between US airlines.

The International Monetary Fund [31] proposes pricing aviation fuel as a highly cost-effective source of revenue compared with broader fiscal instruments in the absence of comprehensive, upstream pricing across all fossil fuel products and countries. It is also observed that the difficulty of taxing the final consumption of international aviation services could point to a carbon charge above that called for by climate considerations. A fully appropriate tax structure would levy both a fuel tax to address environmental concerns and a tax on the final consumption of aviation services. In the absence of such explicit sales taxation, a tax on fuel also plays a role in correcting the potential overconsumption of aviation services and enhancing revenue. This can plausibly call for larger fuel taxes.

International Monetary Fund [31] also studied the case of a distance-based (non-creditable) ticket tax and concluded that, while inferior to a fuel tax, it could also have merit. The air ticket levy in the UK is broadly of this form, being chargeable in an amount that varies with distance (within and outside of the EU) and by travel class. The weaknesses of such a tax—most notably that it discourages emissions only by discouraging travel—are stressed in AGF [1]. The force of these is diminished, however, by the recognition that the absence of sales taxation means that such a tax would at least serve to correct current tax distortions that likely lead to excessive international aviation transport.

Another interesting environmental taxation option called out lately is the introduction of a Frequent Flyer Levy (FFL), which would vary depending on the number of previous flights taken by an individual. At Stephen Devlinand [42] the authors analyse what such a tax regime for the UK should look like to achieve four goals:

- Prevent passenger demand from increasing more than 60% by 2050, as recommended by the Committee on Climate Change
- Be revenue neutral to the exchequer
- Obviate the need for new runway capacity
- Reduce greenhouse gas emissions in line with a low probability of >2 °C warming

It has been found that a progressive tax on frequent flying could play a significant role in restraining demand for flights while at the same time tending to distribute those flights more equally across the income spectrum.

Considering the lack of agreement about the optimum value for such a tax, in a recent study [50], authors constructed marginal curves that represent the effect on demand and CO_2 for different values of a global fuel tax, ranging from 0 to €500/ton of fuel in intervals of €25/tonne. By expressing this information in accumulative percentages, they obtain the marginal CCL curves, which give a straightforward a percentage of reduction for the worldwide demand in the period 2021–2050 for any given tax. This marginal curve can be used as a criteria for design.

Nevertheless, even with all bilateral agreements, studies consider that a carbon tax would only be effective in reducing demand if it is common and equal among countries, as airlines could change their operational behaviour to remain competitive (e.g., changing airports of choice and/or relocating to "low-tax" countries) [24]. Every country would be better off with taxes on aviation in the sense that these would lead to a socially optimal air traffic level [35]. The problem is that there are some incentives for a country to reduce or not introduce taxes in aviation and thus increase its participation in the aviation market. In addition, tax evasion measures taken by tax subjects (e.g., travelling to airports in neighbouring non-tax countries or refuelling abroad) can reduce the positive environmental impact of aviation taxes [35].

Therefore, the main drawback of using taxes as a global climate policy instrument is the need for consensus between states regarding their introduction as well as the detailed policy design, e.g., tax rates and tax bases. Valdés et al. [50], analysed that 70% of the reduction in CO_2 emissions will be produced by aviation with origin in just four regions: Europe, North America, the Middle East, and Southeast Asia. By applying the 80/20 Pareto's law (which states that for many phenomena, 80% of the result comes from 20% of the effort), it would be necessary that at least these regions agree on the implementation of the tax in order to obtain a significant CO_2 saving. A worldwide agreement less than that could lead to insignificant CO_2 savings and, at the same time, produce a negative countereffect on the economy, air transport, and tourism.

Despite all these considerations, the introduction of aviation taxes into an internationally coordinated movement should be possible. For example, the international harmonization of aviation taxes within the European Union or the EU minimum tax rates for energy products used for heating, as a fuel, or as electricity in the Energy Taxation Directive. Regarding energy product taxation, however, many member states apply exemptions or reductions. Moreover, these minimum tax rates are not based on the carbon content of the fuels but mainly on weight or volume. However, some Member States apply an explicit carbon tax on top of the regular energy excise duties. In 2012, energy taxes accounted for 75% of environmental tax revenues in the EU. Vehicle taxes include registration taxes, ownership taxes, and road use charges in the form of vignettes or distance-based toll systems. In 2012, transport taxes accounted for 21% of environmental tax revenues in the EU.

In order to be environmentally effective, air traffic taxes should provide incentives for:

- Reducing fuel consumption and therefore pollution using more efficient aircraft
- Shifting the fuel mix to less emission-intensive sources
- Optimizing aircraft loads and thus decreasing fuel consumption and emissions per passenger, as well as reducing the number of flights [32]
- Avoiding very short and very long distances

Environmental effectiveness also depends on the susceptibility of the tax on the fiscal competition and the possibilities of passengers and carriers to avoid it by

flying from, or by tanking in, third-low or no-tax countries. The various aviation tax choices vary in their ability to set certain incentives.

In terms of exposure to the impacts of climate change, economic development rates, and contributions to international aviation pollution, the different circumstances and respective capacities of developing countries also need to be considered. By introducing a carbon tax on the purchase of airline tickets, it is understood that all passengers can pay such a tax, both passengers in developing countries and those in non-developed countries, which means that countries would have to establish a tax at an appropriate level according to their economic circumstances.

In this context, a carbon fuel tax would reduce flight demand and supply, particularly for very short and very long-haul flights, and it can also be expected to contribute to aircraft load optimization. Carbon-based flight ticket tax offers fewer environmental benefits than the carbon-based fuel tax, as it provides little incentive for airlines to raise aircraft loads and reduce fuel use.

Although there have not been numerous research studies related to carbon taxes in aviation, every factor that can cause a tax to have a negative or positive effect on different variables, such as employment, GDP, or costs and benefits, should be studied prior to implementing any carbon tax [21]. These factors could be the following:

It is likely that there will be lower employment, lower GDP, and an impact on the economy, since it may take time for the economy to adapt to such a change as the imposition of a carbon tax. Despite this, it is difficult to determine how a change in aviation tax will truly impact the economy. On the other hand, carbon taxes in aviation can also result in lower real wages, and this will therefore lead to a reduction in unemployment as well as an increase in GDP and tax benefits.

A tax on aviation would impact both inbound and outbound tourism. When visitors pay more for the goods and services than they cost, taxes account for at least part of that difference. A country will be exporting taxes so that national income and net benefits will increase.

If a country depends on taxation on its goods and services, for example, VAT, all goods and services would usually be taxed. If there is a service like international aviation that is not regulated, welfare will be boosted by imposing on aviation a tax that is not very different from other taxes, resulting in an increase in national income and net benefits [32].

If an externality exists, such as greenhouse gas emissions, taxing the emissions will increase net benefits but not necessarily national income. An aviation tax, regardless of whether it is a tax on fuel or a tax on tickets, will have that positive effect. In the case of emissions, the beneficiary will be the planet rather than the country's population.

The carbon tax will reduce the use of aviation, and if aviation has broader economic benefits, it will reduce national income and affect the economy.

Considering that taxes on aviation jet fuel can only be levied on domestic flights, the following analysis will consider the impact of both existing and potential carbon taxes in European countries, Japan, and the USA. Other countries also tax domestic aviation fuel, but no studies have been found regarding its impact on the country's aviation sector.

7.6.1 Aviation Fuel Tax and the Fit for 55

Through the proposed revisions to the Energy Taxation Directive (ETD), the Commission states that the existing market-based instruments for aviation—the EU ETS and CORSIA—only partially internalize climate externalities. For intra-EEA flights, the climate change impacts are currently not fully internalized through the EU ETS, as a significant proportion (44% in 2019) of total verified emissions are allocated for free to aircraft operators, although this is being reassessed in the proposed EU ETS directive revisions. For extra-EEA flights, the price signal provided by CORSIA clearly falls below the EU ETS carbon price and would only marginally reflect the climate-related external costs generated by extra-EEA flights.

Commission is therefore proposing that the current mandatory EU-wide fuel tax exemption accorded to aviation (and shipping) be removed. The exemption offers these sectors favourable tax treatment in the transport sector, as road transport is not exempted and the exemption of rail transport is optional. Moreover, the present situation substantially weakens the incentives for investing in more energy-efficient and less polluting crafts. The lack of proper differentiation between the different fuels in these sectors covered by the mandatory tax exemptions also does not facilitate reducing the significant price difference between fossil fuels and sustainable fuels. Properly designed taxation measures could support the uptake of sustainable fuels, and at the same time, their production could result in lower prices for these fuels.

Removing the mandatory exemption would allow Member States to unilaterally tax the two sectors if they so wish, but without obliging them to do so. However, with regard to aviation, international flights to third countries outside the EU would be excluded from the scope of the revised tax since legal air services agreements with some third countries do not allow the taxation of fuel uplifted by the carriers of these third countries at EU airports.

Similarly, intra-EU cargo-only flights would also be excluded from the scope due to the special privileges granted to some third countries, for example, the USA through the US-EU Open Skies agreement. Under Open Skies, US carriers have a significant market share of the intra-EU cargo market and have a current exemption from the taxation of aviation fuel uplifted in the EU.

The Commission says a possibility would be for Member States to tax fuel on domestic flights or through bilateral or multilateral agreements between them. Another possibility could, in principle, be to apply a fuel tax to international flights to those third countries that do not have air services agreements with the EU or are not prohibited by air services agreements. In any case, it adds, a passenger ticket tax may be an appropriate alternative to a fuel tax on international flights that would be outside the scope.

The proposal offers several options that could be applied to an aviation fuel tax, which include a transitional period of 10 years (2023–2033), increased in a linear way to the corresponding minimum tax rates applicable to motor fuels used in road transport. A zero rate would be applied for a limited period to advanced biofuels and e-fuels for aviation to help the uptake of these fuels until their production has been

scaled up. This transition would also provide stakeholders with a clear price signal trend in order to adapt investments and technologies.

The Commission acknowledges a potential problem with an EU aviation fuel tax: carbon leakage due to tankering, whereby fuel is uplifted outside EU jurisdiction and used on subsequent intra-EU operations. However, it believes that due to the limited size of the fuel tanks on aircraft, the opportunities for tankering are limited. The collection of an aviation fuel tax is not expected to be a problem from an administrative perspective, as it would be collected as in other transport modes by the fuel supplier and the funds transferred to the relevant tax authority.

An independent study carried out for the Commission and quoted in the ETD proposal modelled a number of tax rates applied to aviation fuel, ranging from €0.17 to €0.50 per litre and a central option of €0.33/litre, with the airline expected to pass the cost on to passengers by raising ticket prices, leading to a reduction in passenger demand and hence fuel consumption. Only to a limited extent does it expect the airline to be incentivized to choose more efficient aircraft for their operations to reduce the fuel consumed. The modelling showed implementing a tax on fuel loaded for intra-EEA flights had noticeable impacts on CO_2 emissions in the long term, with reductions of between 6% and 15% relative to a 2016 baseline.

In terms of tax revenue, the study showed that under the €0.33/litre option, the tax would contribute around €6.7 billion (USD $7.9bn) per year in 2050. The wider impacts on the economy from the reduction in aviation demand would then reduce the rise in total tax revenue over the baseline to €5.4 billion per year.

By 2025, the introduction of a €0.33/litre fuel tax, with no transition period, could lead to a reduction of 10% in the number of flights when compared to the baseline, but the overall flight frequency on most routes would still be higher than it was in 2016, although some variations are expected, and specific regions could see their connectivity reduced. EEA carriers could also suffer negative competitiveness impacts in relation to third-country carriers, which may be subject to a more lenient tax regime in their home countries. The implementation of a fuel tax on intra-EEA flights could also give rise to concerns over ‘hub switching’ as carriers change the connection, or hub, airport from an EEA airport to a non-EEA airport. This is more likely to impact traditional network carriers than low-cost carriers with their mainly direct flights.

Reacting to the fuel tax proposal, the International Air Transport Association (IATA) said a reliance on taxation as a solution to cutting aviation emissions was counterproductive.

In fact, taxes syphon money from the industry that could support emissions-reducing investments in fleet renewal and clean technologies.

IATA acknowledged that market-based measures to manage emissions were required until technology solutions had been fully developed and said it supported the CORSIA scheme to avoid a patchwork of uncoordinated national or regional measures such as the EU ETS. However, they are extremely concerned by the Commission’s proposal that European States would no longer implement CORSIA on all international flights.

Trade body Airlines for Europe (A4E) said the 'Fit for 55' package included proposals that would significantly impact European airlines in the years to come and have a transformative impact on the sector. A4E said that by making flying more expensive, it may shift demand globally and reduce traffic locally, but it will not tackle the source of the emissions. It is necessary to invest in solutions that offer real reductions in CO_2 emissions per aircraft. Increasing costs reduces the capacity to make these investments, whilst CO_2 emissions are potentially shifted to other regions. A4E said unilateral and double pricing of CO_2 under several market-based measures would be economically counterproductive, and if airlines paid for their CO_2 under the EU ETS, for instance, they should not have to pay for it again.

7.7 Revisiting Existing Aviation Subsidies

There is a link between the aviation industry's development and its fiscal framework, and there is not much public data on financial support that airlines and stakeholders receive from governments and institutions. Several papers discuss the effects and design of specific subsidies, such as the provision of air transport services to remote regions, but only a few studies in some countries with a more comprehensive approach quantify subsidies to the aviation industry. Apart from the limited number of academic papers, environmental agencies, non-governmental organizations (NGOs), or aviation lobby organizations have also been addressing the issue of aviation subsidies.

Talking about subsidies, these are a form of economic intervention that can be found in practically all sectors and play an important role globally, even in aviation. Subsidies in aviation have a long history and have been a source of controversy for over half a century. The 1938 Civil Aeronautics Act governed "subsidy payments in support of scheduled domestic and international air services" in the USA. Subsidies were enhanced based on claims of defining aviation as an "infant industry" and in favour of "national defence", although it was noted that aviation could no longer be defined as an "infant industry" by 1950 [23]. Subsidy payments in the USA continued to rise in the following years and exceeded US $60 million in the year 1960, but the Airline Deregulation Act sought to end US airline subsidies by 1978. European flag carriers also received support from governments, with subsidy policies dating back at least until 1919. The European Commission first attempted to define subsidies in the 1980s compared to government measures, as concerns were growing about levels of 'permissible state aid' and 'overcapacity in the European airline industry'. Yet, despite these discussions, governments have continued to provide aid to airlines.

The two main elements of air transport infrastructure are air traffic control services and airports. Air traffic control services are generally financed through user fees or assigned taxes, which allows an almost complete recovery of costs. In addition, the size of an airport is relevant, since most of large airports cover their total costs and even generate profits for their shareholders, while smaller airports are

generally not profitable. In some countries, like Norway and Spain, most airports belong to a group of publicly owned airports. If the group covers its costs, there is no subsidy for the air transport industry, but a cross-subsidy within the industry is possible. In other countries such as Germany, for example, smaller airports are often owned by regional or local municipalities, forcing respective governments to cover losses. However, most passengers depart from large and often profitable airports, generally leading to cost recovery.

Additionally, governments often also pay for airport access infrastructures, including road and railway systems. For example, a large portion of airport financing in the US is based on the Airport Improvement Program, which is funded through taxes on aviation and provides funds for smaller airports.

In some cases, selling duty-free goods at airports or on board an aircraft could be considered a subsidy due to the tax exemption, which allows the airline, the airport operator, or a franchise to sell goods at a higher margin and therefore have higher profits.

On the other hand, fuels for use in international aviation are not subject to taxes. This clause can usually be found in bilateral air services agreements. While Article 24 of the Chicago Convention only prohibits taxes on fuel found on board an aeroplane when it reaches a contracting state, fuel used on domestic flights is usually exempt as well from general fuel tax in many countries.

Governments have been financially involved in the sector's development since the start of commercial aviation, extending a wide range of subsidies to manufacturers, transportation infrastructure providers and airlines, which could have been legal or illegal under various trade agreements and legislation, both local and multinational. A conceptual description of the aviation subsidies is provided in the following table. Airlines were found to receive most types of subsidies. This can be explained by the fact that airlines are the most exposed to competition, and that governments have often tried to protect domestic airlines for reasons of domestic economic growth as well as national priority reasons (Fig. 7.7).

The empty spaces in the table do not necessarily imply that a specific type of subsidy does not exist in practice. Some subsidies are specific by design or definition; for example, only airlines offer frequent flyer programs. Also, some airports can also be expected to receive subsidies related to the provision of airport infrastructure or other programs such as renewable energy generation. And there may be some more subsidies, but they have not been identified due to a lack of transparency.

When subsidies constitute undue economic advantages under competition and trade agreements or national legislation (e.g., open skies agreements; EU rules on state aid), the definition of what constitutes a subsidy is crucial. Temporary subsidies should also be considered, as they are often linked to specific and extraordinary events such as an external crash, the bankruptcy of an airline, or a significant investment in the airport. On the other hand, although subsidies to aircraft manufacturers are often related to specific activities such as the development of a new type of aircraft, these are provided on a regular basis, which makes them permanent and non-temporary examples of state aid.

	Manufacturers		Infrastructure Providers		Airlines
	Aircraft	Suppliers	Air Traffic Control	Airports	
Grants:					
Research and development	X	X	X		
Exports	X	X			
Investments	X	X	X	X	X
Loss coverage				X	X
Equity infusions	X		X	X	X
Loans and loan guarantees	X	X		X	X
Grants to provide air transport services to remote regions				X	X
Dedicated transfers to Residents buying tickets					X
Hidden subsidies:					
Reduced infrastructure fees	X			X	X
Cross-subsidisation				X	X
Monopoly rights				X	X
No or reduced taxes:					
Fuel					X
Value Added Taxes					X
Frequent flyer programs					X

Fig. 7.7 Overview of subsidies extended to aviation. (Source: Gössling et al. [41])

Land grants, low-interest loans, and direct grants to airports and operators are among the many forms of inappropriate state aid that distort competition with other modes, exacerbate aviation's climate effects, and do little to improve connectivity, as the low-cost carrier experience has clearly shown [47]. One policy approach to moderate demand for air travel could be to phase out these direct (passed on to aircraft manufacturers, airports, or airlines) and indirect (provided to customers) subsidies, which also include subsidies for airport expansion and operational aid to airlines or tax exemptions—both energy and CO_2 taxes [7, 41]. The European Green New Deal [15], launched last year, reinforces the need to end global fossil fuel subsidies, phasing out financing by multilateral institutions of fossil fuel infrastructure, strengthening sustainable financing, and phasing out all new coal plant construction, besides putting an end to tax exemptions in the aviation sector.

Subsidies should only be given in the most exceptional cases, i.e., where there is a clear need to maintain appropriate scheduled air services on routes that are vital and/or improve connectivity to remote regions. Public service obligations (PSOs) are designed to do just that. For example, in PSOs, in case no air carrier is interested in operating the route on which the obligations have been imposed, the Member State concerned may restrict access to the route to a single air carrier and compensate its operational losses resulting from the PSO (European Commission [12], 2008). At an MS level, this operational aid given to airlines has the most distorting effect on competition [45].

A report made by the Economic and Social Research Institute analyzed the economic and environmental impacts of the removal of eight different fossil fuel subsidies in Ireland by using the Ireland Economy-Energy-Environment (I3E) model. The model estimated both the impact of the removal of each subsidy alone by

holding the other subsidies constant (as it was made in the CE Delft study) and the impact of the removal of all subsidies in 2020. The base year (for comparison) used in the report was 2014. Also, the model considered a separate set of scenarios in which the removal of each subsidy is accompanied by a gradual increase (of €6 per year starting from 2020) in the level of the carbon tax (which was estimated to increase from €20/ton CO_2 in 2014 to €26/ton CO_2 in 2020) to quantify the combined effects of these different policy instruments.

In the report, the removal of sectoral fossil fuel subsidies is translated into increases in the production tax rates, whereas the elimination of subsidies on the energy commodity is translated as increases in the sales tax. For the aviation sector, the removed subsidies considered were the excise exception on aviation fuel (represented by the indicator PT_ATS and accounted as a fuel production tax rate) and the excise exception on the commodity of kerosene (represented by the indicator ST_KRS and accounted as a sales tax rate). Removing the entire PT_ATS subsidy (€425.9 million in 2014) paid to the aviation sector would increase its production tax rate by 212.07%, and, for the ST_KRS subsidy (€460.1 million in 2014), its removal would correspond to a 31% sales tax rate on kerosene. The higher production rate and introduction of sales tax lead to a reduction in the transportation (TuRbi Propulsion (TRP) indicator) sectoral real value-added (i.e., its contribution to GDP). In 2030, removing the entire PT_ATS subsidy, the aggregate transportation sector's real value-added would shrink by almost 10% (the largest impact in the transportation sector with the removal of one subsidy alone) as the production tax rate of the aviation sector would triple, relative to business-as-usual (BaU), and its value-added constitutes more than half of the aggregate value-added of the transportation sector. In the cases of removing the fossil fuel subsidies on commodities, the impacts are the largest for the transportation and electricity (ELC indicator) production sectors when the subsidy on kerosene is removed. As expected, the sectoral impacts become larger in absolute terms when the subsidy removal process is accompanied by an increase in the level of the carbon tax (scenarios PT_ATS_tax and ST_KRS_tax).

In macroeconomic terms, removing the PT_ATS and ST_KRS subsidies would lead to a reduction in the overall economic activity (measured by real GDP in 2014 prices) by 0.34% and 0.23%, respectively, in 2030, compared to the BaU scenario. Regarding the average price changes of the energy commodities, which are measured by the energy consumer price index (CPI), they would be mainly positive (i.e., the energy inflation would be higher relative to BaU level), with the removal of fossil fuel subsidies on diesel and kerosene having the highest impacts across all scenarios. This is due to the estimated higher production cost that would lead to increased domestic fuel prices and consequently decrease the national firm's profits and competitiveness in international markets.

However, overall, the report concluded that removing the subsidies one-by-one (in 2020) would not generate a significant decrease in the overall economic activity in 2030. Even if all subsidies were removed simultaneously (a scenario represented by the indicator ALL), real GDP would be only 1.3% lower in 2030, compared to its BaU level. With the addition of the carbon tax (scenario ALL_tax), in 2030, the GDP would be 1.83% lower and the level of inflation would be 1% higher, compared to the BaU level.

In Japan, based on the tax reform decided on 2012, the government established the introduction of a Special Tax for Climate Change Mitigation (JPY 289/ton CO_2 or approximately €2.1/ton CO_2), which was added to the existing Petroleum and Coal Tax (that is, JPY 760/KL or approximately €5.6/ton CO_2 for crude oil or oil products) that covers all fossil fuels and was intended to be used for the introduction of renewable energy, among other purposes. An analysis made by the Japanese Ministry of the Environment estimates that the new carbon tax's impact on Japan's GDP was an increase of between 0.04% and 0.1%.

7.8 Fiscal Measures

Besides revising existent aviation subsidies, a greater use of green fiscal measures or instruments is needed to help redirect public investment, consumption, and taxation to green priorities and away from harmful subsidies.

Even though today there are several commercial-size plants in operation producing aviation biofuel, the consumption is very low due to several barriers that are limiting market uptake, such as the cost relative to conventional aviation fuel and the low priority in most MS's national bioenergy policies. In 2017, it was estimated that the EU had a potential bio-based aviation fuel output or production capacity of 0.335 million tons per year, which accounts for approximately 0.6% of the total EU conventional fossil fuel demand [30]. The International Environmental Agency (IEA) [30] estimates that, in 2018, aviation biofuel production was about 15 million litres and accounted for less than 0.1% of total aviation fuel consumption/demand.

Considering this, it is fundamental to create fiscal incentives (subsidies) to increase the global uptake of Sustainable Aviation Fuels (SAFs). For instance, at the international level, ICAO's CORSIA will allow aircraft operators to reduce their offsetting obligations by using SAFs and fossil-based 'lower-carbon aviation fuels', which will decrease the costs resulting from purchasing emissions units and thus the estimated burden for the aircraft operator.

However, countries have more control over policy support for domestic than international aviation, and the introduction of national policy mechanisms to facilitate SAF consumption is also gathering pace. The US, the EU, the Netherlands, the U K, and Norway have all established policy mechanisms that will support the use of aviation biofuels. In the EU, following the European Green New Deal, a review of the European economic governance framework will include a reference to green public investment in the context of the quality of public finance, which allows improving EU fiscal governance.

According to IEA's Sustainable Development Scenario (SDS), which anticipates biofuels reaching around 10% of aviation fuel consumption/demand by 2030 and close to 20% by 2040, subsidizing the consumption of SAF envisaged in the SDS scenario in 2025, around 5% of total aviation jet fuel demand, would require about USD $6.5 billion of subsidy (based on closing a cost premium of USD 0.35 litre between Hydroprocessed Esters and Fatty Acids Synthetic Paraffinic Kerosine (HEFA–SPK) and fossil jet kerosene at USD 70/barrel oil prices). This is far below

the support for renewable power generation in 2017, which reached USD $143 billion.

A report [43] from the UK's Sustainable Aviation (SA) estimates that in 2035 there may be between 14.5 and 30.9 million tons per year of SAFs produced globally. This would correspond to 4–8% of global aviation fuel use. If UK production were to grow in line with global production, SAFs produced in the UK could provide between 3.3% and 7.8% of the UK's 2035 aviation fuel demand. For this, it would be necessary to develop up to 14 full commercial-scale plants until 2035, which corresponds to a new plant every 1.3 years and around £4 billion (approximately €4.4 billion EUR) of total investment. By 2035, the development of a domestic industry to produce SAFs could generate a Gross Value Added (GVA) of up to £742 million (approximately €823 million EUR) annually and support up to 5200 UK jobs. A further 13,600 jobs could be generated from the growing market for SAFs through global exports.

7.9 Conclusion

In recent years, we have seen a higher compromise with the decarbonization objective at the European level. At the same time, data and studies have brought evidence that technological measures included in strategic plans seem not to be enough to achieve these exigent CO_2 decarbonization objectives. In this context, EU policy has reinforced a set of market-based measures of different natures to translate into practice the principle that the polluter pays.

However, experience and a variety of studies revisited through this paper, as well as the opinion of several groups of interests, also brought doubts about those economic strategies being the most adequate for maintaining European aviation leadership while achieving the Fit for 55 goals.

Stakeholders are concerned that inefficient taxation policies could lead to a disproportionate cost burden and hamper aviation's decarbonization plans. As one of the sectors hardest hit by the pandemic and with an essential role in kick-starting societal and economic recovery, future EU policies must guarantee and support the sector's competitiveness.

Risks have been identified of negative impacts on European competitiveness and that of the entire aviation ecosystem, Europe's tourism industry, and the wider EU economy, as well as the risk of carbon leakage. The EU environmental policies and associated costs are being reviewed bearing in mind two other priorities, that may be competing for funding and/or related legislation: (i) the security concerns following the invasion of Ukraine and possibly weaker US commitment to NATO; (ii) the need to support the competitiviness of Europe with regard to the challenges from the US and China.

References

1. AGF, 2010. Work Stream 2: Paper on Potential Revenues from International Maritime and Aviation Sector Policy Measures. Addendum No. 2 to Report of the Secretary-General's High-level Advisory Group on Climate Change Financing, New York: United Nations: s.n.
2. Alberici, S. S. B. P. v. B. Y. D. S. F. A. G. V. v. G. K. G. H. G. D. d. J. E. K. W. P. M. S. E. d. V. T. W. K., 2014. Subsidies and costs of EU energy—Final report, s.l.: Ecofys.
3. Anon., 2003. Decision No. 377/2013/EU derogating temporarily from Directive 2003/87/EC establish-ing a scheme for greenhouse gas emission allowance trading within the Community,, s.l.: s.n.
4. Anon., n.d. [Online] Available at: http://ec.europa.eu/clima/policies/transport/aviation/documentation_en.htm
5. ATAG, 2016. How a global carbon offsetting scheme for aviation would work. [Online] Available at: https://aviationbenefits.org/media/136900/atag_carbon_offset_infographic.pdf
6. Azarova, V. & Mier, M., 2021. Market Stability Reserve under exogenous shock: The case of COVID-19 pandemic. Applied Energy, Volume 283.
7. CAN & ICSA, 2018. Contribution of the Global Aviation Sector to Achieving Paris Agreement Climate Objectives,. [Online] Available at: https://icsa-aviation.org/wp-content/uploads/2018/04/CAN-ICSA-Aviation-TD-submission.pdf
8. CE Delft, 2019. Taxes in the Field of Aviation and their impact. Final Report, s.l.: European Commission.
9. Commission Europea, 2019. The European Green Deal.Communication from the Commission to the European Parliament, the European Council, the Council, the European Economic and Social Committee and the Committee of the Regions. COM(2019) 640 final, 11.12.2019, s.l.: Brussels.
10. EASA, 2020. [Online] Available at: https://www.easa.europa.eu/eaer/topics/market-based-measures
11. European Commission, 2017. Report on the functioning of the European carbon market, COM(2017) 693 final, s.l.: s.n.
12. European Commission (2008), 2008. Regulation (EC) No 1008/2008 of the European Parliament and of the Council of 24 September 2008 on common rules for the operation of air services in the Community, Retrieved from: https://eur-lex.europa.eu/legal-content/EN/TXT/, s.l.: s.n.
13. European Commission, 2003. Consolidated version of Directive 2003/87/EC of the European Parlia-ment and of the Council establishing a scheme for greenhouse gas emission allowance trading within the Community and amending Council Directive 96/61/EC, s.l.: s.n.
14. European Commission, 2013. Regulation (EU) 2018/842 on binding annual greenhouse gas emission reductions by Member States from 2021 to 2030 contributing to climate action to meet commitments un-der the Paris Agreement and amending Regulation (EU) No 525/2013, s.l.: s.n.
15. European Commission, 2019. Communication "United in delivering the Energy Union and Climate Action - Setting the foundations for a successful clean energy transition" COM(2019) 285 , s.l.: s.n.
16. European Commission, 2021. Revision of the EU ETS Directive concerning aviation. [Online] Available at: https://ec.europa.eu/clima/policies/transport/aviation_en
17. European Council, 2014. European Council, 23-24 October 2014. [Online] Available at: http://www.consilium.europa.eu/en/meetings/european-council/2014/10/23-24/
18. European Parliament, 2012. Regulation (EU) No 601/2012 of the European Parliament and of the Council of 21 June 2012 on the monitoring and reporting of greenhouse gas emissions pursuant to Di-rective 2003/87/EC of the European Parliament and of the Council, s.l.: s.n.

19. European Parliament, 2003. Directive 2008/101/EC of the European Parliament and of the Council of 19 November 2008 amending Directive 2003/87/EC so as to include aviation activities in the scheme for greenhouse gas emission allowance trading within the Community, s.l.: s.n.
20. European Union, 2011. Flightpath 2050, 2011. Europe's Vision for Aviation.. ISBN 978-92-79-19724-6, doi https://doi.org/10.2777/50266. ed. Luxembourg : Publications Office of the European Union.
21. Forsyth, P., 2019. "Aviation Taxes – Impacts, Costs and Benefits" (online: http://greenfiscalpolicy.org/wp-content/uploads/2019/06/1560238770_AviationTaxesImpactsCostsandBenefits.pdf), s.l.: s.n.
22. Görlach, B., 2013. What constitutes an optimal climate policy mix? Defining the concept of optimality, including political and legal framework conditions, CECILIA2050 Deliverable 1.1,, s.l.: s.n.
23. Gossling, S., Fichert, F. & Forsyth, P., 2017. Subsidies in Aviation. online: https://www.mdpi.com/journal/sustainability. Sustainability.
24. Hosoda, R. G. &. E. B., 2016. Environmental impact of aircraft emissions and aviation fuel tax in Ja-pan. s.l.:s.n.
25. ICAO, 2010. Resolutions of the 37th Session of the ICAO Assembly (2010), s.l.: s.n.
26. ICAO, 2013. Resolutions of the 38th Session of the ICAO Assembly (2013), s.l.: s.n.
27. ICAO, 2016. Resolutions of the 39th Session of the ICAO Assembly (2016), s.l.: s.n.
28. ICAO, 2019. Resolutions of the 40th Session of the ICAO Assembly (2019), s.l.: s.n.
29. ICAO, 2020. CORSIA-and-Covid-19, ICAO. [Online] Available at: https://www.icao.int/environmental-protection/CORSIA/Pages/CORSIA-and-Covid-19.aspx
30. IEA, 2019. Are aviation biofuels ready for take-off?, Retrieved from: https://www.iea.org/commentaries/are-aviation-biofuels-ready-for-take-off, s.l.: s.n.
31. International Monetary Fund, 2011. Market-Based Instruments for International Aviation and Shipping as a Source of Climate Finance, s.l.: s.n.
32. Jones, B., Keen, M. & Strand, J., 2012. "Fiscal Implications of Climate Change" (online: http://documents.worldbank.org/curated/en/778871468332985467/pdf/WPS5956.pdf), s.l.: s.n.
33. Jörgen Larsson, A. E. T. S. &. J. Å., 2019. International and national climate policies for aviation: a re-view. Climate Policy, 19(6), pp. 787-799.
34. Kelly de Bruin, E. M. &. A. M. Y., 2019. The Impacts of Removing Fossil Fuel Subsidies and Increas-ing Carbon Tax in Ireland, The Economic and Social Research Institute, RS98, DOI: https://doi.org/10.26504/rs98 , s.l.: s.n.
35. Krenek, A. & Schratzenstaller, M., 2016. "Sustainability-oriented EU Taxes: The Example of a Europe-an Carbon-based Flight Ticket Tax" (online: https://ec.europa.eu/budget/mff/Library/hlgor/selected-readings/41-Carbon-basedFlightTicketTax-Schratzenstalle., s.l.: s.n.
36. Ministere de la Transicion Acologique, 2020. Understand and implement the EU Emissions Trading Scheme. [Online] Available at: https://www.ecologique-solidaire.gouv.fr/en/understand-and-implement-eu-emissions-trading-scheme
37. Ministry of the Environment, 2017. Greening of Whole Tax System and Carbon Tax in Japan: Envi-ronment-related Tax in Japan, https://www.env.go.jp/en/policy/tax/20170130_greening.pdf, s.l.: s.n.
38. Mu Huang, A. L. M. M. D. S. &. M. T., 2019. The Potential Financial Impact of CORSIA on Airline Companies & Carbon Markets, s.l.: s.n.
39. Psaraki-Kalouptsidi, I. P. & P., 2018. Econometric supply-and-demand models to analyse carbon pric-ing policies. International Journal of Transportation Science and Technology, 7(4), pp. 274-282.
40. Stagl, M. C. a. S., 2009. Ecological Economics. An Introduction.. s.l.:Cambridge University Press..
41. Stefan Gössling, F. F. P. F. &. H.-M. N., 2017. Subsidies in Aviation. Sustainability, 9(8), p. 1295 https://doi.org/10.3390/su9081295.
42. Stephen Devlinand, S. B., 2015. Managing aviation passenger demand with a frequent flyer levy.. s.l.:managing aviation passenger demand with a frequent flyer levy. Stephen Devlinand Sandra Bernick New Economics Foundation © 2015 The New Economics Foundation.

43. Sustainable Aviation, 2020. SUSTAINABLE AVIATION FUELS ROAD-MAP: Fueling the future of UK aviation. [Online] Available at: https://www.sustainableaviation.co.uk/wp-content/uploads/2020/02/SustainableAviation_FuelReport_20200231.pdf
44. The International Civil Aviation Organization, 2000. "Criterios de la OACI sobre impuestos en la esfera del transporte aéreo internacional" (online: https://www.icao.int/publications/Documents/8632_cons_es.pdf), s.l.: s.n.
45. Transport & Environment, 2018. Roadmap to decarbonising European aviation, Retrieved from: https://www.transportenvironment.org/publications/roadmap-decarbonising-european-aviation, s.l.: s.n.
46. Transport & Environment, 2019. Ryanair joins the club of Europe's top 10 carbon polluters,. [Online] Available at: https://www.transportenvironment.org/press/ryanair-joins-club-europe%E2%80%99s-top-10-carbon-polluters
47. Transport and Environment, 2017. Decarbonisation of aviation: why EU and ICAO action is needed. [Online] Available at: https://www.transportenvironment.org/publications/decarbonisation-aviation-why-eu-and-icao-action-needed
48. Transport and Environment, 2020. Understanding the indirect land use change analysis for Corsia. [Online] Available at: https://www.transportenvironment.org/publications/understanding-indirect-land-use-change-analysis-corsia
49. United Nations High Commissioner for Refugees, 2017. ≏Cuál es el objetivo de los impuestos pigovia-nos? (online: https://eacnur.org/blog/objetivo-los-impuestos-pigouvianos/), s.l.: s.n.
50. Valdés, R., Comendador, V., Campos, L. & Graga Campos, L., 2021. How Much Can Carbon Taxes Contribute to Aviation Decarbonization by 2050. Sustainability, Volume 13, p. 1086.
51. World Bank Group , 2019. State and Trends of Carbon Pricing 2019, Retrieved from: http://documents.worldbank.org/curated/en/191801559846379845/pdf/State-and-Trends-of-Carbon-Pricing-2019.pdf, s.l.: s.n.
52. Zetterberg, M. E. a. L., 2020. Can the EU ETS weather the impact of Covid-19?. CEPS Policy Insights No 2020-14 , June.

Chapter 8
High-Flying Future: Attracting Young Talents to Aeronautics

Ana Costa, Clara Pinho, and Mariana Denis Malta

Contents

8.1 Catching and Retaining the Attention of Children

Children's attention span undergoes significant changes as they grow. Initially, it is driven by their fascination with nearby objects that they can manipulate, easily shifting focus when something more interesting appears. However, as they mature, their attention span becomes more stable, tripling within a few years, from around 30 min at ages 3–4 to about 1.5 h at ages 5–6. This transition is marked by the

A. Costa (✉) · C. Pinho · M. Denis Malta
INOVA+, INNOVATION SERVICES S.A., Matosinhos, Portugal
e-mail: ana.costa@inova.business; clara.pinho@inova.business; mariana.malta@inova.business

L. M. Braga da Costa Campos (ed.), *Accelerating Sustainable Aviation Initiatives*, Sustainable Aviation, https://doi.org/10.1007/978-3-031-73993-4_8

capacity to focus on specific objects and activities consciously and voluntarily, with parents and teachers playing a crucial role in stimulating and shaping their attention span. In this context, aviation-related activities, such as playing with kites, listening to space-themed stories, and watching aviation-related cartoons and movies like Super Wings and Planes, serve as engaging incentives for children. These activities not only entertain them but also stimulate their interest in the aviation sector, potentially inspiring future career choices in the field.

Encouraging children's interest in STEM, specifically in aerospace, involves a variety of engaging activities and resources. Initiatives like those from the American Institute of Aeronautics and Astronautics (AIAA) offer STEM programs that provide teachers, students, parents, and professionals with resources to inspire the next generation of aerospace scientists and engineers. These programs emphasize hands-on learning experiences, classroom grants, and aerospace micro-lessons, making STEM subjects more accessible and enjoyable for young learners.

Additionally, NASA's website offers a wealth of educational resources, including videos, interactive games, and hands-on activities related to science, technology, engineering, and mathematics (STEM). Encouraging children to experiment with simple STEM projects, like testing the strength of folded paper columns or building paper airplanes, can spark their curiosity and interest in these fields. Sports can also serve as a platform to introduce aerodynamics principles, as understanding the science of flight can enhance performance in various sports. Moreover, leveraging cartoons and movies UP! and Leap! can further ignite children's fascination with aviation and aerospace.

Contests, events, and educational initiatives around the world, such as the Science and Engineering Festival in the United States, Fédération Aéronautique Internationale's (FAI) Young Artists Contest in Switzerland, and programs like Cool Aeronautics and Amy's Aviation in the UK, play a vital role in exposing children to the wonders of aerospace and aviation. These efforts aim to inspire young minds and cultivate their interest in STEM fields from an early age, ultimately shaping their future career inclinations in a fun and enjoyable way.

8.2 Promoting STEM Interest in Young Students

Promoting aerospace education and sparking interest in STEM subjects, particularly aerospace engineering, among young students is crucial. Basic teaching of physics, natural science, and mathematics at the secondary school level can lay the foundation for understanding concepts related to flight, space travel, and celestial phenomena. These early experiences are essential as they can influence students' choices of university studies and careers in the aerospace sector.

Initiatives like the ALLIES project partnership focus on providing secondary schools with wind tunnels to facilitate hands-on learning experiences related to aerospace. Wind tunnels have proven effective in generating interest among secondary students in aerospace-related phenomena. Additionally, projects like Learn&Fly

aim to address underachievement in essential skills like math and science by using aeronautics as an inspirational theme. This project not only enhances students' interest in STEM subjects but also provides information about career opportunities in the aerospace field, supporting students in their career decisions.

Furthermore, the transition from secondary to higher education in the aerospace field involves considerations like university choices and career paths. These steps are vital for students pursuing aerospace education, and the collaboration between teachers, parents, and professionals plays a significant role in ensuring a positive environment for the development of the younger generation in the aerospace sector.

For students graduating from secondary school, choosing a university degree can be overwhelming, especially within the realm of STEM. Aerospace engineering presents an appealing option, offering technical knowledge and the prospect of engaging careers in research, manufacturing, or operations with competitive compensation. However, aspiring aerospace engineers must possess strong motivation, solid mathematics, and physics skills and perform well in university admission exams, which typically involve rigorous assessments.

8.3 University Pathways in Aerospace Engineering

The aerospace industry plays a pivotal role in the European Union, driving integration, independence, prosperity, and global competitiveness. This sector, which has seen remarkable growth and collaboration over the past four decades, now stands as a world leader. Aeronautics and air transport contribute significantly to European cohesion and competitiveness, generating 220 billion euros in revenue and supporting 4.5 million jobs in Europe, a number expected to double by 2050, underscoring its crucial role in fostering economic growth, social inclusion, and regional development [2].

A key challenge for the aerospace sector is attracting young talent to sustain European leadership and competitiveness. To achieve this, high-quality university programs are essential, equipping future professionals with the knowledge, skills, and abilities required for effective aerospace work. In Europe, students have a wide range of options for aerospace education, typically starting with a three-year bachelor's degree in aerospace engineering. Following their undergraduate studies, students can pursue specialized two-year master's degrees in various aerospace disciplines.

Many European universities offer aerospace engineering programs, with notable offerings in several countries, including Germany, France, United Kingdom, Italy, Netherlands, Spain, and Portugal. These programs cover diverse areas within the aerospace sector, including aircraft operations, airports, propulsion systems, aerospace materials, management, air navigation, and aircraft systems (Table 8.1). The trend in recent years has been to pursue a bachelor's in aerospace engineering, followed by a specialized master's program, enabling students to focus on their chosen area of expertise within the aerospace field.

Table 8.1 Areas covered by the European degrees within the aerospace sector

University	Aircraft	Airport	Propulsion	Science/ materials	Management	Navigation	Systems
Polytechnic University of Catalunya	x		x	x			
University of Sevilla	x	x				x	
Polytechnic University of Valencia	x	x	x	x			
University of Leon	x	x		x			
University of Cadiz	x		x	x		x	
Polytechnic University of Madrid	x	x	x	x		x	x
European University of Madrid	x			x			
Carlos III University of Madrid	x		x				
University of Rey Juan Carlos	x	x	x			x	
University of Alfonso X el Sabio	x	x	x			x	
Polytechnic University of Milan	x		x	x	x		x
Polytechnic University of Turin	x		x	x	x		
Sapienza University of Rome	x		x	x			x
University of Luigi Vanvitelli	x		x	x		x	
University of Bologna	x					x	x
University of Padova	x			x	x	x	
University of Pisa	x		x				
University of Napoli	x		x				x
Zurich University of Applied Sciences	x	x	x		x		

(continued)

Table 8.1 (continued)

University	Aircraft	Airport	Propulsion	Science/ materials	Management	Navigation	Systems
Delft University of Technology	x		x	x			
Amsterdam University of Applied Sciences	x	x			x		
RWTH Aachen	x	x					
TU Berlin			x	x			
TU Braunschweig	x			x			x
Universitat Stuttgart	x		x			x	x
TU Dresden	x		x				x
Instituto Superior Técnico (IST)	x		x	x	x	x	x
Inholland University of Applied Sciences	x			x			
Institute of Technology Carlow	x			x			x
University of Limerick	x			x	x		x
FH JOANNEUM Graz	x		x		x		
University of Patras	x		x	x	x		
University of Zilina	x	x			x		
Kaunas University of Technology	x			x			x
Vilnius Gediminas Technical University	x		x	x			x
Polytechnic University of Bucharest	x	x	x	x	x		
Transilvania University of Brașov	x			x	x		
Technical University of Košice	x		x		x	x	

(continued)

Table 8.1 (continued)

University	Aircraft	Airport	Propulsion	Science/ materials	Management	Navigation	Systems
BRNO University of Technology	x			x			
Czech Technical University in Prague	x			x		x	
Riga Technical University	x		x	x			
Newport University Centre for Educational Development	x		x	x	x		
University of Belgrade	x	x	x			x	
National Technical University of Ukraine	x			x		x	
Ecole-air de Salon de Provence (military school)	x			x		x	
ENAC Toulouse	x	x				x	x
ENSMA Poitiers	x	x	x	x			
ISAE-SUPAERO Toulouse	x	x		x			
ESTACA (France)	x	x				x	
University of Bristol	x			x		x	x
Queen Mary University of London	x			x			
Loughborough University				x	x		
University of Glasgow	x			x		x	x
Warsaw University of Technology	x			x		x	

Fig. 8.1 Academic Institution members of the PEGASUS Partnership. (Source: Pegasus)

Universities in Europe offer a variety of programs in Aeronautics and Aerospace Engineering, but the specifics of these programs can vary significantly across institutions and countries. To provide insight into the organization of typical university degree programs in this field, we can focus on those participating in specialized networks and partnerships, such as PEGASUS (Partnership of a European Group of Aeronautics and Space Universities), which was established to align academic standards and objectives.

PEGASUS was initiated by leading French Grandes Écoles specializing in aerospace education. It aims to attract top students and offer relevant educational and research programs in aeronautics and aerospace engineering. PEGASUS comprises public and non-profit higher education institutions within the European Union, originally formed by 20 founding members in 1998 and currently expanded to 25 member institutions representing various European countries (Fig. 8.1).

PEGASUS emphasizes coordinated change and innovation in collaboration with aerospace industry stakeholders and government agencies to meet defined objectives. The network is open to all European Union (EU) institutions providing

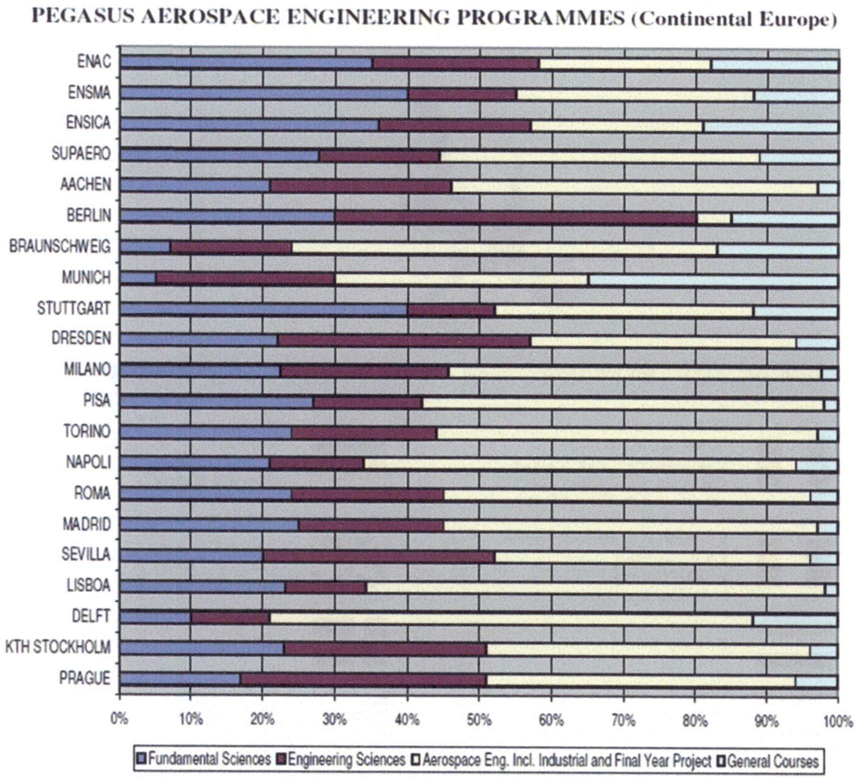

Fig. 8.2 Percent composition (in terms of ECTS) of the Degree Programs in Aeronautic/Aerospace Engineering in the PEGASUS Universities. (Source: Pegasus Aerospace Engineering Programmes, 3rd edition, July 2009)

high-quality aerospace engineering education, including degree-awarding programs, continuing education, research, and international collaborations.

In 2009, PEGASUS introduced standardized curriculum descriptions through the PEGASUS Course Catalogue, which serves to facilitate a clear understanding of the educational standards offered by member institutions. According to the catalog, all the programs have common categories: fundamental sciences, engineering sciences, aerospace engineering sciences, general courses, and industrial training/final year project (Fig. 8.2).

PEGASUS member institutions each bring their unique aerospace engineering specialties to the consortium, offering a diverse range of courses and programs. While some courses are mandatory for all students, others are elective, allowing students to tailor their education. Common topics covered by PEGASUS programs include aerodynamics, aeronautical structures and materials, aircraft design, propulsion, and combustion. Additionally, member institutions often have specific strengths or expertise in areas like aircraft operations, air traffic control, space engineering, and more, contributing to a rich and comprehensive aerospace education network.

The wide range of competences in PEGASUS member universities motivates students in multiple exchange programs including joint academic degrees.

Although the sector has expanded significantly over the past decades, there are several hurdles for educating engineers, including addressing environmental concerns, promoting safety, reducing noise, enhancing fuel efficiency, and adapting to shifts in competitiveness. Aerospace engineering programs must evolve to meet these demands. They should incorporate interdisciplinary education, emphasize global perspectives, and foster a culture of lifelong learning. Employability and adaptability are essential attributes for aerospace engineers, and universities should focus on developing these qualities in their graduates.

8.3.1 Factors Influencing Student Persistence in Aeronautics and Aerospace Education

To support the choice and persistence of students in aerospace disciplines, it's crucial to understand the role of environmental factors and barriers, especially in the context of aeronautics. Research shows that students who start their education in engineering are more likely to continue in the field [3], but academic success and retention of engineering students remain significant challenges.

Models of student success and persistence reveal the importance of both cognitive and noncognitive factors. While cognitive indicators are strong predictors of achievement and persistence among engineering students, noncognitive factors also play a vital role. These factors encompass sociocultural aspects, academic preparation, and personal traits like motivation and self-efficacy [4].

Students choose aerospace engineering for various reasons, including their interest in the field, prior academic preparation, and social influences. Motivated students often exhibit traits like persistence, goal-setting, and resilience. Self-efficacy, a belief in one's ability to perform tasks successfully, is a key factor influencing academic performance and career aspirations in aerospace-related fields [5]. Self-efficacy promotes positive outcome expectations and interest in academic subjects in which students feel confident. Career goals are influenced by self-efficacy, outcome expectations, interests, and environmental supports, with academic satisfaction playing a central role [6]. However, the challenge of attracting and retaining students in STEM fields, particularly women and minorities, can be related to self-efficacy expectations. Women report greater self-efficacy than men for traditionally female fields and lower self-efficacy than men for traditionally male occupations, which may result in women limiting their range of career choices and thus failing to realize their full career capabilities [7]. This has a special impact for aeronautics, where one of the reasons for the problems in attracting and training professionals is the gender bias and the fact that there is only a small number of minorities in the industry. Since gender stereotypes were a factor in career choice, many young women were discouraged from pursuing aeronautics careers [8].

To ensure student persistence in aerospace, it's essential to maintain factors associated with persistence, including goals, satisfaction, social support, and self-efficacy. Interventions should focus on social cognitive factors, such as providing access to support networks, sharing information about occupational rewards, and offering role models from diverse backgrounds.

External factors, like early exposure to aviation careers and support from friends and family in the industry, significantly contribute to student success. Encouraging students to explore aviation careers at a young age through career training programs and activities can enhance career awareness and preparedness. Collaboration with educators and formalized mentoring programs in universities can further support students' persistence in aerospace disciplines.

8.4 Careers in Aeronautics and Space

8.4.1 Evolution of the Aerospace Industry

The aerospace industry is currently undergoing a transformative period marked by intensified global competition, driven by the entry of emerging countries such as China, India, Brazil and Russia. In the most important large airliner market the dominance of the Airbus-Boeing duopoly remains, with historically the largest orderbooks ever, in terms of number and value of aircraft, and efforts to increase production rates. The post-Covid recovery has led to a shortage of personnel in all sectors. To thrive in this changing landscape, aerospace companies are forming strategic clusters to pool resources and expertise. In 2009, the European Aerospace Cluster Partnership (EACP) was established as a platform to facilitate collaboration among European aerospace clusters (Fig. 8.3).

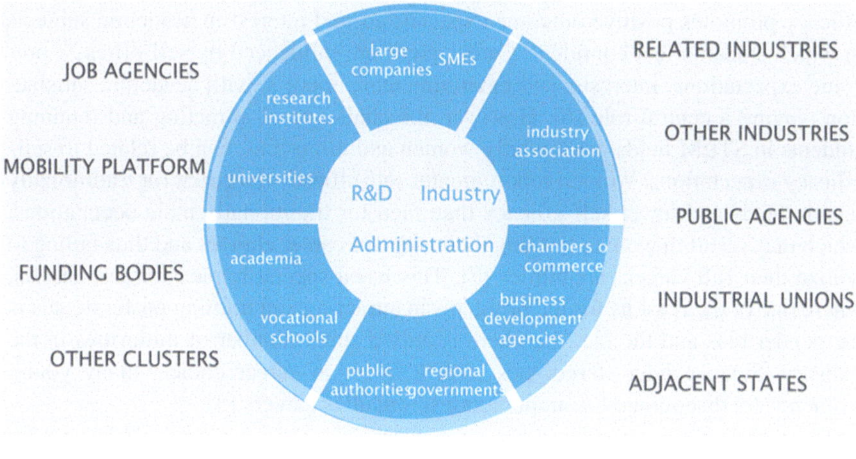

Fig. 8.3 Stakeholders represented within EACP. (Source: EACP)

Fig. 8.4 EACP Cluster map. (Source: EACP)

EACP membership is contingent on representing all facets of the regional aerospace sector, including industry, research and development (R&D), and administrative bodies.

EACP has evolved into a network comprising 42 aerospace clusters from 17 European nations, collectively spanning the entire aerospace value chain (Fig. 8.4).

Guided by core values such as trust and dependability, EACP's primary goal is to bolster Europe's global competitiveness within the aerospace industry. To achieve this, the partnership focuses on three core areas: knowledge exchange, innovation promotion, and strengthening the EU's aerospace positioning. EACP promotes knowledge sharing, best practices, and participation in European Strategic Cluster Partnerships to prepare for future industry developments. It also drives innovation by enhancing the skills of the aerospace workforce, supporting EU projects, and facilitating collaboration between industry and R&D. In parallel, EACP addresses supplier challenges, monitors technological advancements, and collaborates with strategic actors and institutions to reinforce the EU's standing in the aerospace sector. Ultimately, EACP envisions positioning Europe as a preeminent hub for aerospace innovation and competitiveness on the global stage, all while catering to regional needs and characteristics.

8.4.2 The Impact of the Aerospace Sector in Europe

The global aerospace and defense industry reported revenues of USD $741 billion in 2022, an increase of 3.0 percent over 2021 [9]. The sector can be divided into two major areas: aeronautics industry and air transport.

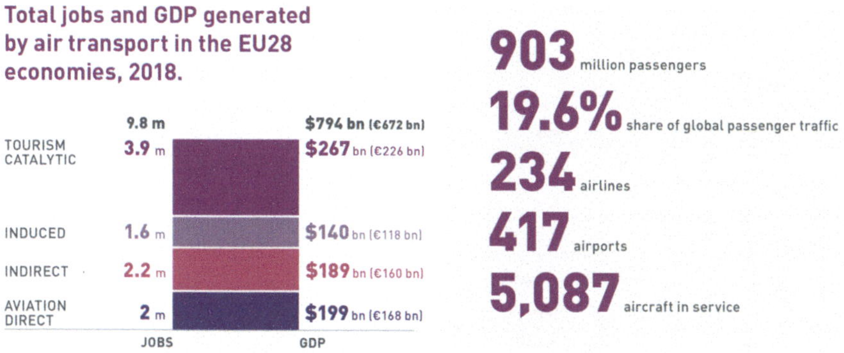

Fig. 8.5 Air transport in Europe. Aviation benefits beyond borders 2020 report

According to the European Commission, the EU is a global leader in the aeronautics sector, excelling in the production of civil aircraft, including helicopters, aircraft engines, parts, and components. This industry also encompasses military aircraft, drones, and aeroengines, with major players like Airbus, British Aerospace, Leonardo, Embraer, and Bombardier designing and manufacturing aircraft, while companies like Rolls-Royce, Safran, General Electric and Pratt & Whitney produce engines. Despite high employment costs, the sector remains profitable, with a significant focus on research and development (R&D) reflected in a growing number of patent applications, reinforcing its position as a high-tech and innovative sector within the EU.

Air transport plays a pivotal role in Europe's economy, supporting 9.8 million jobs and contributing USD $794 billion to its gross domestic product (GDP; Fig. 8.5). In 2018, the industry directly generated approximately 2 million jobs and the depression during the Covid pandemic has been mostly recovered. When considering the broader impact, including the direct, indirect, and induced jobs, the air transport sector was responsible for 5.8 million jobs and a GDP contribution of USD $528 billion in Europe. Furthermore, air travel significantly benefits regional economies by driving tourist spending, adding an estimated 3.9 million jobs and USD $266 billion to GDP in 2018.

8.4.3 Human Dimension of the Sector

Employment in aerospace and defense industry has been increasing and, in 2015, reached its maximum by employing 847.700 people. Employment in the aerospace sector is particularly significant in Europe, mainly in the United Kingdom, France, Germany, Italy, Spain, Poland, and Sweden, with increasing demand for civil and military aviation production and development.

In 2019, 408 thousand people were employed in the air transport sector in the EU, representing an 11% increase compared with 2018. Among EU member states, Luxembourg recorded the highest share of persons employed in the air transport

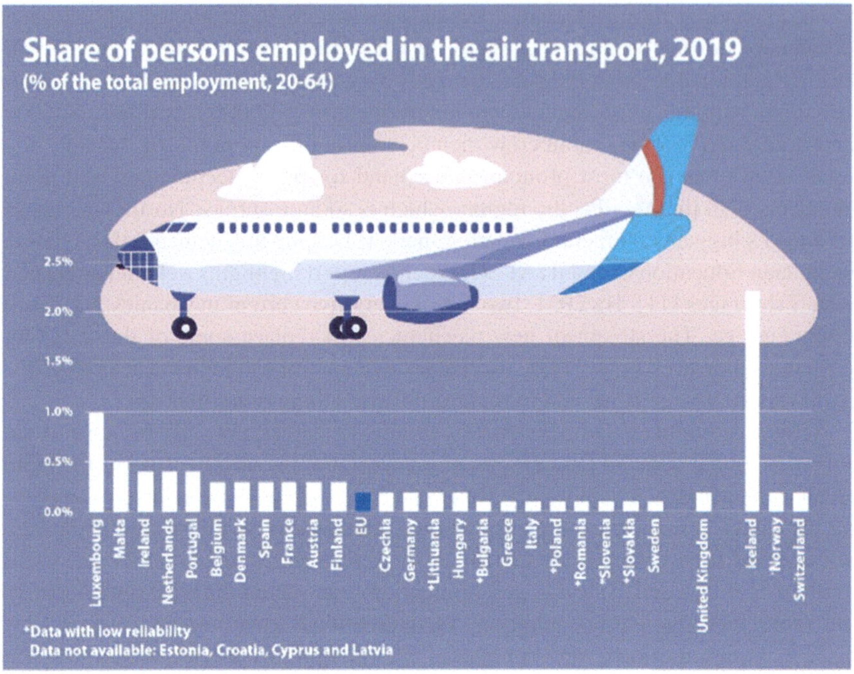

Fig. 8.6 Employment in air transport by country. (Source: Eurostat)

sector, followed by Malta, Ireland, the Netherlands, and Portugal (Fig. 8.6). With regard to the gender balance, a small majority of air transport workers were men (56%). The highest share of men was recorded in Sweden (77%), followed by Denmark (74%), and Malta (73%). By contrast, Finland was the only EU Member State where women were predominant in the air transport sector (57% of air transport workers were women).

8.4.4 Needs and Tendencies

The commercial passenger aircraft fleet is on a significant growth trajectory, with Airbus forecasting its expansion from approximately 19,000 aircraft to 40,000 in the next 20 years [10]. This growth is set to have a profound impact on three critical aspects of the aerospace industry: the Maintenance, Repair, and Overhaul (MRO) market, the demand for new pilots, and the need for technicians for production and maintenance.

As the fleet grows, the MRO business is expected to mirror this trend, with Airbus projecting a doubling in size from USD $60 billion to over USD $120 billion annually by 2036. The Asia-Pacific region is set to play a substantial role in this growth, accounting for 36% of the MRO market's value, exceeding USD $660 billion over the next two decades [10].

The expanding aircraft fleet also demands an increase in the number of pilots and technicians. Airbus predicts that more than a million aviation professionals will need to be trained to meet industry demands over the next 20 years. Boeing's forecast aligns with this need, indicating a requirement for 637,000 new pilots, 648,000 mechanics, and 839,000 cabin crew members during this period [10]. Notably, the report shows that the most pronounced demand for pilots, technicians, and flight attendants is in the Asia-Pacific region, which is soon to surpass North America as the world's biggest commercial aviation market. Despite this demand, the Aviation Technician Education Council (ATEC) Pipeline report highlights a challenge related to staff shortages [11]. By 2031, two out of every five current mechanics will reach retirement age. The stream of new mechanics will replace some of those exiting mechanics; however, only 79,000 mechanics will enter the field in the next 10 years, 11,000 short of what is needed to replace those reaching retirement age.

Aerospace engineers will be another promising career, not only because of the estimated global demand for new passenger airplanes but also because of the rapid advances in aerospace technology. R&D projects, such as those related to improving the safety, efficiency, and environmental friendliness of aircraft, will help sustain demand for workers in this occupation. Aerospace engineers in all areas will continue to be needed as design and production focus on aircraft that are less noisy and more fuel-efficient. In addition, as international governments refocus their space exploration efforts, new companies are emerging to provide access to space beyond the access afforded by standard governmental space agencies. The growing use of unmanned aerial vehicles will create more opportunities for aerospace engineers as authorities find domestic uses for them, such as finding missing persons lost in large tracts of forest or measuring snowpack and other water resources. The increased geopolitical tensions are also expanding demanding the markets for military aviation. Employment opportunities should be favorable for those trained in software or with education and experience in stress and structural engineering.

Entering the aerospace industry often begins with internships, which serve as a valuable entry point. Graduate programs and direct entry positions are also viable avenues for starting a career. Companies in this sector typically provide extensive training and opportunities for professional development, enabling individuals to transition into various roles within the industry. Carrie Lambert, who served as a capability manager in engine noise at Rolls-Royce in 2016, emphasized that the aerospace industry actively seeks graduates from diverse disciplines, including aerospace/aeronautical, chemical, control, electrical, electronics, environmental, instruments, manufacturing, materials, mathematics, mechanical, physics, power systems, and software. Emerging areas of interest encompass unmanned aerial vehicles, supersonic aircraft, distributed propulsion, electric aircraft, and advanced engine technologies, such as geared fans and open rotor systems. Alongside a strong technical foundation, effective communication skills are deemed critically important in this industry. Aerospace engineers are required to think logically, employ both forensic and creative thinking as problem solvers, and possess excellent organizational skills to manage tasks efficiently in this dynamic and demanding field.

8.5 Motivating and Rewarding the Workforce

High-performance working practices (HPWP) are identified as integral components of strategic business development, with a key role played by human resources. HPWPs encompass various aspects, including recruitment, employee integration, involvement, internal communication, training, rewards, and commitment. Strategic human resource management theory suggests that these practices enhance employees' knowledge, skills, and abilities (KSAs), empowering them to utilize these assets for the organization's benefit while increasing their motivation to do so [12]. This leads to greater job satisfaction, reduced turnover rates, heightened productivity, and improved decision-making, all of which contribute to enhanced organizational performance. Consequently, HPWP positions the HR function as a strategic element in formulating and executing company strategies, emphasizing not only the attraction and selection but also the retention, motivation, and development of the workforce.

Several practices identified in the HIPAir Project that prove effective in motivating the workforce include:

- Reward and commitment practices that foster a stronger sense of belonging within the organization through financial and non-financial incentives, promoting long-term positive effects on employees.
- Employee involvement practices focus on increasing employee engagement and commitment, aiding in performance evaluation, improvement, and achieving excellence in specific areas.
- Training practices aim to enhance the skills and competencies of employees, aligning them with company goals.

Additionally, the motivation of individuals can be influenced by environmental conditions, including job enrichment factors such as autonomy, feedback on performance, skill variety, task identity, and task significance. Organizational justice is another crucial factor in enhancing motivation, as individuals often compare their treatment with that of relevant others, and equitable reward allocation plays a significant role in this context [13]. Nonetheless, it's important to recognize that individual differences in motivation and the interaction between individuals and organizations are complex and influenced by various factors, including cultural contexts and organizational conditions. This interplay between individual motivations and organizational dynamics contributes to the establishment of a psychological contract within the workplace, making it crucial for managers to consider the diverse motivations and needs of their employees.

The new challenge is to create innovative methods to reward employees to boost productivity and competitiveness in global markets. Global literature suggests a link between systematically and comprehensively implementing HPWP and performance indicators like productivity and profitability. This connection may stem from

Table 8.2 High-performance practices used in European companies

Recruitment and integration	Interviews, theoretical tests. Induction programs, internship programs. Search for talented people in collaboration universities and other training institutions. Preparation of job descriptions and selection procedures.
Employees involvement	Functional flexibility program (mobility inter- and intra-department). Surveys: organizational climate and satisfaction; employee engagement. Individual development plans; employee action plans; performance appraisal. Annual objectives for each employee, individual and/or global; KPI for teams. Continuous improvement system. Excellence achievement program.
Internal Communication	Communication packages for employees. Intranet. Webinars. Communication meetings (involving all the workers who may present their opinions and suggestions to improve company's procedures).
Training (learning and education)	Scholastic program; post-graduate studies; language courses. Training technical and non-technical (internal and external courses); mentoring, coaching, tutoring. Talent identification and development plan. Talent development program for women leadership in the aeronautics industry. Specific training plans based on competences/skills matrix to answer identified gaps and to support life-long learning processes.
Reward and commitment	Compensation plans (production, administrative, management, etc.). Flexible working conditions. Rooms for nursing mothers; retention policy; standardized job roles Compensation, bonuses related with objectives achievement. Salary increment related with objective review.

the impact of HPWP on employee attitudes and behaviors, including job satisfaction, organizational commitment, and employee empowerment. HPWP can improve working conditions, influence employee perceptions toward their employer, elevate satisfaction levels, and enhance employees' willingness to contribute more effectively to the workplace. While integrated sets of HPWP are not widely adopted at the European level, some isolated practices, such as leadership development programs, job rotation, and talent retention initiatives, are considered high-performance practices within the industry (Table 8.2). Companies that adopted high-performance practices are resumed in Table 8.3.

Table 8.3 Examples of companies that use High-Performance Practices and its characteristics

Company	FerroNATS	Fokker Elmo Turkey	Groundforce	Global Training and Aviation	FTB Lisi Aerospace	MTU Aero Engines Polska	Pratt and Whitney Rzeszów S.A.	Southwest Airlines
Employees	+ 130	302	2335	+150	555	580	3500	46000
Products	Air traffic control services	Electrical Wiring Interconnection Systems	Auxiliary activities to Air Transport	Aviation Training Solutions	Fasteners for airplanes	Aviation business	Aircraft engine manufacturers	Airline industry
Location	Spain	Turkey	Portugal	Spain & Indonesia	Turkey	Poland	Poland	USA
Established	2011	2007	2003	2002	2001	2008	1937	1967
High Performance Practices	Training Reward and commitment	Employees involvement	Recruitment and integration Training	Internal Communication Training	Employees involvement	Internal Communication	Employees involvement	Recruitment and integration

8.5.1 Employer Branding Strategy

The so-called "war for talent," driven by global competition, international mobility, technological changes, and the recognition of human capital as a valuable organizational asset, has made attracting and retaining talent a significant challenge in human resource management [14]. Talent management, involving the systematic attraction, identification, development, and retention of individuals with high potential, has become crucial for organizations [15].

Employer branding, which communicates an organization as a desirable and distinctive workplace, emerges as a powerful strategy in this talent war [16]. It encompasses long-term efforts to manage stakeholders' knowledge and perceptions and is influenced by factors like culture, values, recruitment, remuneration, training, and leadership [17].

A well-crafted employer brand positively impacts employee attraction, retention, talent acquisition, recruitment cost reduction, performance optimization, and increased employee satisfaction and commitment [18]. Thus, investing in employer branding provides a solution to the challenges of attracting and retaining talent. These challenges can be costly for organizations due to expenses related to recruitment, training, and development, often resulting in high turnover rates. It's crucial to adapt this strategy to the prevailing socio-economic conditions, considering the specific needs and factors that make an organization attractive to potential employees [19].

In the aerospace sector, attracting young talent is particularly challenging, making employer branding a central strategy for human resources departments worldwide. An illustrative example of successful employer branding in aerospace is TAP (Transportes Aéreos Portugueses; Portuguese Air Transport in English) Portugal. TAP has received recognition for its employer branding in recent years, aligning personal and organizational contexts to balance employee professional success with competitiveness and sustainability. The company's participatory management model, personal recognition, and emphasis on health and well-being contribute to sustainable development and employee performance. TAP's human resources practices align with the company's strategy, enhancing the credibility of the brand and emphasizing the importance of specialized and competent human capital. Practices include special health conditions, nutrition, physical exercise, training, and social benefits to promote employees' work-life balance and well-being, reflecting the significance of employer branding in attracting and retaining talent in the aerospace industry.

8.5.2 Organizational Culture

The study of culture in organizations is a complex subject with multiple interpretations and theoretical perspectives [20]. In a broad sense, culture can be defined as a set of values and practices established by an organization. It forms the basis for a

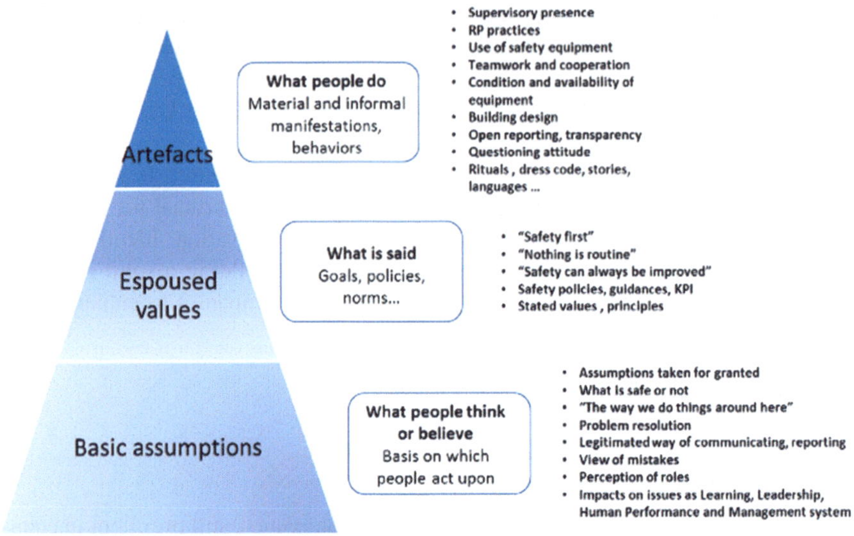

Fig. 8.7 Schein's layered model. (Bernard [23])

system of beliefs, norms, and expectations that shape the thoughts and behaviors of individuals [21]. Schein's model has been instrumental in providing a deeper understanding of organizational culture. According to this model, culture encompasses shared implicit assumptions within a group that influence how it perceives, thinks, and acts in different contexts [22]. These assumptions are developed over time as the group addresses external and internal challenges and are transmitted to new members as the accepted way to approach problems. Schein's model further categorizes culture into three levels, as summarized in Fig. 8.7 [23].

Organizational culture is a powerful force in attracting and motivating a workforce, particularly in industries like aerospace. A strong organizational culture can positively impact both employee and customer satisfaction, as well as profitability. The safety culture is a specific example of organizational culture in the aviation industry, playing a critical role in ensuring the safety of airlines. It consists of various individuals and subcultures within an organization and is reflected in consistent approaches to safety issues [24].

The 8 core components of a total safety culture include [25]:

- Management commitment to safety
- Performance management
- Organizational commitment
- Worker involvement
- Job satisfaction
- Training, equipment, and physical environment
- Co-worker support
- Personal accountability

Furthermore, organizational culture significantly affects innovation, a vital factor in the aeronautics industry [26]. As aeronautics companies operate in highly competitive global markets, fostering an innovative culture becomes paramount. As an example the recent series of quality issues at Boeing and Spirit Aero Systems are being addressed as mainly a question of organizational culture.

In conclusion, safety and innovation are essential in aeronautics. Beyond technical measures, maintaining a strong organizational culture is crucial for preventing accidents, sustaining competitiveness, and promoting innovation. Identifying key variables related to safety culture, innovativeness, and building a solid organizational culture should be a top priority for aeronautics companies to enhance their results and sustainability.

8.6 Retaining the Fidelity of Employees

The traditional model of lifelong employment in companies, still prevalent in countries like Japan, is sometimes criticized as lacking mobility compared to the US and Europe. However, the mobility of employment is more influenced by cultural traditions than by company efficiency. For example, Japanese companies with stable employment can be just as efficient, if not more so, than American companies with more volatile employment practices. Factories of Japanese companies in the US sometimes achieve higher productivities than the home-based competitors possibly due to closer management-worker contact.

Worker mobility is essential for individuals to find jobs that align better with their skills. Nevertheless, there is merit in companies cultivating employee loyalty by providing good working conditions, competitive wages, job stability, and retirement prospects.

8.6.1 Satisfaction of Employees

Employee satisfaction is a measure of how happy workers are with their job and working environment and can be seen as an emotional state that results from their evaluation of their job experience.

In the journey towards a company's consolidation, employees and collaborators play a pivotal role in working towards a common goal. It's common for some employees to become dissatisfied with their work when their conditions within the company are not optimal, so they end up leaving the company, which causes a constant rotation in the organizations. This is detrimental to the company, causing distrust, instability, and increased costs in recruiting new personnel. Therefore, retaining employees is essential for a company's success.

Often, employee dissatisfaction arises from wages that don't align with working hours, workload, or job quality. An employee's perception of being undervalued is highly subjective, and solely increasing wages may not always solve the issue.

However, offering fair and competitive wages that match the job's requirements and market standards is crucial.

In a work environment where employees typically spend around 8 hours a day, flexibility can greatly improve job satisfaction. Companies that offer flexible schedules and other supportive facilities tend to retain their employees better, as it demonstrates understanding and empathy towards their personal needs. Furthermore, employees highly value feeling appreciated by their superiors. Recognizing and valuing the experience and commitment of employees over the years strengthens their loyalty to the company. In fact, good relationships with colleagues, competitive salaries, favorable working conditions, opportunities for training and education, career advancement prospects, and other benefits are all factors that can contribute to increased employee satisfaction [27].

Thus, to improve employees' satisfaction, some changes can be made:

- Organizational development
- Policies of compensation and benefit
- Promotion and career development
- Job satisfaction, including job challenge, autonomy.
- Job security
- Working environment and condition
- Good relationship with supervisor
- Work group
- Leadership styles

Various studies have been showing that employee satisfaction is positively correlated with motivation, job involvement, organizational citizenship behavior, organizational commitment, life satisfaction, mental health, and job performance and negatively related to absenteeism, turnover, and perceived stress [28]. Moreover, satisfied employees tend to care about the quality of their work, leading to increased productivity and, thus, contributing more and better for the organization.

In conclusion, practices that prioritize employee well-being and focus on fostering satisfaction, happiness, and loyalty contribute to worker retention and workforce stability.

8.6.2 Stability of the Workforce

Employee fidelity can be defined as employees who are devoted to the success of their organization and believe that being an employee of this organization is in their best interest. Not only do they plan to remain with the organization, but they do not actively seek alternative employment opportunities.

Not only do they plan to remain with the organization, but they do not actively seek alternative employment opportunities.

Workforce-related challenges are among the top concerns for the aerospace industry, since a considerable portion of aerospace workers is nearing retirement eligibility, leading to uncertainty regarding workforce retirement and future

planning [29]. Despite these challenges, companies must maintain existing expertise while anticipating future hiring needs, including what skill sets will be needed and when to hire for them.

In the competition for talent, aerospace companies face unique industry-specific hurdles. Attracting and retaining talent, particularly individuals with STEM skills, that are so highly prized in many different industries, is a priority. Ensuring the transfer of knowledge and skills from older to younger employees is of increasing importance as older employees retire [29]. This demographic imbalance, coupled with high attrition rates and increased labor mobility, poses significant risks for the industry. Companies can mitigate these risks through employee retention and succession planning. As retirement rates rise, systematic succession planning that includes leaders, management, and key technical employees becomes increasingly vital.

The loyalty of the aerospace workforce is a critical factor in the quality of its output, with pay being the first determinant. The working conditions are the second determinant, and an "esprit de corps" (a feeling of pride and mutual loyalty) is the third. To further enhance employee fidelity, factors like perceived job security, a robust pension system, effective health insurance, and optional retirement age extensions could increase attractiveness and employee retention, especially among those with STEM skills. In conclusion, retaining the aerospace workforce should be a top priority. Employee intentions to leave and actual departures can trigger a domino effect on other human resource concerns, affecting service quality, productivity, and overall organizational success. Newly recruited staff may lack the experience level of retirees and need training to reach similar levels of competence and efficiency ('learning curve' effect).

8.7 Conclusions and Future Considerations

Overall, to promote youth educational pursuits in the STEM and aerospace sectors, it is essential to develop comprehensive policies that address several key areas:

- Enhancing learning experiences: this can be achieved through curriculum and pedagogical changes aimed at enabling students to master STEM subjects effectively. These changes should also incorporate social persuasion, which involves providing active verbal support and constructive feedback to influence students' self-efficacy and outcome expectations.
- Promoting contextual support: this includes fostering a supportive environment within families, school communities, among classmates, and with the help of role models. These supports have a particularly strong influence on engineering and STEM self-efficacy.
- Investing in vocational guidance: interventions should be based on the four sources of efficacy: mastery experiences, learning from others, social persuasion, and physiological reactions. These interventions can help students build confidence and motivation in pursuing STEM subjects and aerospace careers.

- Building awareness and employer branding: creating awareness of STEM disciplines and the aerospace sector among youth is crucial. Additionally, employer branding efforts by aerospace companies can make these fields more appealing to young talent.
- Youth persistence: once students are attracted to aerospace-related courses, it's essential to address the issue of youth persistence in their educational pursuits in this sector. Social support from parents, peers, and faculty members plays a critical role in this regard. More research is needed to understand and enhance youth persistence in the sector.
- European STEM policy: a European STEM policy should provide a coherent framework for identifying and articulating STEM-specific strategies and programs across all levels of education, including primary and secondary schools, vocational institutions, higher education, and research. This approach ensures a coordinated effort to promote STEM education and careers.
- Teacher professional development: teacher professional development is essential for the successful implementation of integrated STEM education. Implementable frameworks and curriculum resources should be made available to educators to facilitate effective STEM teaching.
- Promotion of aerospace courses and careers: specific attention should be given to the promotion of aerospace-related courses and jobs, which has strategic importance and requires targeted efforts to attract and retain talent.

In conclusion, to promote youth engagement in STEM disciplines and the aerospace sector in alignment with national and European agendas, a multifaceted approach is required. This approach encompasses curriculum improvements, social support networks, vocational guidance, awareness building, policy coordination, and a focus on engineering education. Continuous research and collaboration are essential to drive these initiatives forward effectively.

References

1. Lent RW, Brown SD. Social cognitive career theory and adult career development. In: Adult career development: Concepts, issues and practices, 3rd ed. Columbus, OH, US: National Career Development Association; 2002: 76–97.
2. European Commission. Directorate General for Mobility and Transport., European Commission. Directorate-General for Research and Innovation. Flightpath 2050 : Europe's vision for aviation : maintaining global leadership and serving society's needs. Publications Office; 2011.
3. Orr MK, Ramirez NM, Lord SM, Layton RA, Ohland MW. Student choice and persistence in aerospace engineering. Journal of Aerospace Information Systems. 2015; 12: 365–73.
4. Jackson AP, Smith SA, Hill CL. Academic Persistence among Native American College Students. Vol. 44, Journal of College Student Development. Johns Hopkins University Press; 2003.
5. Burtner J. The use of discriminant analysis to investigate the influence of non-cognitive factors on engineering school persistence. Journal of Engineering Education. 2005; 94: 335–8.
6. Lent RW, Ezeofor I, Morrison MA, Penn LT, Ireland GW. Applying the social cognitive model of career self-management to career exploration and decision-making. J Vocat Behav

[Internet]. 2016; 93: 47–57. Available at: https://www.sciencedirect.com/science/article/pii/S0001879115300245
7. Clement S. The self-efficacy expectations and occupational preferences of females and males. Journal of Occupational Psychology. 1987; 60: 257–65.
8. Fowler R, Lundberg A, Siao D, Smith C. Women and Minorities in Commercial Aviation: A Quantitative Analysis of Data from the United States Bureau of Labor Statistics. International Journal of Aviation, Aeronautics, and Aerospace. 2023; 10.
9. PwC's Global aerospace and defense industry performance and outlook 2023 [Internet]. Available at: www.pwc.com/us/aerospaceanddefense
10. Airbus Global Market Forecast. Growing Horizons 2017–2036. 2017.
11. Aviation Technician Education Council. 2022 Pipeline Report. 2022.
12. Combs J, Liu Y, Hall A, Ketchen D. How Much Do High-Performance Work Practices Matter? A Meta-Analysis Of Their Effects On Organizational Performance. Vol. 59, PERSONNEL PSYCHOLOGY. 2006.
13. Mubashar T, Musharraf S, Khan S, Butt TA. Impact of organizational justice on employee engagement: The mediating role of organizational trust. Cogent Psychol. 2022; 9.
14. Michaels E, Handfield-Jones H, Axelrod B, Eds. The War for Talent. Harvard Business Press; 2001.
15. Tansley C. What do we mean by the term "talent" in talent management? Industrial and Commercial Training. 2011; 43: 266–74.
16. Berthon P, Ewing M, Hah LL. Captivating company: Dimensions of attractiveness in employer branding. Int J Advert. 2005; 24: 151–72.
17. Sullivan J. The 8 Elements of a Successful Employment Brand. ER Daily. 2004.
18. Alves P, Santos V, Reis I, et al. Strategic talent management: The impact of employer branding on the affective commitment of employees. Sustainability (Switzerland). 2020; 12: 1–21.
19. Alnıaçık E, Alnıaçık Ü, Erat S, Akçin K. Attracting Talented Employees to the Company: Do We Need Different Employer Branding Strategies in Different Cultures? Procedia Soc Behav Sci. 2014; 150: 336–44.
20. Chambel M, Curral L. Psicologia organizacional. Da estrutura à cultura. 2008.
21. Balthazard PA, Cooke RA, Potter RE. Dysfunctional culture, dysfunctional organization: Capturing the behavioral norms that form organizational culture and drive performance. Journal of Managerial Psychology. 2006; 21: 709–32.
22. Hogan SJ, Coote L V. Organizational culture, innovation, and performance: A test of Schein's model. J Bus Res. 2014; 67: 1609–21.
23. Bernard B. Safety culture oversight: An intangible concept for tangible issues within nuclear installations. Safety. 2018; 4.
24. Hudson T. Safety Culture-Theory and Practice. 2001;
25. Saffle TR, Evans RB, Mitchell RG, Martin DB, Markham D, Jonker B. Idaho National Engineering and Environmental Laboratory Site Environmental Report for Calendar Year 1998. 2000.
26. Duygulu E, Ozeren E. The effects of leadership styles and organizational culture on firm's innovativeness. African journal of business management. 2009; 3.
27. Sageer A, Rafat S, Agawal P. Identification of variables affecting employee satisfaction and their impact on the organization. IOSR Journal of Business and Management. 2012; 5: 32–9.
28. Lee M, Kim B. Effect of Employee Experience on Organizational Commitment: Case of South Korea. Behavioral Sciences. 2023; 13: 521.
29. Feeko P, Fuller S. 2021 A&D Workforce Study: how to prepare now for the work of the future. EY. 2021.

Chapter 9
From Rockets to Careers: Women's Progress in Aerospace

Ana Costa, Clara Pinho, and Mariana Denis Malta

Contents

9.1 Generating Interest in Aerospace and Building Confidence

9.1.1 Influences That Can Shape Girls and Young Women in STEM

Women's underrepresentation in STEM-related fields can be attributed to several key factors, including the absence of encouragement from their social circles (friends, family, and teachers), the lack of awareness about STEM as a viable career option, and lower self-confidence among young women to pursue roles in STEM.

A. Costa (✉) · C. Pinho · M. Denis Malta
INOVA+, INNOVATION SERVICES S.A., Matosinhos, Portugal
e-mail: ana.costa@inova.business; clara.pinho@inova.business;
mariana.malta@inova.business

L. M. Braga da Costa Campos (ed.), *Accelerating Sustainable Aviation Initiatives*, Sustainable Aviation, https://doi.org/10.1007/978-3-031-73993-4_9

In 2017, Microsoft commissioned Europe-focused research involving 11500 schoolgirls and young women from 12 European countries, aiming to understand the reasons why girls and young women in Europe were not actively pursuing STEM subjects and careers [1]. This study revealed that most European girls develop an interest in STEM subjects between the ages of 11 and 12, which significantly declines during high school (15–16 years old), highlighting the importance of cultivating their enthusiasm for STEM during this four-year period. The Engineering UK Report 2018 [2] supports this conclusion, stating that interest in engineering drops off for girls as they get older, particularly after the age of 16. This means there is a critical four-year period to cultivate girls' passion for STEM subjects before they start to lose interest.

The Microsoft study also assessed the extent to which teachers were actively involving their students in STEM subjects and whether parents were encouraging their daughters to pursue STEM. The results vary widely from country to country, highlighting the substantial influence of the girls' cultural and social surroundings on their attitudes toward STEM, both positively and negatively. In certain countries, the primary barrier to STEM engagement was a lack of self-confidence, while in others, it revolved around peer approval or the absence of female role models. Importantly, the study demonstrated that across Europe, schoolgirls and young women expressed confidence in their own STEM capabilities, and between 46% and 68% of them rejected the notion that they couldn't excel in STEM subjects as well as boys. Nonetheless, all of them acknowledged the existence of differential treatment between men and women in STEM-related roles, and this perceived inequality was identified as a significant constraint to further pursuing STEM studies and careers among girls, with 59% of them admitting they would feel more confident pursuing a career in STEM if they knew that men and women had equal opportunities. Considering these results, it is vital for teachers and family members to actively encourage any girl with an early interest in STEM before it starts to disappear. The five biggest factors identified in the report that can sustain a girl's interest in STEM subjects and careers are presented in order of importance in Fig. 9.1.

Empowering Girls and Young Women: STEM as a career option
A study commissioned by the EC's Directorate-General for Mobility and Transport (DG MOVE) [3] assessed the appeal of the transport sector as a workplace for young people and identified ways to assist the recruitment of young people to the sector, particularly young women. This study involved a web-based survey across all 28 Member States (MS), targeting young individuals aged 16–25. Results revealed that both young women and men had limited knowledge about the sector, leading to complex and confused perceptions. While both groups found the idea of working in the transport sector appealing, their preferences for specific modes (air, road, rail, and maritime) varied. Air transport was viewed positively due to travel opportunities, good wages, and the social aspect of the job, while road and rail were viewed more negatively, with road being criticized for noise and dirt based on motorway experiences, and rail seen as unattractive due to strikes and poor industrial relations. The knowledge about maritime transport among participants was very limited, and therefore they had no opinion regarding its attractiveness.

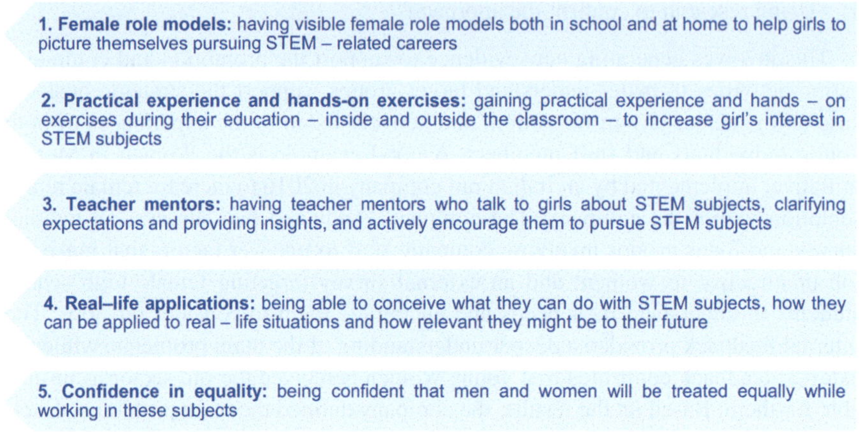

Fig. 9.1 Five main factors that can sustain girls' interest in STEM subjects and careers. (Source: Why Europe's girls aren't studying STEM. Report 2017)

Gender bias played a role in terms of numbers and job preferences, with young women preferring roles involving people rather than machinery/equipment. Only 15% of them responded positively to the prospect of driving a truck, train, bus, or tram, while air transport is seen by the majority as a prestigious and exciting sector. Intriguingly, interest in a pilot position was less gender-related, although under 20% of young females and over 30% of young men responded positively.

Initiatives to make the transport sector equally attractive to both genders received stronger support from young women; approximately 70% supported proposals to, for example, (1) set up promotional schemes; (2) present successful examples of women; (3) improve the working environment; and (4) make changes in education and training. Nonetheless, less than half that number of young men supported such actions, highlighting the need to ensure that men understand the gender imbalances in the workplace and engaging them in becoming part of the solution.

Currently, across the European Union (EU), several communication campaigns aiming at attracting youth to the sector have been attempted but have been ineffective mainly due to unsuccessful promotion of jobs being now crucial to support and reinforce communication activities in this area. In fact, the EC's DG MOVE study indicated that this lack of effectiveness may arise from the existing initiatives' inability to offer the requisite level of tailoring and face-to-face engagement required to attract young people to pursue careers in this sector.

9.1.2 Strategies to Attract Women to the Transportation Sector

DG MOVE's study identified 10 specific communication good practices and strategies from a shortlist of 25 communication initiatives implemented in recent years in the EU to effectively attract and retain women in the transport sector.

1. "Using research to confirm the approach"

This involves generating new evidence to support the promotion and communication activities to better understand target groups, support the strategic business case and generate buy-in for action, and actively listen to the experiences of both young individuals and staff members. A good example is the Women in Motion initiative, implemented by an Italian rail company in 2016 to increase female representation within the company. The company conducted two surveys: an internal survey and focus groups involving company staff to uncover factors that make the job unattractive to women; and an external survey targeting female high school students to understand their needs and encourage them to consider rail jobs. The internal feedback provided a deeper understanding of the main problems, while the external feedback confirmed that young women perceived the rail sector as unsuitable for them. Based on the results, the company defined the main problems, developed a strategy to challenge beliefs, and selected 80 successful women working in the rail sector as inspirational examples.

2. "Taking a strategic approach (a long-term plan with specific targets)"

Define SMART (specific, measurable, achievable, realistic, and timely) targets to steer communication efforts and generate the desired results.

3. "Going into schools, colleges, and universities"

Generate opportunities for face-to-face contact between students and companies to provide detailed information to young people regarding career choices in the transport sector, including developing a plan of events in several schools, working with career advisors, bringing apprentices to talk to schoolchildren, offering work experience to schoolchildren, and helping schools to develop employability skills training/curriculum.

4. "Providing opportunities to experience the job"

Enable young people to experience a job of their interest, with the aim of giving them a better understanding of the diverse range of roles available across the different transport sectors.

5. "Showcasing real people as role models"

Providing young people with concrete examples of successful employees working in the transport sector, with the goal of influencing them on the prospects of jobs in transport.

6. "Working with men to engage women"

Engaging male employees to better understand the focus of recruitment promotion and fostering a workplace culture that attracts women.

7. "Communicating with young people on their terms"
   Using the channels and tools that young people use, e.g., social media, and trying to replicate the world that they live in in order to attract them.

In the Women in Motion initiative, the company identified social media, namely Facebook, LinkedIn, and Instagram, as a crucial communication channel. However, just posting information on social media does not guarantee that it will reach the consciousness of young people. The following campaigns took approaches with impact:

- Invest in Shipping: a Swedish campaign used YouTube personalities as brand ambassadors to raise awareness about career opportunities in shipping, drawing inspiration from a successful technology sector campaign ('Teknikföretagen'). This campaign focused on encouraging girls to pursue vocational and technical training through videos produced by a YouTube star active in fashion, with the key metric for impact being the number of applicants generated rather than video views.
- 'Üstra rockt' ('Üstra rocks')': implemented in 2015 by the German bus operator Üstra with the goal of attracting women to potential jobs in public transport and raising awareness of gender issues among men currently working there. Six videos were disseminated via social media showing bus drivers dancing in skirts, and by putting an emphasis on humor, the videos went viral on Facebook and Twitter, making the campaign successful in attracting young people.

8. "Building in careers advice provision to promotional strategies"

Including a career advice aspect in the organization promotional drive with the goal of raising young people awareness of the professional opportunities offered by the sector. For example, in 2017, Transport for London (TfL) took advisors and apprentices to career events to explain what the work is all about, leading to an increase in the number of apprenticeships in this area.

9. "Using networks and mentoring to support female retention"

   Using networks and mentoring to support female employees in technical jobs, particularly in contexts where female representation is limited and they often feel isolated, experience sexist attitudes or intangible/invisible barriers when it comes to promotion or wage increases.

   The initiatives that took measures to support female retention are:

   - 'Au féminin': network created in 2012 by around 250 employees (mostly women and a few men) of the French national railway company SNCF. The initiative combines a support network for the career development of women who are already inside the company.
   - Women in Rail (WR): UK networking group that connects women and men and supports mentoring to improve diversity in the country's rail industry. WR provided, in June 2007, 300 mentorships. Mentees are women who are matched with a senior professional (male or female) outside the company who they would not otherwise be in contact with. Participating in the mentoring gave women a confidence boost and inspiration that is crucial to support the vast network of women working in the sector.

- Women in Logistics (WiL): UK networking group providing a platform for women to be part of the wider debate about logistics issues, focusing on networking, mentoring, and showcasing. WiL runs an annual awards event, which showcases very successful women, putting them in a position where they can be role models for other women in the sector, who can see what can be achieved.

10. "Using existing resources and networks to increase cost effectiveness."
    Making the most of what each organization has in-house to help reduce the costs of the activities undertaken within the initiative. Many of the teams behind the following initiatives were small, but they worked in a focused way to harness the insights, ideas, and efforts of their colleagues and partners:
    - Invest in Shipping: the whole sector joined forces to address the common challenge of recruitment. The close relationships formed between the industry, authorities, and schools allow the campaign organizers good insights into the challenges and opportunities in the sector.
    - 'Üstra rockt' ('Üstra rocks')': the initiative involved the company's own employees in developing the campaign solutions. Employees designed all the campaign's concepts and made the dissemination videos themselves.
    - Transport for London (TfL): the small team at TfL responsible for developing the early careers' offering couldn't be as successful without linking different TfL groups, including Women in Transport and Youth Ambassadors to specific schools.

Altogether, the EC's DG MOVE study concluded that promotional efforts that engage young people on a face-to-face basis work well, particularly when they (1) provide opportunities to explain what is involved in the range of jobs available; (2) help young people to understand the fit between their skills and transport jobs; and (3) give young people opportunities to meet others who are employed in these jobs. Regarding women's attraction, promoting an increase in the number of female recruits only addresses part of the problem, being fundamental to generating male employees buy-in, supporting female career progression, recognizing the contribution of female employees, and facilitating more flexible working for careers [4]. There are successful examples oriented not to the transport sector but to STEM in general. In the UK, Women Into Science and Engineering (WISE) introduces career opportunities for girls and women in STEM with the goal of achieving gender parity in the UK's scientific, technological, and engineering fields.

Importantly, role models from STEM industries can inspire the next generation of women by helping girls improve their self-confidence and picture themselves pursuing a career in these areas. Self-confidence significantly impacts engagement with science, particularly in STEM fields seen as demanding and rigorous. Gender stereotypes associate science with masculinity, causing women to doubt their abilities and demotivate them to aspire to STEM careers. A study of 1327 Swedish secondary school students reveals that social belongingness and self-efficacy are key factors influencing gender disparities in subject choice. Boys tend to avoid subjects with a higher female presence due to perceived lower societal value and rewards,

while girls often underestimate their capabilities in STEM despite, on average, outperforming boys in school, reflecting the enduring influence of gender stereotypes even in gender-neutral environments.

Gender Stereotypes in STEM

According to the Organisation for Economic Cooperation and Development (OECD) [5], in OECD countries,[1] less than 1/3 engineering graduates and 1/5 computer graduates are girls, most likely due to stereotypes and expectations rather than performance differences in mathematics and science. For example, at age 15, only 4.7% of girls (compared to 18% of boys) reported expecting to have a career in engineering or computing.

European Institute for Gender Equality (EIGE) defines gender stereotypes as "preconceived ideas whereby females and males are arbitrarily assigned characteristics and roles determined and limited by their gender." Gender stereotypes develop due to complex socio-cultural factors and can have a direct and indirect impact on gender segregation. They also impact the choice of study fields or occupations that women and men take by driving interest towards specific subjects that are deemed "appropriate." People's choices can, therefore, be interpreted as a sign of what the culture accepts and enforces as appropriate gender behavior in relation to a specific field of study or occupation, which highlights the influence of the cultural context in the creation of gender stereotypes.

UNESCO Cracking the Code report [6] stated that stereotyped ideas about gender roles develop very early in life. It is found that girls and boys often have different toy preferences by the end of the first year of their lives; they understand gender stereotypes and want to behave like others of the same sex by as early as age two; and they learn to adjust their behavior according to internalized gender stereotypes by age four. Gender stereotypes about STEM specifically are prevalent throughout the socialization process, during which girls learn and develop gender roles. Therefore, preconceptions or assumptions about the roles and activities of men and women should be countered as soon as possible.

9.1.3 Pre-conceptions Induced by Society

A study from Leiden University [7] indicated that when addressing gender stereotypes in science education, it is important to consider stereotypic education materials (e.g., presence of only a male version in science schoolbooks instead of both) and gender-biased teaching (e.g., assumption that "girls are diligent, neat, and calm while boys are loud, laidback, and naturally clever"). In fact, the EIGE Study and work in the EU Report defend that participation in STEM is traditionally associated

[1] Austria, Belgium, Czech Republic, Denmark, Estonia, Finland, France, Germany, Greece, Hungary, Ireland, Italy, Latvia, Lithuania, Luxembourg, Netherlands, Poland, Portugal, Slovakia, Slovenia, Spain, Sweden, United Kingdom

with various stereotypically masculine identity traits and roles and is enforced by the long-lasting historical and cultural idea that science is male-gendered. This stereotypical association stems from the association of men with objectivity and rationality, while irrationality and emotionality are allegedly women's traits.

There are two predominant stereotypes with relation to gender and STEM: "boys are better at math and science than girls" and "science and engineering careers are masculine domains."

This dominant association of science as masculine and the fact that scientists are usually described/drawn as males makes it particularly challenging for girls to see STEM as a potential career choice. Women are also found to be less likely to aspire to STEM careers due to expectations of feeling less good in contexts with unfavorable gender stereotypes. Overall, the EIGE report defends that gender expectations are stricter for boys than girls and that cross-gendered pathways are currently more acceptable for girls than for boys.

Within the Microsoft research, schoolgirls were asked if they agreed or disagreed that STEM subjects are for boys, being that all examples that were ever given to them about STEM were mostly about what boys liked. The results were similar across the participant countries, with more girls disagreeing, except for Germany, in which a slightly higher percentage of girls believe that all STEM-related examples in school are crafted towards boys' interests. Also, in Finland and Russia, there was a perception of STEM subjects as gender-neutral since there was a bigger difference between the percentage of girls that agreed and the ones that disagreed.

Stereotypes can also make it challenging for individuals to remain in their chosen career pathway. Students who differ from what is considered normal within their field experience more challenges in being academically and socially accepted as well as in developing a feeling of belonging to the discipline. Likewise, in the labor market, deviation from the "norm" is not tolerated, e.g., women working part-time is typically viewed as an enabling factor in terms of work and family balance. The equivalent choice among men, however, often encounters a strong resistance as it deviates even further from the "norm."

Mass media and social media are also considered to play an important role in the socialization process, influencing opinions, interests, and behaviors. Gender stereotypes portrayed in the media are internalized by children and adults and affect the way they see themselves and others. Media images of STEM professionals may be particularly salient for girls during adolescence as they actively consider future professional paths. Particularly in television and movies, the underrepresentation of female scientists leads to girls losing interest in STEM subjects as they get older. For example, some studies have found that when women are shown in television advertisements that allege sex-based abilities in math, they report being less interested in pursuing careers involving technical or quantitative skills.

Gender stereotypes on social media platforms can also have a harmful effect. For instance, a recent study of social media users found that gender stereotypes and negative messages about STEM were prevalent and often transmitted by girls and young women themselves. Female social media users were more likely than male users to post or support posts promoting negative views about STEM subjects,

especially mathematics. In this study, 75% of all self-mocking mathematics messages were posted by girls, and one-third of students' social media shares about women and girls in STEM were sexist.

As seen before, gender stereotyping and gender imbalance affect women's self-confidence in STEM. There is also a correlation between both factors, being that gender stereotypes can have an impact on gender segregation, and gender segregation by itself is considered, according to the EIGE Study and Work in the EU report, to further reinforce gender stereotypes.

Gender Imbalance in STEM Roles

EIGE defines gender parity as the relative equality in terms of numbers and proportions of women and men and is often calculated as the ratio of female-to-male values for a given indicator. It is an essential issue to understand whether and how economies and societies thrive and was a great influence on the growth, competitiveness, and future-readiness of economies and businesses worldwide.

For instance, EIGE commissioned a study on the "economic benefits of gender equality" [8] in the EU that found that closing the gender gap in STEM would contribute to an increase of 2.2–3.0% in EU GDP per capita and would increase total employment in the EU by 850 000 to 1 200 000 jobs by 2050. In monetary terms, closing the STEM gap leads to an improvement in GDP by EUR 610–820 billion in 2050.

An EIGE report on gender segregation [9] showed that the share of women in STEM occupations in EU-27 was less than 15% in 2014 and in the ten-year period (from 2004 to 2015) only increased 1%. Analyzing by country, Bulgaria (26% share of women), Lithuania (21%), and Portugal (21%) had the most balanced STEM workforce, while in Luxembourg (10%), the Netherlands (9%), and Austria (10%), the gender segregation was the highest.

In the UK, according to a 2016 [10], women made up just 14.4% of individuals working in STEM occupations, and in 2013, only 30% of women with STEM qualifications were working in STEM-related industries. This report concluded that more women typically work in jobs where the importance of "soft skills" is high but where technical or "hard skills" are not as important. Men, on the other hand, typically work in occupations where there is a more even blend of cognitive, social, and technical, or "hard skills."

Gender segregation also varies across STEM occupations within a country (Fig. 9.2), being that gender segregation among stationary plant and machine operators is the most varied, ranging from being predominantly men—dominated in Luxembourg to predominantly women—dominated in Bulgaria and Lithuania. On the other hand, men-dominated occupations across EU countries included building and related trades, electrical and electronic trades, and metal, machinery, and related trades. Regarding science and engineering professionals and associated ones, the highest share of women is observed for Latvia (44%), while Bulgaria has the highest share (33%) of women among Information and Communications Technology (ICT) professionals and Romania has among ICT technicians (36%). Nonetheless, the latter occupations are still men-dominated in all EU members.

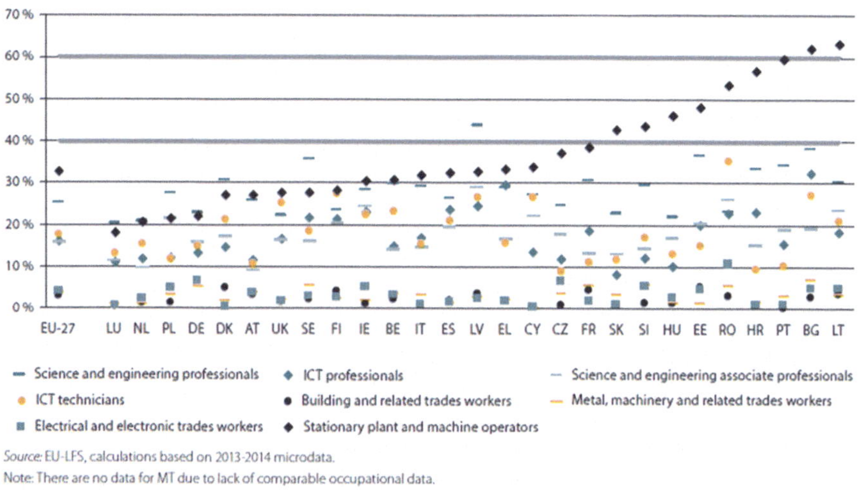

Fig. 9.2 Gender segregation across STEM (share of women) occupations (% 2013–2014). (Source: Study and work in the EU Report, 2018)

According to the EIGE Study and Work in the EU report, gender segregation narrows life choices, education, and employment options, leads to unequal pay, and limits access to certain jobs while also perpetuating unequal gender power relations in the public and private spheres. Additionally, this contributes to skill shortages and surpluses, thereby exerting substantial influence on various policy initiatives, such as efforts to stimulate economic growth and combat long-term unemployment.

Policymakers have long recognized the importance of eliminating occupational segregation and promoting women's participation in male-dominated fields. Achieving gender equality in STEM can be achieved through effective gender mainstreaming efforts.

9.1.4 Importance of Gender Mainstreaming for Gender Equality

According to the EIGE [11], gender mainstreaming is a strategy towards realizing gender equality and combating discrimination that involves the integration of a gender perspective into every stage of policy development, from planning and design to implementation, monitoring, and evaluation of policies, regulations, and spending programs. As policies focus on the general public, they often impact women and men differently. Therefore, this approach is essential to enhance the quality of policymaking and legislative efforts, ensuring they effectively address the needs of all citizens.

Top of Form

Gender mainstreaming consists of two dimensions: (1) integration of a gender perspective into the content of each policy, and (2) addressing the representation of women and men within a given policy area. Both dimensions should be considered throughout all four phases of the policymaking process: define, plan, act, and check.

According to the 2019 Report on equality between women and men in the EU [12], the European Commission (EC) remains dedicated to its efforts to promote gender equality and combat stereotypes through mainstreaming activities. One noteworthy initiative within its research and innovation funding program—Horizon 2020—was the end of the HYPATIA project in 2018. This project focused on building partnerships among schools, science museums and centers, and industries aiming to provide gender-inclusive STEM education to young people, especially girls.

The EU platform "Women in Transport—EU Platform for Change" [13] launched in November 2017 has continued with the addition of new members in 2018. These new members include representatives from prominent employers' and workers' organizations within the transport sector, MS' representatives, the EU Agency for Railways and the Shift2Rail joint undertaking. The platform serves as a forum for discussing and sharing best practices aimed at enhancing women's employment opportunities and promoting gender equality in the transport sector, with members being encouraged to take concrete actions to meet this objective. In the context of air transport, the EC is actively encouraging aviation stakeholders to bring concrete actions to the platform to improve employment and working conditions to attract and retain women in aircrew professions.

According to the EIGE 2019 Report on equality between men and women in the EU, EIGE has prepared a report on 'Gender equality and youth: opportunities and risks of digitalization,' which recommends that the EU institutions and MS should integrate a gender perspective into all digital initiatives focused on young people and recognizes that digital media offers a powerful tool for mobilization in advancing gender equality. As for Microsoft Europe's research focus, the report offers the recommendations for policymakers summarized in Fig. 9.3.

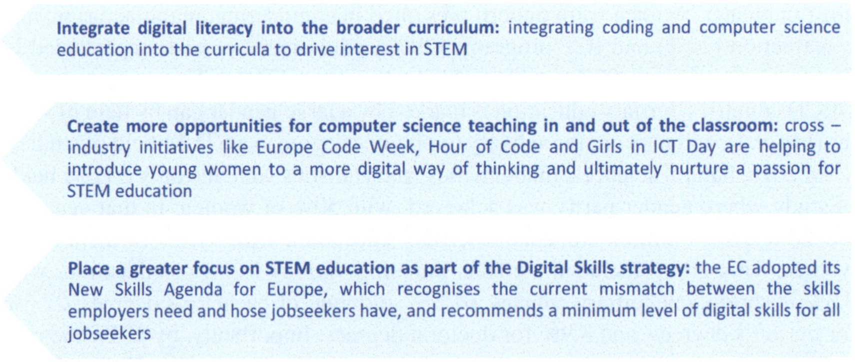

Fig. 9.3 Recommendations for policymakers. (Source: Why Europe's girls aren't studying STEM. Report 2017)

9.2 Changes in the Educational Context

Gender segregation within STEM fields is closely linked to, and often a result of, gender segregation within STEM education. Therefore, to comprehend the segregation present in the labor market, it is essential to gain a profound understanding of the factors influencing the educational disparities between girls and boys during their school and college years. Research indicates that there is little or no difference in boys' and girls' average ability at STEM subjects, which means that to attract more girls to STEM subjects at schools and universities, the solution is to tackle the stereotypes that they are exposed to from primary and secondary education until university graduation rather than addressing performance differences in STEM subjects.

Addressing gender stereotypes in education is essential to combating gender segregation, which creates inequalities in the labor market and restricts women's choices and opportunities. To tackle this issue, educational reforms should incorporate appropriate role models, consider peer influence, improve recruitment and retention strategies of students, and enhance classroom activities through hands-on and project-based learning.

Gender imbalance in STEM education

According to Eurostat Statistics, in the EU-28, there were 29.1 million pupils in primary education in 2016, of whom approximately 49% were girls and young women, nearly achieving gender parity. In secondary education, there were 20.5 million pupils in lower secondary and 22 million in upper secondary, with a balanced gender distribution across MS. In post-secondary non-tertiary education, which broadens the knowledge and skills acquired in upper secondary education, there were 1.6 million pupils in the EU-28 in the same year. On average, there were more female than male graduates, which varied between EU MS. In tertiary education, according to the 2015 OECD's "Education at a Glance 2017" report, in OECD countries, a higher percentage of women (57%) graduated from tertiary education compared to men, with an average of 57% of first-time graduates. Despite being most graduates, women were underrepresented in engineering, manufacturing, and construction (24%) and ICT programs (29%), while men are underrepresented in degrees in education (22%), and health and welfare (24%). This means that, in OECD countries, tertiary education is marked by a large gender gap by field of studies (Fig. 9.4), and men are more likely to obtain a degree in a STEM field of studies.

In 2016, natural sciences, mathematics, and statistics were the only STEM fields of study where gender parity was achieved, with 50% of women. In that year, the EU-28 had 19.6 million students in tertiary education, with 54.1% of them being women. This group included students pursuing various levels of education, with 7.3% in short-cycle tertiary courses, 61.3% studying for bachelor's degrees, 27.6% for master's degrees, and 3.9% for doctoral degrees. Importantly, by 2017, the proportion of women among all tertiary graduates had increased to 57.6%. When examining fields of study, nearly 32% of all tertiary students in the EU were enrolled in social sciences, journalism, information, business, administration, or law, with

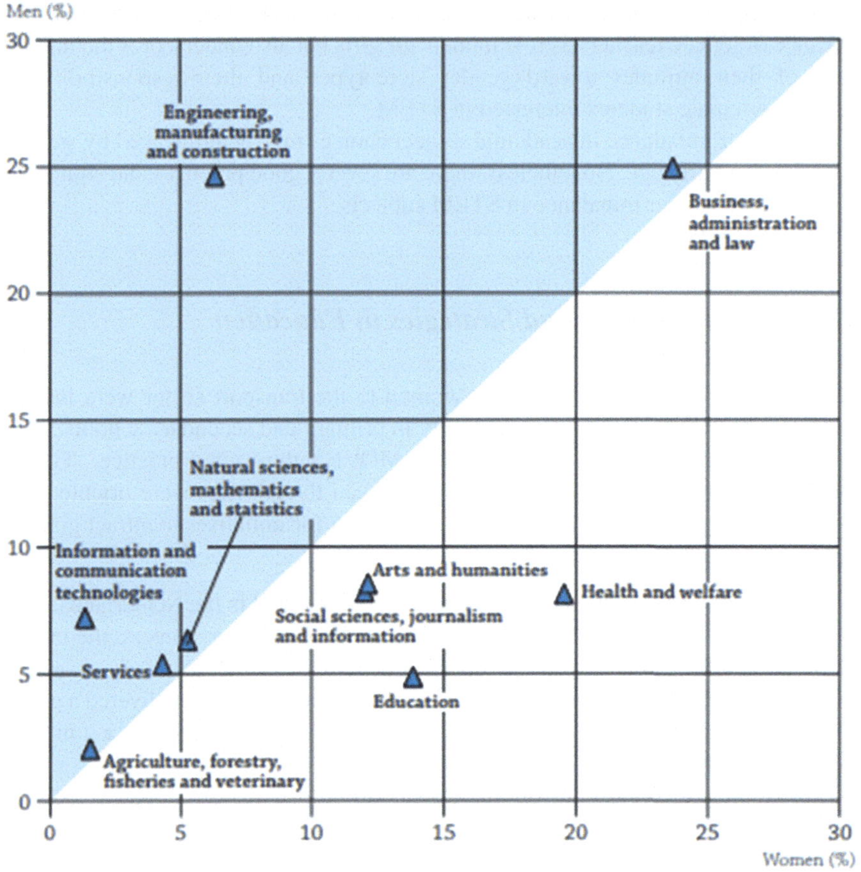

Fig. 9.4 Distribution of tertiary graduates, by gender and field of study (2016). (Source: OECD's "Education at a Glance 2018" report)

women being overrepresented (57.6%). The next most common field was engineering, manufacturing, and construction-related studies, with 25.9% women, slightly higher than the OECD average. In fields like education, 78% of students were women, and in arts and humanities, women accounted for nearly two-thirds (64.5%) of all students. In contrast, natural sciences, mathematics, statistics, and information and communication technologies had a higher proportion of male students, at 61.1%.

In the EU, STEM fields are much more prevalent among men, while social sciences, health, and humanities are more popular among women. Following the importance of educators in influencing girls and young women in studying and pursuing STEM fields, teachers' gender in the several levels of education was also analyzed. In 2017, primary and secondary education in the EU had a clear dominance of female over male teachers. Conversely, in tertiary education, including aerospace academic careers, men held most teaching positions, although the gender

scenario was more balanced than in other education levels. This indicates the significance of female teachers as role models for girls but also underscores the importance of their attitudes toward gender stereotypes and their responsibility in supporting female students interested in STEM.

The gender imbalance in academic subject choices may be influenced by stereotypes from an early age. Nonetheless, there are several good practices and strategies to address the gender imbalance in STEM subjects.

9.2.1 Good Practices and Strategies in Education

Several initiatives aimed at attracting women to the transport sector were implemented in educational settings, particularly in primary and secondary schools.

The following initiatives originated DG MOVE's third good practice, "Going into schools, colleges, and universities," and, even though they were oriented for children in general, they can serve as an inspiration for initiatives to attract girls to STEM subjects:

- 'Zeebenen in de Klas' ('Sea legs in class'): implemented in the Netherlands with the goal of inspiring children's interest in maritime transport careers, the initiative successfully conducted more than 300 school visits to 200 schools, reaching approximately 10,000 children. During each visit, a shipworker delivered a guest lesson that aimed to convey his/her experience to the children in a fun and inspiring way.
- Five employee volunteers from the Raben Group in Poland conducted visits to kindergartens and schools and, collaborating with the local police, talked about safety and the importance of transport. The idea was to "put transport into pupils and children DNA" to be able to draw on their knowledge and potential interest later in their lives.

The initiative 'Women in Motion' (DG MOVE's fifth good practice—"Showcasing real people as role models") featured 80 successful female employees from an Italian rail company. These women, all aged below 45, visited high schools to share their personal stories and experiences with female students, trying to overcome stereotypes about girls and promote the rail industry as a viable career choice. In these visits, girls had the opportunity to establish personal contact, ask questions, and get more insights into the various career paths available to them in the rail industry. Another example is Speakezee [14], a web platform that connects academics and non-academic audiences, such as universities and schools. As part of a campaign for International Women's Day, Speakezee, in partnership with the Institute of Physics and the Association of State Girls' Schools (ASGS), is sending young female STEM students to schools to talk to and inspire girls to consider pursuing STEM topics at A-level. These are some measures that have been adopted in Europe to reduce gender disparities and enhance the participation, advancement, and success of girls and women in STEM fields. Other examples include [15]:

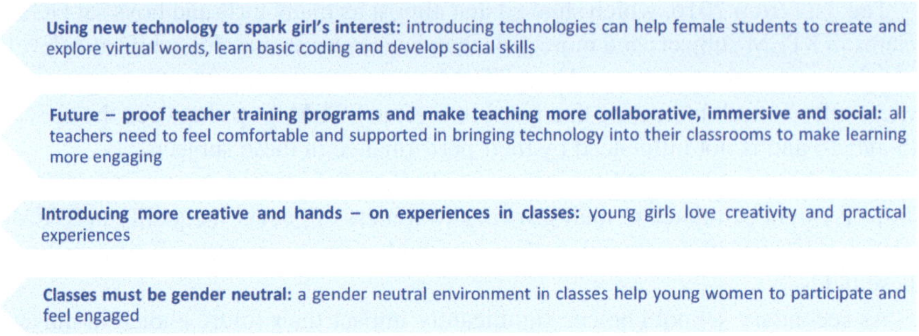

Fig. 9.5 Recommendations for educators. (Source: Why Europe's girls aren't studying STEM. Report 2017)

- Provide online access to stories and cartoons where girls drive cars and fly airplanes, as much as boys do, and let them play with vehicle models or ask for them as presents.
- Create a toolkit for primary and secondary school teachers to fight gender stereotypes and promote awareness of transportation careers among young people, including gender-neutral activities such as flight experiments.
- Identify and disseminate good practices for creating family-friendly work rosters to be made available to all stakeholders;
- Reinforce and accelerate visits to universities and industry, featuring successful role models and captivating technologies.

To further diminish gender imbalances in STEM-related subjects, educators should consider the recommendations outlined in Fig. 9.5 from Microsoft Europe-focused research.

9.2.2 *Performance and Choices of Girls Concerning STEM*

According to the Deloitte UK report [10], in 2016, nearly as many girls as boys took the General Certificate of Secondary Education (GCSE) in STEM subjects. However, three times more boys than girls took computing, and 50% more boys than girls took design and technology. Mathematics, additional science, and biology were the only STEM subjects, with slightly more girls than boys. Despite being underrepresented, the number of girls awarded A*-C grades was 20 percentage points higher than for boys. Girls generally performed better than boys in all subjects, including most STEM subjects, except for mathematics, where boys slightly outperformed girls. At the A-level, where students who pass their GCSE exam go to, gender differences in subject choices become more apparent, with boys being 40% more likely to pursue STEM subjects than girls in 2016, although girls tend to perform better in these subjects.

The data from 2016, which showed that almost as many girls and boys sat GCSE exams in STEM subjects and more girls than boys were successful, while at A-level, a higher proportion of boys chose STEM subjects, supports the findings of the Microsoft research indicating that girls' interest in STEM drops between the ages of 15 and 16 and is not influenced by their performance in these subjects.

As highlighted in the Microsoft research, this drop in girls' interest in STEM subjects might be related to their growing awareness of gender inequality in STEM-related roles and the perception that men and women are treated differently in these fields.

As secondary school choices significantly impact their future choice in university, and being those studying STEM subjects at the A-level are more likely to pursue STEM-related degrees, girls tend to select university programs that exclude science and mathematics.

Choices and performance of women in university

Historically, most women in engineering chose chemical engineering, while few girls opted for fields like civil or mechanical engineering. The idea that chemical engineering is like cooking lacks substance and is another form of biased judgement. Instead, women might have felt uncomfortable in certain engineering environments, as, e.g., a lady mechanical engineer might not be at ease in a factory shop full of rough workers. A better prospect for women inclined towards science would be to work in a chemical laboratory or medical profession less influenced by societal biases. Nonetheless, the situation has seen significant improvement with more engineering branches now prioritizing office and laboratory work, such as engineering design and informatics, and the societal attitudes have evolved towards greater gender equality and respectful treatment, although some barriers persist.

A real-life example illustrates this. A distinguished male university professor leading a campaign to increase women's interest in STEM and engineering programs encountered resistance, particularly from female secondary school teachers. This resistance might arise from their own negative experiences or perceptions in STEM fields, and perception of unequal treatment can perpetuate itself from one generation to the next, creating a "snowball effect." It is crucial to reverse both the actual disparities and the perception of gender inequality in STEM fields. This requires a commitment to fostering inclusivity, support, and showcasing successful outcomes to break the cycle of negative experiences.

9.2.3 Performance and Choices of Women Concerning Stem Subjects

According to the EIGE Study and work in the EU Report, in ten years (2004–2015), women's share among STEM graduates in the EU has fallen from 23% to 22%. Notably, gender segregation is more pronounced in vocational (technical) education than in tertiary education across most EU countries, with only 13% of graduates

from STEM vocational education being women, compared to 32% in STEM tertiary education. Additionally, a significant gender disparity exists in STEM occupations, where only one-third of female STEM graduates work in STEM fields, while half of male graduates do. Among vocational education graduates, the gap is even wider, with only 10% of women compared to 41% of men working in STEM occupations. Among those moving away from STEM, 21% of women at the tertiary education level work as teaching professionals, and 20% of women with vocational STEM education work in sales.

The EIGE Study and work in the EU Report also reveal that women graduating from male-dominated fields face significantly lower employment rates than men. In 2014, the employment rate for female STEM graduates at the tertiary level was 76%, over 10 percentage points lower than their male with the same qualification and 3 percentage points below the average for women with tertiary education. According to the Engineering UK Report 2018, in the academic year 2015/2016, women comprised only 16.0% of first-degree entrants in engineering and technology entrants, being the STEM field with the second lower proportion of female first-degree entrants; only computer science had a lower proportion. However, these female entrants performed well academically, with higher proportions than men continuing or qualifying at the first-degree level. This suggests that women must remain a priority target group in terms of potential recruits to the engineering workforce, making it important to prioritize efforts to increase their representation in the field.

Data from the WISE campaign [16] demonstrated a notable increase in the number of women in the UK graduating in a core STEM subject, with an overall increase of over 1000 women graduating in these fields between 2017 and 2018. In each specific core STEM subject, the number of female graduates increased, except in mathematical sciences. In 2018, women made up 26% of graduates in core STEM subjects. However, analyzing by core STEM subject, the percentage for the academic year 2017/2018 was as follows: 42% in physical sciences, 39% in computer science, 15% in engineering and technology, 15% in architecture, building, and planning, and 36% in mathematical sciences. Furthermore, the Deloitte UK report concludes that, since girls are significantly underrepresented in STEM subjects at school, at university, and consequently in working life, there is potential talent among girls not being fully exploited, which could help fill skills shortages existing in STEM-related occupations.

University environment influence on women persistence in STEM fields

The study "Girl Power" [17] indicates that women in engineering majors enter college with the same interest and intent to persist in the major as men but are less likely to complete STEM degrees and pursue related careers. Overall, research shows that positive experiences, including support from faculty and peers, research involvement, and participation in STEM organizations, significantly enhance women's probability of continuing in STEM studies and careers [18–25]. Additionally, many women in science and engineering report experiences with discrimination and bias that make it difficult to persist and succeed in their majors. Over time, women

in these disciplines often report feeling marginalized, isolated, and vulnerable to stereotype threats, contributing to a sense of growing discomfort and exclusion [24, 25]. Favouritism and differential treatment can delay women's persistence and academic success, as male and female faculty in science and engineering have been found to favor and be more responsive to male students in comparison to females [26, 27]. Furthermore, undergraduate women in science and engineering frequently report bias in the form of sexual harassment from both their classmates and faculty members [28]. The report also defends that differential treatment and bias faced by women also have negative underlying consequences for women's perceptions of their own abilities. Another study suggests that women tend to underestimate their performance on engineering tasks when compared to men [29].

This phenomenon is intensified by unwelcoming environments in science and computing, which contribute to decreased self-confidence, reduced self-efficacy, and self-limiting behaviors among women that negatively impact overall success and persistence in these fields [28, 30], with gender discrimination also implicated in lower academic performance [31].

9.3 Improvement of the Employment Context

In the male-dominated aviation industry, women's participation, especially in technical roles, remains low despite ongoing efforts. To oppose this, it is necessary to increase awareness of career opportunities in aviation, ensure equal treatment and abuse prevention to guarantee the protection of family, maternity, and parenthood, recognize female employees' achievements, and address gender disparities. Ultimately, having more women in aviation should be seen as a broadening of talent, not just a numerical increase.

According to the "*Aviation: Benefits Beyond Borders*" report [32] for the Air Transport Action Group (ATAG), the aviation sector in Europe directly employed an estimated 2.6 million people in 2016. This sector encompasses five segments, including airport operators, on-airport services, airlines, civil aerospace, and air navigation service providers, each offering a diverse range of employment opportunities (Fig. 9.6).

In Europe's aviation industry, women constitute 41% of the workforce, but this percentage does not provide insights into skill distribution or female representation in senior roles. For instance, while there is a high share of female cabin crew, only about 4–5% of the world's commercial airline pilots are women [33]. The same applies to technical positions, which require STEM skills, which are also likely to be skewed toward men. In airport operational roles and ground handling, women represent 25.8% of employees, with over 32% in the under-30 age group. According to a 2017 survey, among air navigation service providers, 25% of air traffic controllers are women. In airline executive positions, the gender gap is significant, with women comprising just 3% of the top 100 airline chief executive officers, 8% of chief financial officers, and 3% of chief operating officers. In contrast, women make up 32% of human resources directors.

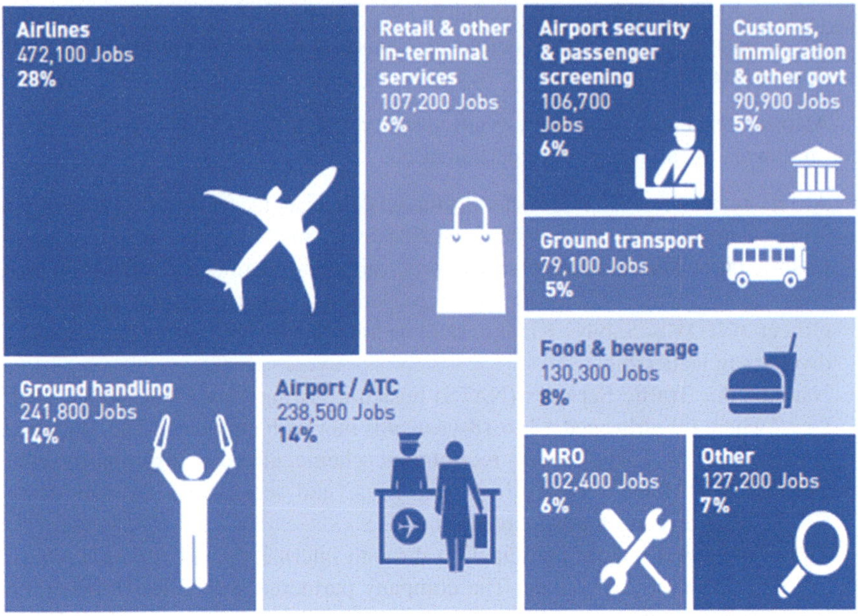

Fig. 9.6 Overview of the types of jobs at a typical European airport. (Source: ATAG Report 2018)

9.3.1 *Communication: Good Practices and Strategies in Employment*

DG MOVE's sixth good practice—*"Working with Men to Engage Women"*

In the *Üstra rockt' ('Üstra rocks')'* campaign, bus and tram male drivers wore skirts as part of their work uniforms, conveying the idea that "if men can wear skirts, women can be drivers," to challenge traditional gender stereotypes.

DG MOVE's fourth good practice—"Providing opportunities to experience the job"

Lufthansa's "#Nolimitsforwomen" initiative in Germany invited women to participate in a career day in typically male-dominated areas such as logistics, aircraft maintenance, and air traffic control. The winner also received a trip to Lufthansa headquarters in Frankfurt, making it a valuable opportunity for women to explore different career paths.

DG MOVE's second best practice—"Taking a strategic approach (a long-term plan with specific targets)"

- Air France—32.2% of the company's senior executives and 48% of managers are women, and it aims at increasing the share of women in management by 5 percentage points by 2030.
- 'Üstra rockt' ('Üstra rocks')—implemented in 2015 by the German bus operator Üstra with the goal of increasing the number of women employees from 16% to

22%, including in driving and technical jobs. The number of female applications for all positions increased threefold in 2 years, and their proportion went from 20% to 60%.

Also, several companies in the aviation sector are currently implementing different initiatives to enhance female participation:

- Airbus invites female high school students to visit its facilities and meet its female employees to discuss their experiences via the "Elles du Futur" and similar initiatives. The company also supports women's networks and mentoring programs to encourage balanced leadership. Notably, the current Chief Technology Officer (CTO) is Sabine Klauke, demonstrating their commitment to gender diversity in leadership.
- National Air Traffic Services (NATS) hosted a "Bring Your Daughter to Work Day," where 60 girls aged 13 to 18-year-olds met with engineers. This initiative is part of NATS' Early Careers recruitment scheme, aimed at inspiring the next generation of female controllers, engineers, and technicians by promoting STEM-related career paths in aviation.
- Thales promotes female participation through internships and apprenticeships, offering hands-on experience. The company partnered with Elles Bougent, an association that provides networking opportunities for young women, and over 150 Thales women acted as mentors in 2018, sharing their passion for engineering. Additionally, Thales also actively engages in events like the annual Réseaux et Carrières au Féminin forum, the Journée des Sciences de l'Ingénieur au Féminin event, and various initiatives to encourage careers in science and technology among young females.

In aerospace academic careers, women remain underrepresented, despite performing as well as or better than men. A real-life example of an assistant professor position received 56 applications, with only 3 from female candidates. Ultimately, the chosen candidate was initially reluctant to accept due to personal concerns, but with encouragement, she excelled in the role and is very likely to have tenure at the 5-year completion. This example underscores the importance of men mentoring women against self-doubt, as they often face unnecessary barriers to their success in male-dominated fields.

Guaranteeing equal treatment and preventing abuse
Ensuring gender equality in all aspects of job recruitment, from the announcements to interviews and benefits (including salary), is crucial. Equally important is addressing gender abuse, treating it as seriously as gross incompetence or financial misconduct. Clear policies and consequences should be implemented to discourage and prevent any form of gender abuse in the workplace.

- Job Recruitment and Retainment

While aviation employers aim to encourage young women to enter the industry, the representation of women in technical roles remains low. To address this issue, it's important to make career information and support activities more appealing and

effective for girls. Additionally, organizational support for women should exist at the leadership level, creating a positive work environment that includes mentoring and networking opportunities. Training for human resource personnel and formal practices can further contribute to an inclusive and equitable recruitment and retention process.

- Gender Pay Gap

Women are disproportionately more likely to go into jobs in industries or sectors where pay levels are lower, being the UK's average gender pay gap for full-time workers at 9.4% [34]. Interestingly, the gender pay gap for individuals with STEM qualifications who enter careers in those fields is smaller compared to other academic subjects. Thus, encouraging more women to pursue STEM careers not only diversifies their skills but also helps reduce the gender pay gap, particularly in the early stages of their professional journey.

The gender pay gap has multiple underlying causes, making it a complex issue to address. One contributing factor is the tendency for many women to take breaks from their careers for family-related reasons and, upon returning to work, often opting for part-time positions, which generally offer lower hourly wages compared to full-time roles. The age of workers also influences the gender pay gap, with a wider gap typically observed between older men and women. Additionally, the gender segregation present in STEM fields is considered a significant contributor to the gender pay gap, emphasizing the importance of addressing these factors to achieve greater pay equity.

The Deloitte UK report [10] highlights the need to address the gender pay gap by focusing on the differences in skills that women acquire and the fields they choose to study. Encouraging more women to pursue STEM subjects tailored to their cognitive preferences not only prepares them for a changing workforce but also narrows the pay gap. This effort requires collaboration between businesses, educators, and policymakers, with businesses taking a proactive role in supporting women's skills development and career opportunities. Recommendations for business comprise providing educators and policymakers with practical career insight, providing more support for women returning to work, and publishing detailed information on the gender pay gap.

- Gender Abuses

Supporting women in the workplace involves recognizing their diverse skill sets, providing opportunities for career advancement, and maintaining a zero-tolerance policy towards sexist behavior. Companies typically take serious action against two forms of inappropriate conduct: professional incompetence and misconduct that harms the organization. It is essential to add a third category, addressing gender abuse, with similar sanction to incompetence or dishonesty, because all three represent improper forms of conduct. Moreover, many institutions, including large corporations, do not apply individual rights equally. Disparities in rewards and consequences exist, where top executives may receive substantial rewards, even in failure, while lower-level employees may face dismissal. Addressing gender abuse

should involve a focus on those in positions of power, holding them accountable for their actions, and promoting ethical behavior. This approach aims to create a more equitable and inclusive workplace for all employees, irrespective of their position in the organizational hierarchy.

9.3.2 *Integration of Women in the Aerospace Sector*

When it comes to on-the-job treatment and preventing abuse in the workplace, there are three main factors to take into consideration: mentoring programs, job satisfaction determinants, and networking. Women mentoring programs are particularly important and can contribute to industry transparency, knowledge exchange, career advancement, and woman self-confidence. Evidence shows that the assistance of a mentor is important for women in all career stages [35–38], particularly in terms of career advancement [39–42]. Of particular significance, mentoring acts as a channel for the exchange of tacit knowledge and information that is often linked with promotion opportunities [40, 43], helping them to break through the 'glass ceiling' [41, 42]. Also, mentoring increases women's visibility within organizations [44] and contributes to raising aspirations and levels of self-confidence [45]. However, women often face challenges in finding mentors, especially in male-dominated industries [39, 40]. The Alta project, an Economic and Social Research Council (ESRC), is designed to address these challenges in the aviation and aerospace sectors. The project was based on a knowledge-exchange partnership between the University of West England, the Royal Aeronautical Society (RaeS), the Royal Air Force (RAF), and Airbus seeking to empower female professionals through a formal mentoring program. The project provides one-to-one mentoring from successful women and a network that's dedicated to ensuring that talented women reach their full career potential, for the benefit not only of individuals but also the industry.

Women's job satisfaction is also an important factor to have in consideration when integrating women in the aerospace sector. For instance, in 2015, independence at work, being part of a larger team, valorization of work by the supervisor, flex time, and job stability were pointed out as crucial to women's workforce in aviation [46].

Establishing networks of women to share experiences and promote opportunities is a key element in improving working conditions, particularly in environments where gender disparities and challenges persist. In the aviation sector, several organizations aim to help women through networking opportunities. Women in Aviation International is a non-profit organization that supports women in the aviation industry through educational resources, scholarships, outreach programs, an annual conference and job fair, "Girls in Aviation Day" for girls ages 8–17, and Girl Scout Aviation Patch. Women in Aerospace (WIA) is committed to boosting women's leadership roles and visibility in the aerospace field. Networking is a significant benefit of WIA, connecting individuals with diverse backgrounds and technical expertise through various events throughout the year. The Royal Aeronautical Society's Women in Aviation & Aerospace Committee (WAAC), founded in 2009,

aims to inspire young women to pursue careers in aviation and aerospace and offers support to women already working in these fields. The International Aviation Women Association (IAWA) is a global nonprofit founded in 1988 that promotes women's advancement in aviation and aerospace. IAWA hosts conferences, networking events, and informative meetings, connecting accomplished women in the industry and keeping members informed about industry developments.

Protection of the Family, Maternity, and Parenthood
The protection of the family and children is a fundamental value of society that the law tries to ensure in all situations, especially in employment. However, some dismissals based on irrelevant arguments are hidden violations of the prevailing law. Decisions about how parenthood allowances are shared should remain a family matter, with room for special circumstances. In some cases, a change of type of work or a different assignment can make more compatible company priorities and family needs by adjusting schedules or timetables. Some aviation sector companies are actively implementing initiatives to protect their employees' family, maternity, and parenthood rights. NATS fosters a supportive environment for women to advance their careers while accommodating work-time flexibility. Aeropuertos Españoles y Navegación Aérea (ENAIRE) prioritizes recruiting female talent and promotes gender equality through the Flight Plan 2020, featuring work-life balance programs and women leading sustainability efforts. The German Aerospace Center (DLR) is dedicated to achieving gender balance in technical and leadership roles, employing target quotas, flexible work arrangements, childcare assistance, and structured career re-entry for those who've taken family-related breaks. Additionally, Frequentis, as a family-owned company, emphasizes work-life balance, offering various parental leave options and childcare programs for its employees.

Recognizing achievements and compensating eventual gender differences
Recognizing professional achievements should always be based on objective and unbiased criteria, regardless of gender, age, or beliefs. However, fairness also means equal opportunities, and while applying the same final criteria, all should have the same opportunities to attain those objectives.

The recognition of job achievements should also include a reasonable allowance for special circumstances, such as adjusting working conditions or responsibilities, to create an equitable workplace, fostering task distribution based on skills rather than biases. It's essential to avoid both direct and reverse discrimination while promoting the utilization of the best talents and skills from all genders for increased creativity and efficiency. Achieving gender equality should focus on providing equal opportunities and fair treatment rather than favoritism, with the same principles applying to other potential forms of discrimination. Despite limited recognition, women have achieved remarkable success in aviation, suggesting even greater potential under more favorable conditions.

9.3.3 Achievements of Women in Aviation

Women were present in aviation activities since the first flights. However, until the 1970s their presence was, with rare exceptions, mainly as passengers or flight attendants onboard of male-piloted vehicles (Fig. 9.7). Prejudices formed in the general social life were exacerbated in aviation, and still today gender imbalances continue to persist in various aviation roles, underscoring the ongoing need for gender equality efforts in the industry.

Enlargement of a workforce with broader talent

Women and men can have different sensibilities, distinct approaches to the same problem, and complementary abilities that can be of benefit to the balanced and efficient performance of many tasks. Choosing which skills fit best each task is part of the efficient management of human resources in a company. Gender-based discrimination is not only unfair but also a loss of valuable talent.

Greater participation of women in aeronautics is not only an enlargement of the workforce in numbers, but it is also an enrichment in quality and talent, which are the foundations of inventiveness and competitiveness, on which depends the continuing European leadership in an ever more competitive world with new

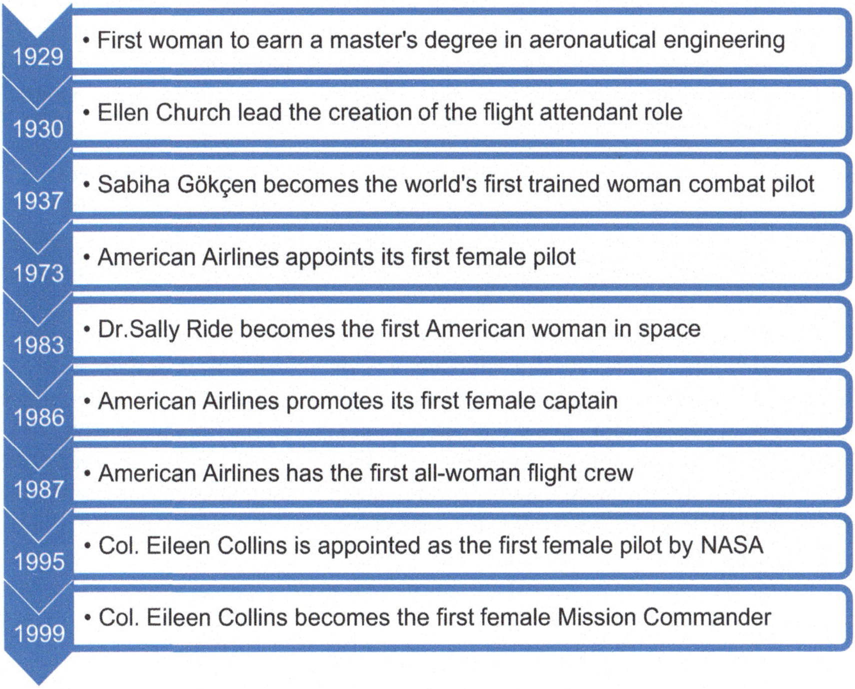

Fig. 9.7 Milestones for Women in Aviation. (Source: PARE Report 2020)

challengers. Europe's leadership in the competitive global aerospace market relies on a smaller but highly skilled workforce that excels in science, engineering, and technology. While emerging competitors like BRIC (Brazil, Russia, China, and India) nations are increasing their presence, Europe's strength lies in innovation. This makes the STEM-able professionals the key actors of the future prosperity of the aerospace sector in Europe, forming a community in which both genders can contribute equally.

9.3.4 Women in STEM: Role Models

María Moliner
Born in 1900, María Moliner was a Spanish lexicographer and librarian who played a pivotal role in promoting women's participation and recognition in the field of science and is a good example of prejudices against women outside the aviation sector. By compiling her groundbreaking dictionary, "Diccionario de uso del español," her work not only enriched the Spanish language but also served as a symbol of the potential for women to excel in traditionally male-dominated fields, ultimately paving the way for greater gender equality and inclusivity in science. Maria Moliner was the first female teacher at the University of Murcia, and her legacy serves as a reminder of the importance of empowering women in science and celebrating their contributions to the advancement of knowledge and society.

Amelia Earhart
Amelia Earhart, an iconic figure in aviation history, led a life defined by courage, determination, and a relentless pursuit of adventure. Born in 1897 in the United States of America, she developed a passion for flying in her early twenties and quickly rose to prominence as one of the world's pioneering aviators. She made history as the first woman to fly solo across the Atlantic Ocean in 1932 and continued to break records with her transcontinental and transoceanic flights. Her adventurous spirit and fearless determination inspired generations of women to challenge gender stereotypes and pursue careers in aviation and science. Her legacy endures as a symbol of empowerment and an example for women in aviation.

Valentina Tereshkova
Born in 1937 in Russia, Valentina Tereshkova made history as the first woman to go into space. She pioneered spaceflight, marking a significant milestone in space exploration, breaking gender barriers, and inspiring women worldwide to pursue careers in science and space. Her contributions continued post-flight as she became a prominent figure in Soviet space program administration and politics, advocating for space exploration and scientific research. Her life and achievements remain a proof of the power of determination and dedication in the pursuit of scientific and space exploration goals.

Dana Newman
Dana Newman, born in 1964, is a prominent figure whose career spans academia and space exploration. She has held key roles, including Deputy Administrator of NASA and Apollo Program Professor at MIT. Dana's research focuses on human performance in different gravity environments and includes four spaceflight missions. She authored "Interactive Aerospace Engineering and Design," published over 300 papers, and holds numerous compression technology patents, leaving an enduring impact on academia, space science, and human adaptation research.

9.4 Conclusions and Future Considerations

In summary, there's a noticeable underrepresentation of women in some areas of work, especially those where some knowledge of STEM subjects is required. More women typically work in jobs valuing soft skills and men in roles requiring a mix of cognitive, social, and technical abilities. This clear division in skills between genders needs to be addressed so that all students are provided with an equal foundation upon which they can build the career of their choice. To do so, it is important to actively promote women's participation in STEM fields to increase the diversity of ideas in the related workplaces, thus reducing the gender gap in these fields and encouraging teamwork among women and men. To achieve this goal, it is mandatory to have greater equality in the support provided to students, acknowledge that teaching approaches for males and females may need to be different, and increase the scope of career guidance from schools and parents. Only through these measures can women make well-informed decisions about their future careers. In summary, the suggested strategies and changes include:

- Remove existing barriers and impediments: identify impediments to progress and work toward eliminating them to achieve gender equality in the aviation field. This also includes examining potential mechanisms to improve work-life balance.
- Increase visibility and outreach to younger girls through the existence of role models: having role models who resemble them is crucial for young girls to consider pursuing a career in the field.
- Provide support for women while they are students, trainees, and employees: social support plays a crucial role in building confidence and fostering a sense of belonging, which provides women with the assurance they require to pursue STEM education and careers in the aerospace sector.
- Address retention in addition to recruitment: further analysis is necessary to understand the reasons behind the drop in women's participation after training in the aerospace sector (e.g., difficult schedule and lifestyle);
- Reinforce leadership and organizational support: ensuring the presence of mentoring systems and promoting diversity training that addresses attitudes, prac-

tices, and approaches when working with individuals who are different than yourself.
- Change cultural perceptions: if a culture has been traditionally male-oriented, achieving diversity can be challenging. Cultural attitudes need to evolve to open more paths for the career development of previously underrepresented occupations.

References

1. Trotman A. Why don't European girls like science or technology? [Internet]. 2017. Available at: https://news.microsoft.com/europe/features/dont-european-girls-like-science-technology/
2. Neave S, Wood G, Research TM, et al. Engineering UK 2018: The state of engineering. 2018.
3. Directorate-General for Mobility and Transport (DG MOVE). Study on a Pilot project: Making the EU transport sector attractive to future generations. 2017.
4. Giannelos I, Smit G, Lakamp R, et al. Women in European transport [Internet]. 2018. Available at: http://www.europa.eu
5. Organisation for Economic Co-operation and Development (OECD). Where are tomorrow's female scientists? [Internet]. 2017. Available at: https://www.oecd.org/gender/data/whereare-tomorrowsfemalescientists.htm
6. UNESCO. Cracking the code: girls' and women's education in science, technology, engineering and mathematics (STEM). 2017.
7. Kerkhoven AH, Russo P, Land-Zandstra AM, Saxena A, Rodenburg FJ. Gender stereotypes in science education resources: A visual content analysis. PLoS One. 2016; 11.
8. European Institute for Gender Equality. How gender equality in STEM education leads to economic growth. 2017.
9. European Institute for Gender Equality. Study and work in the EU: set apart by gender. 2018; Available at: http://eurogender.eige.europa.eu
10. Deloitte UK. Women in STEM Technology, career pathways and the gender pay gap. 2016.
11. European Institute for Gender Equality. What is gender mainstreaming? 2016; Available at: www.facebook.com/eige.europa.eu
12. Commission E, Consumers D-G for J and. 2019 report on equality between women and men in the EU. Publications Office; 2019.
13. https://transport.ec.europa.eu/transport-themes/social-issues-equality-and-attractiveness-transport-sector/equality/women-transport-eu-platform-change_en. Women in Transport – EU Platform for change. 2017.
14. Gjersoe N. Bridging the gender gap: why do so few girls study Stem subjects? [Internet]. 2018. Available at: https://www.theguardian.com/science/head-quarters/2018/mar/08/bridging-the-gender-gap-why-do-so-few-girls-study-stem-subjects
15. Opengart R, Ison D. A strategy for alleviating aviation shortages through the recruitment of women. Vol. 3, Int. J. Aviation Management. 2016.
16. Wise. Core STEM Graduates 2018 [Internet]. 2018. Available at: https://www.wisecampaign.org.uk/statistics/core-stem-graduates-2018/
17. Smith KN, Gayles JG. 'Girl Power': Gendered academic and workplace experiences of College Women in Engineering. Soc Sci. 2018; 7.
18. Kyoung Ro H, Fernandez F, Ramon EJ. Gender Equity in STEM in Higher Education; International Perspectives on Policy, Institutional Culture, and Individual Choice; First Edition. PlosOne. 2016;

19. Kezar AJ, Holcombe EM. Voices from the Field Viewpoints Shared Leadership in Higher Education: Important Lessons from Research and Practice. 2017.
20. Hughes RM. Keeping University Women in STEM Fields. Int J Gend Sci Technol [Internet]. 2010; 2. Available at: http://genderandset.open.ac.uk
21. Litzler E, Young J. Understanding the risk of attrition in undergraduate engineering: Results from the project to assess climate in engineering. Journal of Engineering Education. 2012; 101: 319–45.
22. Gayles JG, Ampaw F. To stay or leave: Factors that impact undergraduate women's persistence in science majors. NASPA Journal About Women in Higher Education. 2016; 9: 133–51.
23. Beyer S. Why are women underrepresented in Computer Science? Gender differences in stereotypes, self-efficacy, values, and interests and predictors of future CS course-taking and grades. Computer Science Education. 2014; 24: 153–92.
24. Reilly D, Neumann DL, Andrews G. Sex differences in mathematics and science achievement: A meta-analysis of national assessment of educational progress assessments. J Educ Psychol. 2015; 107: 645–62.
25. Marra RM, Rodgers KA, Shen D, Bogue B. Women engineering students and self-efficacy: A multi-year, multi-institution study of women engineering student self-efficacy. Journal of Engineering Education. 2009; 98: 27–38.
26. Moss-Racusin CA, Dovidio JF, Brescoll VL, Graham MJ, Handelsman J. Science faculty's subtle gender biases favor male students. Proc Natl Acad Sci U S A. 2012; 109: 16474–9.
27. Milkman KL, Akinola M, Chugh D. What Happens Before? A Field Experiment Exploring How Pay and Representation Differentially Shape Bias on the Pathway Into Organizations. Journal of Applied Psychology. 2015;
28. Morris LK, Daniel LG. Perceptions of a chilly climate: Differences in traditional and non-traditional majors for women. Res High Educ. 2008; 49: 256–73.
29. Woodcock A, Bairaktarova D. Gender-biased self-evaluations of first-year engineering students. J Women Minor Sci Eng. 2015; 21: 255–69.
30. Haines C, Stevens H. International women in science: a biographical dictionary to 1950. 2001.
31. Beyer S. Predictors of Female and Male Computer Science Students's Grades. J Women Minor Sci Eng. 2008; 377–409.
32. Air Transport Action Group (ATAG). Aviation Benefits 2017. 2017.
33. SESAR. Gender equality at full throttle. 2017.
34. Annual Survey of Hours and Earnings: 2015 Provisional Results. 2016.
35. Hill CA, Corbett C. Why So Few? Women in Science, Technology, Engineering, and Mathematics. 2010; Available at: https://www.researchgate.net/publication/234647120
36. Vinnicombe S, Singh V. Women-only management training: An essential part of women's leadership development. Journal of Change Management [Internet]. 2002; 3: 294–306. Available at: https://doi.org/10.1080/714023846
37. Singh V, Bains D, Vinnicombe S. Informal Mentoring as an Organisational Resource. Long Range Plann [Internet]. 2002; 35: 389–405. Available at: https://www.sciencedirect.com/science/article/pii/S002463010200064X
38. Ehrich LC. Mentoring and women managers: Another look at the field. Vol. 23, Gender in Management. 2008.
39. Durbin S, Tomlinson J. Female Part-Time Managers: Careers, Mentors and Role Models. Gend Work Organ [Internet]. 2014; 21: 308–20. Available at: https://doi.org/10.1111/gwao.12038
40. Durbin S, Tomlinson J. Female part-time managers: networks and career mobility. Work, Employment and Society [Internet]. 2010; 24: 621–40. Available at: https://doi.org/10.1177/0950017010380631
41. Linehan M, Walsh JS. Mentoring relationships and the female managerial career. Career Development International [Internet]. 1999; 4: 348–52. Available at: https://doi.org/10.1108/13620439910295718
42. Ragins BR, Cotton JL. Mentor Functions and Outcomes: A Comparison of Men and Women in Formal and Informal Mentoring Relationships. Vol. 84, Journal of Applied Psychology. 1999.

43. Swap WC, Leonard DA, Shields M, Abrams L. Using Mentoring and Storytelling to Transfer Knowledge in the Workplace. In: Managing Knowledge Assets, Creativity and Innovation [Internet]. WORLD SCIENTIFIC; 2001: 137–58. Available at: https://doi.org/10.1142/9789814295505_0006
44. Hersby M, Ryan M, Jetten J. Getting Together to Get Ahead: The Impact of Social Structure on Women's Networking. British Journal of Management. 2009; 20.
45. Flynn J, Heath K, Holt MD. Four Ways Women Stunt Their Careers Unintentionally. 2011.
46. Hedden CR, Sands C. Aviation Week 2015 Workforce Study: A Reality Check as Competition for Talent Increases. 2015.

Chapter 10
Key Findings and Actions as Challenges for Future Aviation

Luiz Manuel Braga da Costa Campos and Joana Rodrigues

Contents

L. M. Braga da Costa Campos (✉)
Instituto Superior Técnico da Universidade de Lisboa, Lisbon, Portugal
e-mail: luis.campos@tecnico.ulisboa.pt

J. Rodrigues
INOVA+, Porto, Portugal

L. M. Braga da Costa Campos (ed.), *Accelerating Sustainable Aviation Initiatives*, Sustainable Aviation, https://doi.org/10.1007/978-3-031-73993-4_10

10.1 Introduction

The present conclusion chapter is an overview of the current status of aeronautics, with minimal passing reference to the past only as necessary and instead main focus on needed actions and future challenges. It summarizes the conclusion [1–3] of the project PARE (Perspectives for Aeronautical Research in Europe) concerning the ACARE (Advisory Council for Aeronautical Research and Innovation in Europe) goals for aviation in the 2050-time frame [4–6], including: (i) current status and additional progress needed to reach the 24 ACARE goals; (ii) 32 supporting PARE objectives underlying the achievement of those goals. The outcome is summarized

into a set of 58 recommendations for aeronautical research in Europe [7], based on key findings of the PARE project detailed in its three yearly reports [1–3].

The broad scope of aeronautics is covered in two sets of subjects:

1. Five vertically integrated objectives formerly identified by ACARE:
   - Meeting societal and market needs
   - Maintaining and extending industrial leadership
   - Protecting the environment and energy supply
   - Ensuring safety and security
   - Prioritizing research, testing capabilities and education
2. Five horizontally transversing cross-cutting supporting activities were subsequently added in the PARE project:
   - Aircraft Markets
   - Emerging aviation technologies
   - International cooperation
   - Attracting young talent to aeronautics
   - Increasing the participation of women in aerospace

Each of these 5 vertical and 5 horizontal subjects is addressed in a Sects. 10.3, 10.4, 10.5, 10.6, 10.7, 10.8, 10.9, 10.10, 10.11, and 10.12, in two parts:

- An overview of key findings suggesting a set of key actions
- A set of specific recommendations on how to implement those actions

There are 68 recommendations, all with the same structure consisting of 6 points:

1. **Recommendation(s):** a brief statement of the actions to be taken
2. **Rationale**: current situation and future prospects motivating the recommendation
3. **Stakeholders**: institutions that could contribute to implementation: European Union (EU), Member States (MS), Aerospace Industry (AI), Research Centres (RC), Universities and Academia (UA), Certification Authorities (CA), Accident Assessment Organizations (AA), Air Navigation Service Providers (AN), Airlines (AR), Airports (AP), Professional Associations (PA)
4. **Relevance**: potential impact of the initiative
5. **Priority**: justification of the priority rating on a scale from 3 asterisks (top) to zero asterisks
6. **Justification**: reference to the section of the PARE Report containing more detailed information

Thus, the present chapter addresses the five vertical and five horizontal subjects, based for each on a set of key findings and actions leading to specific recommendations.

10.2 Meeting Societal and Market Needs

By 2050, passengers and freight experience must be improved in order to meet the increasing demand for travel and to handle more easily unforeseeable events. Travel services should be affordable, quick, reliable, efficient, seamless and sustainable, based on a resilient air transport system and capable of automatically reconfiguring the journey, including transfers to other transport modes if necessary.

To guarantee this improvement, ACARE established the Flightpath 2050 goals. The second chapter of PARE's report, entitled "Meeting Social and Market Needs", addresses Flightpath 2050's goals 1 to 5, which concern air traffic capacity, ground infrastructure, mobility, speed and punctuality, respectively.

10.2.1 Air Traffic Capacity

Currently, the European part of world airline traffic is around 10 million flights per year of all types of vehicles and is expected to rise to 25 million by 2050, including unmanned and autonomous vehicles. This forecasted growth of air transport puts increasing demand on air traffic capacity with undiminished safety; airspace capacity involves runway, airways terminal area and route capacity, and its evolution is closely related to Air Traffic Management (ATM). The limiting factor in the number of flights is the smallest of these three capacity limits, at present the runway capacity.

Key Findings

- In 2007, in Europe, there were about 45 main airports (large and medium hubs) and about 450 country and regional airports (commercial service airports). Three years later, the five major European airport hubs were at saturation—operating at full capacity.
- In 2016, the total numbers of passengers travelling by air in the EU could be established at 973 million, an increase of 5.9% compared to 2015.
- Over a European network of more than 2100 airports, 528 airports account for just 25% of arrivals but 98% of the departures. Also, the 25 largest airports in Europe generate 44% of all flights, and 90% of all traffic comes from the largest 250 airports.
- There is a geographical concentration of airports in the region London–Amsterdam–Munich–Milan, which creates dense air traffic with large numbers of climbing and descending aircraft: a significant challenge for terminal area and en route capacity.
- The cities closest to Europe's busiest airports have between 4 and 46 airfields within 100 kilometres (km) from the city centre. For 8 of the 10 cities close to Europe's biggest airports, a single airport handles 80% or more of all the departures within 100 km.
- By 2030, it is expected that no fewer than 19 airports will be operating at full capacity 8 hours a day, every day of the year, which means they will be highly

congested and 50% of all flights will be affected by delays upon departure or arrival, or both.

During the past years, there has been a growing gap between capacity and demand at a number of busy European Union (EU) hubs, leading to the forecast that Europe will not be in a position to meet a large part of the expected demand due to a shortage of airport capacity. It is estimated that in 2050, up to 36% of flight demand will not be accommodated at European airports.

Key actions: it is recommended that a broad and deep research effort be maintained concerning all aspects of ATM that can contribute to increasing airspace capacity, which is the purpose of the 1st Flightpath 2050 goal, with equal or greater safety. Additionally, there are proposed projects and measures that could improve air traffic capacity:

1. Eurocontrol measures to increase capacity
2. Innovative technologies and procedures researched in SESAR
3. Cooperation in ATM between Europe and other regions of the world

10.2.2 Ground Infrastructure and Multimodal Transport

Nowadays, new airports to serve major cities tend to be built farther, requiring transport to reduce access time to the airport, which affects passenger convenience. Vertiports and heliports can be sited much closer to city centres, providing an alternative with faster access than airports, if noise and community issues can be resolved. By 2050, the air transport ground infrastructure should comprise major hubs, secondary airports, vertiports and heliports, all seamlessly connected within a multimodal transport system, and should include interfaces with other modes of transport.

Key Findings

- In the USA, of the existent 5664 heliports in 2016, both for private and public use, most of them were essentially unused and have been declared, throughout the years, inactive and for emergency use only. In Europe, unconfirmed reports indicate less than 100 civilian-type heliports.
- The most frequent distance between European airport pairs is related to approximately 1000 km, while there are only a few potential links above 3000 km. At smaller airports, departures most often travel less than 300 km, and at large and very large airports, the 400 km distance bracket is the most common, even though they have the largest share of over 3500 km flights.
- Very high-speed train point-to-point connections between city centers (travelling up to 250 km/hour) where they exist can be more time-efficient than air transport over a distance up to about 600 km, although load factors are lower than in aviation (85% on average).

- Over 50 city-pairs will be connected by new or improved links between 2019 and 2035, such as high-speed trains that can offer comparable transport times for distances up to 800 km. Passengers opting for rail will reduce the demand for flights by a little over 0.5% (estimated 0.7%) in 2035.
- The advantage of air travel between being able to link any two airports remains relative to fixed rail or surface high-speed links.

A network of small, traditional or electric aircraft that take off and land vertically, called Vertical Take-off and Landing (VTOL), would enable rapid, reliable transportation between suburbs and cities and, ultimately, within cities. On a different perspective, shifting short-haul flights to high-speed trains would reduce, even slowly, the unaccommodated demand for flights by reducing the demand for flights.

Key actions: it is recommended that urban and land planning methodologies are developed to optimize, on a regional basis, the location of airports, vertiports and heliports to simultaneously provide convenient links to other transportation nodes and minimize environmental impact and disturbance of populations. Some of the proposed measures mentioned before could also improve the connectivity of (or the mobility between) airports, which is the aim of the 2nd Flightpath 2050 goal, as well as the new initiative, named ONE Order, proposed by the International Air Transport Association (IATA).

10.2.3 Choice of the Most Efficient Mobility Solutions

The progress in mobile communications and availability of information should ensure that, in 2050, citizens can make informed mobility choices among several available travel options and have affordable access to one another, taking into account economy, speed and level of service (that can be tailored to the individual customer). Additionally, continuous, secure and robust bandwidth communications should be provided for added value applications.

Key Findings

- In 2009, according to a survey conducted within the ModAir project, no airport website provided enough information for the customer to be able to plan the entire trip. Customers needed to visit several websites to do that, and still, websites didn't provide information about transit times between different travel modes or real-time information about delays in ground transportation.
- According to the ModAir project, a clearly important requirement for passengers is to easily access reliable, impartial and real-time information, both for pre-trip planning and to be kept informed of relevant developments during the journey.
- Passengers are mainly wanting better information related to intermodality, comprehensibility of the reservation systems (including better prices when booked air and rail are together), flexibility on their bookings, and a secure framework with clear operators 'liability conditions.

- Automation will enable passengers to be informed about the current status of their journey and alternative options, periodically or on demand, using smart phones or interactive panels/screens situated along the intermodal transport network.

In conclusion, nowadays there are many sources of travel information and services with different objectives and priorities, making the choice confusing and often non-comparable. Moreover, there is the necessity to develop and/or implement solutions to provide passengers with real-time information during their entire journey.

Key actions: to achieve the 3rd Flightpath 2050 goal, it is recommended that a one-shop centralized travel information site is promoted where the citizen can readily find the alternative options for connecting any two locations, including costs and timetables, with links to reliable booking.

10.2.4 Overall Ground Plus Air Travel Time

By 2050, the interfaces of the airport with other modes of transport must allow 90% of passengers within Europe to be able to complete their journey door-to-door (D2D) within 4 hours. This D2D time comprises the origin airport access time, the time inside the origin airport to go through airport services, the air travel time, the time inside the destination airport, and the time from the destination airport to the final destination place.

Key Findings

- In 2016, the average departure delay per flight ranged from 8 to 16 minutes, with an annual average all-cause departure delay per flight of 11.3 minutes. The aircraft turnaround-related processes have been the main factor that have induced delays in the air transport operation over the years.
- For almost 80% of the European cities, the nearest airport is situated at 20 km. Such a short distance reflects that the general accessibility of the European airports is high.
- Passengers spend, on average, using Madrid as an example, more than 30 minutes to get to the airport. Therefore, researching new ways to get to the airports (e.g., the use of VSTOLs) from the city centres would be the first key to actually reducing the travel duration.
- The second key to actually reducing travel duration would be the improvement of the current processes used at the airports or even the design of a new system regarding both passengers and luggage.
- Air travel times can vary significantly in Europe, from 1 hour in central Europe to 4 hours between extremities of the continent. Besides depending on flight distance, which can go up to 3000 km within the EU, travel times also depend on the aircraft engine type, since jet-powered aircraft are faster than one propeller-driven aircraft.

In conclusion, key enablers to reduce overall travel times are a reduction in airport access times, a higher predictability of times accessing the airport, and process times inside the terminal. However, as the aviation factor cannot be responsible for what happens outside the aircraft and airports and cannot influence travel time to or from the airport, the 4th Flightpath 2050 goal should only consider travel times and process times inside the terminal.

Key actions: it is recommended that the 4th Flightpath 2050 goal is revised to take into account the distance and duration of flight and cover the time period from arrival at the departure airport to exit from the destination airport. The project DATASET2050 can support this revision process since it addresses the EU passenger mobility in the context of the D2D objectives defined in the Flightpath 2050 vision.

10.2.5 ATM and Weather

In 2050, flights should arrive within 1 minute of the planned arrival time, regardless of weather conditions, thanks to a resilient transport system capable of automatically and dynamically reconfiguring the journey within the network to meet the needs of the traveller if disruption occurs. Likewise, special mission flights should be able to be completed in the majority of weather, atmospheric conditions and operational environments.

This punctuality of air transport in adverse conditions depends on the availability of meteorological data sufficiently in advance for efficient rerouting. Disruptive events and special flights that require reconfiguration of multiple flight paths can be made efficiently if supported by fast and reliable simulation tools.

Key Findings

- A flight is considered to be delayed when it is 15 minutes later than its scheduled time. In 2016, yearly airline arrival punctuality decreased, with 81% of flights arriving within this time, compared to 82% in 2015. Weather delays slightly increased, compared with 2015, to 0.57 minutes per flight.
- The basic issue is overall ATM capacity, not only at airports and in terminal areas but also en route, with spare capacity to cope with special missions, disruptions and weather hazards (weather conditions such as icing, strong wind, low visibility, snow, etc.).
- A high average weather-related airport arrival delays is usually the result of a notable capacity reduction in bad weather combined with a high level of demand.
- At US airports, the higher frequency of instrument meteorological conditions (IMC) combined with scheduling closer to visual meteorological conditions (VMC) are key elements to reduce winter delays. Weather-dependent delays are more relevant during the winter months.
- Both in the USA and Europe, weather is the predominant element affecting the airport throughput and as a consequence of ATM-related departure restrictions.

In Europe, weather-related constraints represent a smaller share of delays than in the USA, even though the weather in Europe is less favourable.
- Runway throughput rates depend on visibility conditions and are reduced significantly when Low Visibility Procedures (LVPs) have to be adopted since they require an increased spacing between aircraft.

In conclusion, ATM performance depends on a number of factors and is affected by meteorological conditions, such as visibility, wind and convective weather, and can vary significantly in different airports according to the airport equipment, runway configurations (wind conditions) and approved rules and procedures. Therefore, additional efforts are required to relate weather conditions to airport and ATM performance and to develop a more comprehensive assessment of weather impact.

Key actions: it is recommended that a more comprehensive weather data is made available to ATM and airlines to assist in achieving punctuality targets and that a rapid near-real-time simulation capability is developed for ATMs to accommodate special emerging flights and adjust to major disruptive events.

For more information about these topics, please access the full chapter here.

(a) ***ACARE GOAL 1:*** *An air traffic management system is in place that provides a range of services to handle at least 25 million flights a year of all types of vehicles, including unmanned and autonomous systems that are integrated into and interoperable with the overall air transport system with 24-hour efficient operation of airports. European airspace is used flexibly to facilitate reduced environmental impact from aircraft operations.* **Recommendation 1:** A broad and deep research effort must be maintained concerning all aspects of Air Traffic Management (ATM) that can contribute to increasing airspace capacity with equal or greater safety. **Rationale:** The growth of air transport puts increasing demands on air traffic capacity with undiminished safety. The foreseen operation of UAVs in manned airspace will increase the demand for capacity. As capacity limits are approached, there are more delays that cause inconvenience to passengers and increase emissions and fuel costs. In combination, airport noise and air traffic capacity could become the two main bottlenecks for the growth of aviation, together with local and global emissions. **Stakeholders:** EU, MS, AN, AP, AR, AI, RC, UA. **Relevance:** The air traffic capacity must increase with undiminished or improved safety to accommodate traffic growth and UAV's without incurring major delays. **Priority:** Air traffic capacity could potentially become an obstacle to the growth of aviation, and past experience shows that approaching capacity limits can cause major disruption in terms of flight delays and operating costs and emissions. **Justification:** PARE report Section 2.1 and Topics T2.1 and T2.2.

(b) ***ACARE GOAL 2:*** *A coherent ground infrastructure is developed, including: airports, vertiports and heliports with the relevant servicing and connecting facilities, as well as other modes.* **Recommendation 2:** Urban and land planning methodologies need to be developed to optimise on a regional basis the location of airports, vertiports and heliports to simultaneously provide convenient links to other transportation nodes and minimise environmental impact

and disturbance of populations. **Rationale:** The integration of airports into regional planning must include the interfaces with other modes of transport and also the compatibility with other land users. **Stakeholders:** EU, MS, AP, AR, AN. **Relevance:** The airport network should cover population needs at regional, national and European and intercontinental levels. **Priority:** There is the need to harmonize intermodal transport at regional level and fast air transport on a European and intercontinental scale. **Justification:** PARE report Section 2.2 and Topic T2.3.

(c) ***ACARE GOAL 3:*** *European citizens are able to make informed mobility choices and have affordable access to one another, taking into account: economy, speed, and level of service (which can be tailored to the individual customer). Continuous, secure and robust high-bandwidth communications are provided for added value customer applications.* **Recommendation 3:** Promote a one-shop centralised travel information site where the citizen can readily find the alternative options for connecting any two locations, including costs and time-tables, with links to reliable booking. **Rationale:** There are many sources of travel information and services with different objectives and priorities, making a choice confusing and often non-comparable. **Stakeholders:** EU, MS, AR, AP, CA, PA. **Relevance:** An independent, reliable and comprehensive source of comparable and reliable travel information would serve the interests of the citizen and fair competition to offer the highest quality and most efficient air travel, promoting the growth of air transportation. **Priority:** The information is available, although in scattered and inconsistent form, with plenty of scope for improvement. **Justification:** PARE report Section 2.3 and Topic T2.4.

(d) ***ACARE GOAL 4:*** *90% of travellers within Europe are able to complete their journey door-to-door within 4 hours. Passengers and freight are able to transfer seamlessly between transport modes to reach the final destination smoothly, predictably and on time.* **Recommendation 4:** Revise the goal to take into account (a) the distance and duration of flight and (b) cover the time period from arrival at the departure airport to exit from the destination airport. **Rationale:** The aviation sector (b) cannot be responsible for what happens outside the aircraft and airports and cannot influence travel time between home/work and the airport. The flight distances within the EU can go up to 3000 km, and some aircraft are slower (propeller-driven) than others (jet-powered), so travel times can differ. **Stakeholders:** AR, AP, MS, EU. **Relevance:** The realistic information that can be provided to a traveller consists of (i) flight time that airlines can provide and (ii) time for departure and arrival procedures that the airport can manage. Travel beyond the airport is out of control of the aviation sector. **Priority:** The information is available or predictable but not as reliable or accurate as the business traveller needs and can disrupt leisure travel connections and plans. **Justification:** PARE report Section 2.4 and Topics T2.5, T2.6 and T2.7.

(e) ***ACARE GOAL 5:*** *Flights arrive within 1 minute of the planned arrival time, regardless of weather conditions. The transport system is resilient against disruptive events and is capable of automatically and dynamically reconfiguring*

the journey within the network to meet the needs of the traveller if disruption occurs. Special mission flights can be completed in the majority of weather, atmospheric conditions and operational environments. **Recommendation 5.1:** More comprehensive weather data is made available to ATM and airlines to assist in achieving punctuality targets (see recommendations 13 and 15.1). **Recommendation 5.2:** A rapid near-real-time simulation capability is developed for ATM to (a) accommodate special emergency flights and (b) adjust to major disruptive events. **Rationale:** The punctuality of air transport in adverse weather conditions depends on the availability of meteorological data sufficiently in advance for efficient rerouting. Disruptive events and special flights that require reconfiguration of multiple flight paths can be made efficiently if supported by fast and reliable simulation tools. **Stakeholders:** EU, MS, AN, AC, AR, AI, RS, NA. **Relevance:** Being able to maintain a steady flow of air traffic is particularly relevant, and more difficult, during major disruptive weather events, like ash clouds or freezing conditions. **Priority:** This is an area of gradual progress where steady effort should be maintained, but nothing revolutionary is expected. **Justification:** PARE report Section 2.5 and Topic T2.8.

10.3 Maintaining and Extending Industrial Leadership

By 2050, the innovative, sustainable and highly competitive aviation industry must cement its place as the world leader and be recognized globally for its vehicles, engines, services and a large range of very cost-effective and energy-efficient products. This leading position should be secured through a seamless research and innovation system that assures continuity through blue-sky research, applied research, development, demonstration and innovation in products and services.

To assure that this overall goal is met, the ACARE established the specific Flightpath 2050 goals 6 to 8, which concern the maintenance and extension of a leading position in the aeronautical sector, the mastering of a wide range of technologies and the integration of these in an aircraft design and development program, respectively. The 3rd chapter of the PARE report, entitled "Maintaining and Extending Industrial Leadership", addresses these goals.

10.3.1 Retaining and Strengthening Market Share

Nowadays, the aeronautical industry covers a wide range of aerospace applications: large civil jet aircraft, regional aircraft, helicopters, military aircraft, missiles, satellites and launchers, engines and equipment. In 2050, it must be equally competitive, deliver the best products and services worldwide and satisfy environmental requirements.

Key Findings

- Airbus has a share of 50% of the world market for jet airliners with more than 100 seats, and the Airbus–Boeing 'duopoly' dominates the market for jet airlines with more than 120 seats, with a full range of narrow and wide body aircraft.
- ATR is the leading supplier in the regional aircraft market, which has multiple suppliers.
- Airbus Helicopters (formerly Eurocopter) and Augusta–Westland are among the market leaders in helicopters.
- Dual use and specific technologies ensure an equally strong position in the world market for military aircraft, missiles, space launchers and satellites.
- Safran and Rolls–Royce rival Pratt & Whitney and General Electric in aero-engines and in the equipment sector, Liebherr, Safran, GKN and others are major suppliers of European and non-European aircraft.

These impressive achievements across a full range of aeronautical products depend on: leading-edge technologies in all the sectors contributing to the design of aeronautical vehicles and the integration of all these cutting-edge technologies in efficient aircraft production, certification and service support programmes.

Key Actions: it is recommended that a broad-based application-oriented research and development activity is maintained, covering all sectors relevant to the global competitiveness of the European aircraft industry, which is part of the 6th Flightpath 2050 goal.

10.3.2 Cutting-Edge at the Full Range of Technologies

The modern world is facing several challenges, among which is the introduction of innovative technological solutions to the European aviation market. The competitiveness of the aerospace industry depends on mastering cutting-edge technologies over a wide range of 11 technological areas. Therefore, in order to remain competitive, it is necessary to retain leading-edge design, manufacturing and system integration capabilities and jobs supported by high-profile, strategic flagship projects and programmes that cover the whole innovation process from basic research to flight demonstrators by 2050.

Key Findings

- The necessity to introduce new technological solutions results from the needs of society, new technologies that have emerged, and new types of transport means and air transport systems.
- There are many challenges for the aviation industry, especially from the environmental side, such as the amount of greenhouse gases generated by the aviation industry, for example 2% of the total CO_2 generated in the world.
- There are currently two aviation programs being implemented in Europe that address Europe's challenges, namely Clean Sky 2 and SESAR 2020. Many other

smaller programs contribute to the development of innovative, cutting edge solutions.
- Patents in the mentioned technological areas are an indicator of innovation in aviation. Since 1969, the peak in the number of patents in aviation per year was reached in 2016, with approximately 3500 patents in aviation.

Although the aerospace industry is currently quite competitive, there are a number of emerging technologies that could be used by current and new competitors to change the balance, which need to be monitored and supported to ensure Europe remains a leader in the aeronautical sector.

Key Actions: to achieve the 7th Flightpath Goal, it is recommended that a stable independent observatory of citizen needs, global trends in aviation and technological advances that could meet them be supported to ensure that major breakthroughs that occur are matched without delay in reaching the market.

10.3.3 Efficient Development and Life-Cycle Management

The growing capability (related to certification) and complexity (frequently supported by new technologies) of modern aircraft increases the relevance of life-cycle analysis that needs to be considered also at the component level, such as batteries. Taking this into account, by 2050, streamlined systems engineering, design, manufacturing, certification and upgrade processes should have addressed complexity and significantly decreased development costs (including a 50% reduction in the cost of certification). Also, a leading new generation of standards should have been created.

Key Findings
- Rise in the cost of development (including certification) is correlated with the increased complexity of aircraft.
- Two measures of aeronautical development efficacy are: specific development cost (SDC) and specific development period (SDP), both per number of model's passenger seats.
- Significant reductions on the metrics mentioned before can be achieved by:
  - Intensive use of modelling and simulation instead of physical testing and experiment.
  - More specific, flexible and adaptive regulatory requirements (standards) for certification, including the involvement of airworthiness authority in virtual design and.
  - A fully integrated multi-physics and multiscale model of the complete aircraft should be coupled with aerodynamic and thermal models, eliminating ground test rigs completely.

To reduce CO_2 emissions considerably, which is a challenge for the aviation industry, the interest in hybrid and/or electric aircraft is increasing worldwide. This

would require an efficient power source for the electric engine and, in view of its high energy density, long life, and rate capability, the lithium-ion battery is the main current candidate for this purpose. However, batteries also introduce weight limitations and new safety concerns to aircraft that would have to be managed.

Key Actions: it is recommended that the architecture of industrial aviation programmes be analysed in order to identify best practices in matching design, development, certification, production, operations and maintenance in the most cost-effective and time-efficient manner. Also, the introduction of new technologies and stricter safety requirements should be accompanied by more efficient testing and validation to minimize time and cost.

For more information about these topics, please access the full chapter here.

(a) ***ACARE GOAL 6:*** *"The whole European aviation industry is strongly competitive, delivers the best products and services worldwide and has a share of more than 40% of the world market".* **Recommendation 6:** Maintain a broad-based application-oriented research and development activity covering all sectors relevant to the global competitiveness of the European aircraft industry. **Rationale:** The importance of the aeronautical industry to the prosperity of Europe is well documented. Since aeronautics is a synthesis of advanced technologies, it requires mastery of all of them to remain competitive. **Stakeholders:** EU, MS, AI, RC, UA, AN, AA, AP, PA. **Relevance:** taking as example the market for airliners with more than 100 seats, maintaining the Airbus share of 50% of the world market will require technological leadership in a broad range of technologies. **Priority:** This is the core of the aircraft market worldwide. The problems of the Airbus A380 with the passenger infotainment system and those of the Boeing 787 with the lithium-ion batteries show that even seemingly secondary aspects can cause major disruption. **Justification:** PARE report Section 2.5 and Topic T2.8.

(b) ***ACARE GOAL 7:*** *Retain leading edge design, manufacturing and system integration capabilities and jobs supported by high-profile, strategic, flagship projects and programmes that cover the whole innovation process from basic research to full-scale demonstrators.* **Recommendation 7:** Support an observatory of global trends in aviation to ensure that major breakthroughs that occur are matched without a delay in reaching the market. **Rationale:** The industry does by itself a good job of ensuring its own competitiveness, although shorter-term needs may prevail over longer-term prospects. The EC and MS programs aim at long-term competitiveness on the basis of professional recommendations but could still benefit from a stable independent observatory of citizen needs, market trends and technological advances that could meet them. **Stakeholders:** EU, MS, AI, AR, RC, UA. **Relevance:** Although the aerospace industry is currently quite competitive, there are a number of emerging technologies that could be used by current and new competitors to change the balance; these need to be monitored and supported to ensure leadership in the aeronautical sector. **Priority:** Current mechanisms to remain competitive though still relevant in the future may not be sufficient. **Justification:** PARE report Section 3.3 and Topics T3.4 and T3.5.

(c) ***ACARE GOAL 8:*** *Streamlined systems engineering, design, manufacturing, certification and upgrade processes have addressed complexity and significantly decreased development costs (including a 50% reduction in the cost of certification). A leading new generation of standards is created.* **Recommendation 8:** Analyse the architecture of industrial aviation programmes to identify the best practices in matching design, development, certification, production, operations and maintenance in the most cost-effective and time-efficient manner. The introduction of new technologies and stricter safety requirements should be accompanied by more efficient testing and validation to minimise time and cost. **Rationale:** Emerging technologies like 4.0, artificial intelligence and big data offer the prospect of more efficient integration of the aircraft lifecycle from design through production to operations but require large investments in complex systems that should preferably be phased in a try-and-prove approach. **Stakeholders:** EU, MS, AI, RC, UA, AR, CA. **Relevance:** The 4.0 factory is the next big trend in industry, and aeronautics is no exception, adding operations and maintenance to designs and production. **Priority:** Developments in this area are widespread in industry, and aerospace requires specific adaptations. **Justification:** PARE report Section 3.3 and Topics T3.4 and T3.5.

10.4 Protecting the Environment and the Energy Supply

In 2050, the effect of aviation on the atmosphere must be fully understood by the general public, and it must be shown that the aviation sector has made the utmost progress in mitigating environmental impacts and therefore air travel is environmentally sustainable. For this, a combination of measures, including technology development, operational procedures and market-based incentives, should be taken into account to mitigate environmental impacts at a rate outweighing the effects of increasing traffic levels.

To ensure that this overall goal is met, ACARE established the specific Flightpath 2050 goals 9 to 13, concerning the reduction of noise and emissions, emissions-free taxying, recycling enabled by design, alternative fuels and atmospheric research, respectively. The 4th chapter of the PARE report, entitled "Protecting the Environment and the Energy Supply", addresses this set of 5 goals.

10.4.1 Reduction of Noise and Emissions

The growth of air transport at a rate of 3–7% per year leads to flights increasing to double by 2030 and triple by 2050. In order to avoid increased noise exposure near airports and emissions in cruise, the corresponding reductions must be made per flight. Thus, by 2050, the technologies and procedures available must allow a 75%

reduction in Carbon Dioxide (CO_2) emissions per passenger kilometre (km), a 90% reduction in Nitrogen Oxides (NOx) emissions, and a 65% reduction in the perceived noise of flying aircraft. These reductions are relative to the capabilities of typical new aircraft in 2000.

Key Findings

- Overall noise reduction at airports requires consideration of two classes of noise sources: engine noise sources and aerodynamic noise sources. Noise is dominated by the engine at high thrust at takeoff and by aerodynamics at approach with the engine at idle.
- The major contributor to the reduction of engine noise has been the increase in the by-pass ratio of turbofan engines, which also decreases fuel consumption, leading both to lower emissions and more favourable economics.
- The overall noise exposure of near airport residents can be reduced by land planning and by operational measures, such as noise abatement procedures (NAP), which can be broken down into three broad categories: noise abatement flight procedures, spatial management and ground management.
- Aviation emissions of CO_2 and NOx are produced by aircraft, support vehicles and ground transportation predominantly. The emissions from these sources fall into two categories: emissions that cause deterioration in local air quality (LAQ) and emissions that cause climate change.
- In 2012, aviation represented 13% of all European Union (EU) transport CO_2 emissions and 3% of the total EU CO_2 emissions. European aviation specifically represented 22% of global aviation's CO_2 emissions. Similarly, aviation now comprises 14% of all EU transport NOx emissions and 7% of the total EU NOx emissions.
- As a result of technological improvements, the noise footprint (85 decibels dB (A) maximum sound pressure level contour) of new aircraft is at least 15% (up to 50%) smaller than that of the aircraft they replace. Further design improvements offer the potential to reduce perceived noise from aircraft by 65% by 2050.
- Among the projects related to environmentally friendly aircraft are mentioned: Aviation Noise Research Network and Coordination (X-Noise EV) and Forum on Aviation and Emissions (Forum AE), related to aviation noise research and emissions research, respectively.

Key Actions: it is recommended that:

1. A broad research effort is supported to reduce aircraft noise at the source through technology advances and operating procedures also taking into account psychoacoustic effects.
2. A modest effort is made towards a long-term, definitive solution: aircraft inaudible outside airport boundaries.
3. A set of trade-offs is formulated between different types of emissions (CO_2, NOx, particles, and water vapour) in local airports and global cruise flights.
4. Research on propulsion efficiency makes the best trade-offs with noise and emissions.

10.4.2 Emissions-Free Taxying at Airports

The taxying of aircraft on engine power and the use of auxiliary power units (APU) on the ground can be significant contributors to emissions at airports and also generate noise. By 2050, aircraft movements must be emissions-free when taxying, which can be achieved with electric towing vehicles and power supplies, especially batteries and fuel cells. Therefore, the feasibility and economics of emissions-free taxying critically depend on the available electric technology.

Key Findings

- The current preferred battery technologies for ground movements at the airport or on the airfield and in the aircraft itself are the lead-acid and the nickel-cadmium (Ni–Cd) batteries, increasingly replaced by Lithium-ion (Li-ion) batteries with solid-state batteries in the future. These batteries have development potential, and improvements in terms of energy density, cycle life, calendar life, etc. are expected.
- The use of hydrogen fuel cells as an alternative to batteries should be considered, in particular in the context of future hydrogen-powered aircraft.
- Although electric towing and power supplies are feasible, the investment in vehicles and support infrastructure must be assessed, as well as how costs are covered.

Key Actions: it is recommended that a methodology to comprehensively assess the implications of electric taxying and electrical energy supply is developed in terms of requirements, costs, land and environmental impact for a variety of airport configurations.

10.4.3 Design and Manufacture Bearing in Mind Recycling

Competition in the aircraft industry market and global warming have driven the industry to think along economic and environmental lines. For instance, in 2050, air vehicles must be designed and manufactured to be recyclable, preventing depletion of limited resources and making better and repeated use of materials already available. This goalis consistent with the emergence of a more electric aircraft (MEA) concept, providing the utilization of electric power for all non-propulsion systems.

Key Findings

- Recycling of aircraft parts depends mostly on the materials used and also on the fabrication process. The choice of materials for an aircraft is subject to a considerable set of constraints related to performance, weight, availability, cost, ease of manufacture and maintenance, durability and resistance to hostile environments. Adding the recycling ability is an additional constraint that can bring benefits in several other areas.
- Recent technological advances in the fields of power electronics, fault-tolerant architecture, electro-hydrostatic actuators, flight control systems, high-density

electric motors, power generation and conversion systems have ushered in the era of the MEA.

Key Actions: it is recommended that a comprehensive assessment of materials used in aircraft production is made and that recyclable alternatives and related issues of availability, ease of use, certification, maintenance and cost are assessed.

10.4.4 Sustainable Alternative Fuel Sources

There is a strong need to reduce the dependence on fossil fuels, which affects all modes of transport. Even though aviation is not the largest user, it should try to improve its position and contribution to the whole by finding sustainable alternatives: less polluting fuels and possibly also safer handling with undiminished energy density per unit weight and volume. Aviation specifically must guarantee that, in 2050, it is established as a centre of excellence for sustainable alternative fuels, including those for aviation, based on a strong energy policy.

Key Findings

- The European Commission (EC), Airbus, and high-level representatives of the aviation and synthetic producer's industries launched in 2011 the European Advanced Synthetic Flightpath. This action is scheduled to achieve two million tons of sustainable biofuels used in the EU civil aviation sector by the year 2020.
- By now, several alternative synthetic fuels are under scrutiny or are already approved: synthetic Fischer–Tropsch (FT); hydrogenated esters and fatty acids (HEFA); heavy pyrolysis oils (HPO); and alcohol to jet (ATJ). Most promising are e-fuels combining corbon capture from the armosphere with clean hydrogen produced by solar panels or wind turbines. Other sources of fuel include biomass and algae. Although there are several potential alternatives, it is not easy to match the energy density, usability and cost of kerosene/paraffin in large quantities.
- The production of biofuels should not interfere with food production on agricultural land, and other synthetic fuels are needed to satisfy the quantities needed.
- Current certification limits use of SAFs to 50%, and Airbus and Boeing have committed to making all aircraft 100% SAF compatible by 2030.
- Although airlines have been willing to test new fuels, a coordinated effort must be done far upstream to: consider a variety of sources of fuel that do not interfere with food production and whose environmental impact is neutral or positive (waste disposal); establish the technical feasibility to meet all applicable quality and safety standards and certification requirements; and assess the economic and environmental feasibility of large-scale sustained production, distribution and use.

Key Actions: it is recommended that a comparative study in performed on potential alternative fuels, their availability in the required large quantities and the feasibility and cost of large-scale production, distribution and use.

10.4.5 Atmospheric Research, Weather and the Environment

Aviation is one of the most climate/weather sensitive industries: it is affected by changes to visibility, storms, temperature, icing events, etc. Therefore, one of the most important activities is to assess the atmosphere and the environmental state in order to predict future climate issues and develop mitigation strategies, which would help to reduce possible disturbances in the airspace and allow an increase in airport capacity. Taking this into account, in 2050, aviation should be at the forefront of atmospheric research and monitoring and take the lead in the formulation of a prioritized environmental action plan and establishment of global environmental standards.

Key Findings

- The monitoring of the atmosphere is performed by a vast array of earth and satellite sensors, plus specialized weather aircraft used to fly through tropical storms and collect in-situ atmospheric data. It is possible to obtain much more comprehensive weather data both in time and locations aboard aircraft on regular flights.
- There are several methods to monitor the atmosphere, such as routine ground-based measurements (made by ground-based sensors and land-based buoys); systematic aircraft measurements (made by aircraft and balloons); and satellite measurements (made by space-borne sensors).
- Changes to temperature, precipitation and storm patterns are all expected in the near term, certainly by 2030. The impacts of sea level rise are more gradual and not expected until later in the century. However, more frequent and intense storm surges will have an earlier impact on European aviation, reducing capacity and increasing delays.
- Currently, there are several European initiatives and projects that have as their main objective monitoring the atmosphere through satellite and airborne instrumentation, such as:
  - Copernicus project, previously known as the Global Monitoring for Environment and Security (GMES)
  - Constellation Observing System for Meteorology, Ionosphere and Climate (COSMIC) project
  - Advanced Satellite Aviation Weather Products (ASAP) initiative
  - European Organisation for the Exploitation of Meteorological Satellites (EUMESAT)
  - In-service Aircraft for a Global Observing System (IAGOS) project

Key Actions: it is recommended that regular airliner flights are used to collect in-situ atmospheric data, which should further be processed to have near real-time knowledge of conditions along flight routes. This data could be supplemented by drones—unmanned aircraft systems (UAS)—specifically designed to fly in more remote regions of the atmosphere.

For more information about these topics, please access the full chapter here.

(a) ***ACARE GOAL 9:*** *"In 2050, the technologies and procedures available allow a 75% reduction in CO_2 emissions per passenger kilometre and a 90% reduction in NOx emissions. The perceived noise emission of flying aircraft is reduced by 65%. These are relative to the capabilities of typical new aircraft in 2000".* **Recommendation 9.1:** Support a broad research effort to reduce aircraft noise (a) at the source, (b) through operating procedures and (c) taking into account psychoacoustic effects. **Recommendation 9.2:** Besides struggling with short-term solutions to an increasingly pressing noise problem, a modest effort should be made towards a long-term, definitive solution: aircraft inaudible outside airport boundaries. **Recommendation 9.3:** Formulate a set of trade-offs between (a) different types of emissions (CO_2, NOx, SOx, particles, aerosols, water vapour and contrails) in (b) local airports and global cruise flights. **Recommendation 9.4:** Besides struggling with short-term emissions problems, put an effort towards better long-term definitive solutions, such as hydrogen-powered aircraft that are carbon free or electric propulsion that is emissions free. **Rationale:** The growth of air transport at a rate of 3 to 7% per year leads to flights increasing to double by 2030 and triple by 2050; to avoid increased noise exposure near airports and emissions in cruise, the corresponding reductions must be made per flight. Noise is dominated by the engine at high thrust at takeoff and by aerodynamics at approach with the engine at idle; thus, the full range of noise sources needs to be tackled, the operating procedures optimized and psychoacoustic effects accounted for to succeed in this major challenge. The requirements for low emissions of CO_2, NOx, SOx, particles, aerosols, water vapour and contrails near airports and in cruise are sometimes contradictory, and a reasonable compromise needs to be defined to guide engine design. The 'definitive' solutions to aircraft noise and emissions, such as aircraft inaudible outside airports and hydrogen propulsion that emits mainly water vapour, and electric propulsion that is emissions free, are challenging but deserve an effort to make them viable for a wider range of platforms. **Stakeholders:** EU, MS, AI, RC, UA, AN, AR, AP. **Relevance:** Tolerance to airport noise is reducing, and court or other actions to limit airport operations are likely to increase if overall noise exposure cannot be contained. Aviation should have a non-increasing and preferably decreasing role in global emissions aiming towards Net Zero 2050. **Priority:** It is very challenging to contain total noise exposure at airports, and failure to do so could limit airport operations and become a bottleneck for the growth of aviation. Emissions are a major local and global environmental concern, and aviation should be an example of positive action. Beyond the pressing short-term issues of noise and emissions, an effort should be made to assess and mature out-of-the-box long-term solutions. **Justification:** PARE report Section 4.1 and Topics T4.1 and T4.2.

(b) ***ACARE GOAL 10:*** *Aircraft movements are emission-free when taxiing.* **Recommendation 10:** Develop a methodology to comprehensively assess the implications of electric aircraft taxying and electrical energy supply in terms of requirements, costs, land and environmental impact for a variety of airport configurations. **Rationale:** Aircraft taxying under the power of their own engines

and, to a lesser extent, auxiliary power unit (APU) running contribute to airport emissions that could be avoided by electric towing and power supplies. **Stakeholders:** EU, MS, AP, AR, CA. **Relevance:** Although electric towing and power supplies are feasible, the investment in vehicles and support infrastructure must be assessed, as well as how costs are covered. **Priority:** This is mainly an implementation issue that will be supported by progress in electric power, especially batteries and fuel cells. **Justification:** PARE report Section 4.2 and Topic T4.3.

(c) ***ACARE GOAL 11:*** *Air vehicles are designed and manufactured to be recyclable.* **Recommendation 11:** Make a comprehensive assessment of materials used in aircraft production and assess recyclable alternatives and related issues of availability, ease of use, certification, maintenance and cost. **Rationale:** The trend towards recycling is industry-wide, and in the case of aeronautics as others can be integrated as one of the objectives of goal 8. **Stakeholders:** EU, MS, AI, RC, AU, AR, CA. **Relevance:** Aviation must share the goal of preventing depletion of limited resources and making better and repeated use of materials already available. **Priority:** This is a wide effort among most industries, which aviation must share, taking into account its specific needs. **Justification:** PARE report Section 4.3 and Topic T4.4.

(d) ***ACARE GOAL 12:*** *There is a need to establish a centre of excellence on sustainable alternative fuels, including those for aviation, based on a strong energy policy.* **Recommendation 12:** Perform a comparative study of potential alternative fuels, their availability in the required large quantities and the feasibility and cost of large-scale production, distribution and use. **Rationale:** There is a strong need to reduce the dependence on fossil fuels and although there are several potential alternatives, it is not easy to match the energy density, usability and cost of kerosene in large quantities. **Stakeholders:** EU, MS, energy industry; AI, RC, UA, AR, CA. **Relevance:** Avoiding fast depletion of finite oil resources, finding alternative less polluting fuels and possibly also safer handling with undiminished energy density per unit weight and volume: demanding and necessary. **Priority:** This is an issue affecting all modes of transport, among which aviation is a major user, and should try to improve its position and contribution to the whole. **Justification:** PARE report Section 4.4 and Topic T4.5.

(e) ***ACARE GOAL 13:*** *Atmospheric monitoring is at the forefront of atmospheric research and takes the lead in the formulation of a prioritised environmental action plan and establishment of global environmental standards.* **Recommendation 13:** Use the regular airliner flights can collect in-situ atmospheric data and processing this large amount of information can provide near-real-time knowledge of conditions along flight routes. This data could be supplemented by drones specially designed to fly in more remote regions of the atmosphere. **Rationale:** It is possible to obtain much more comprehensive weather data both in time and location by using sensors aboard aircraft in regular flights. The collection, organization and processing of this data with quite a homogeneous structure should be feasible by big data methods. **Stakeholders:**

EU, MS, AN, AR, CA, AI, RC, UA. **Relevance:** Detailed weather data with higher spatial and temporal resolution is the key to achieve several objectives of timeliness and safety of aviation, like ACARE goals 5 and 15. **Priority:** This major improvement in geographical and temporal coverage of weather data can be of use to other sectors as well as aviation. **Justification:** PARE report Section 4.5 and Topic T4.6.

10.5 Ensuring Safety and Security

In 2050, aviation must have achieved unprecedented levels of safety and security and continue to improve. The accident rate in commercial flights must be less than one per million flights, and all types of aircraft and rotorcraft must safely operate in the same airspace and in most weather conditions. On the other hand, security processes for passengers and cargo must allow seamless and non-intrusive security, air vehicles must be resilient to internal and external threats and air transport data networks must be hardened against and resilient to cyberattacks.

To ensure that all of these goals are met, the ACARE established the specific Flightpath 2050 goals 14 to 19, which are addressed in the 5th chapter of the PARE report, entitled "Ensuring Safety and Security".

10.5.1 Ultra-Low Accident Rate in Commercial Flight

The aviation accident rate has been declining throughout the years; nevertheless, the rate of decline has slowed markedly since 2004 and, at the same time, there is a continued growth in the number of flights, which are set almost to triple by 2050. In 2015, the accident rate of the European Union Aviation Safety Agency (EASA) Member States (MS) operators of Commercial Air Transport (CAT) aeroplanes was approximately three accidents per million flights. By 2050, the European air transport system must have reduced this rate to less than one accident per million commercial aircraft flights.

Key Findings

- From 2010 to 2015, on EU's territory and with EU-registered aircraft, there were 188 fatalities (155 fatalities in 2015 from the 3 accidents) in CAT aeroplanes, and, in 2016, there was only one fatal accident involving any EASA MS operator of CAT aeroplanes.
- The Key Risk Areas identified by EASA MS operators for accidents and serious incidents are: (a) system/technical failure, (b) airborne collision/conflict-collision of two aircraft in the air, (c) movement area collision or ground collision/handling, (d) fire, (e) runway excursion or abnormal runway contact, (f) runway

collision, (g) aircraft upset-full range of loss of control situations, (h) terrain collision/conflict-aircraft collision with terrain, and (i) obstacle collision.
- The origins of the causal and contributory factors behind the accidents and serious incidents involving EASA MS operators between 2007 and 2016 are: flight operators (56.64%), technical (21.30%), human factors (8.53%), Air Traffic Management (ATM) (5.87%), aerodrome operations (4.49%), organisational (2.40%) and maintenance (0.77%).

Key Actions: Aviation as one of the safest modes of transport can only benefit from being made even safer, which requires investigating accident classes, finding preventive and corrective actions and proving that they can be implemented. It is recommended that accident causes are considered by order of statistical occurrence and that appropriate safeguards for each class are identified and implemented.

10.5.2 Weather-Hazards and Risk Mitigation

The aviation system is highly sensitive to disturbing weather effects that can become hazardous for the aircraft operation by producing setbacks such as delays and accidents/incidents. Moreover, due to the expected growth in aviation, an increasing number of airports will operate near their capacity limit and hence will be more sensitive to disturbances by weather phenomena. To ensure safety, by 2050, weather and other hazards from the environment must have been precisely evaluated and risks properly mitigated.

Key Findings
- Weather is responsible for 13% of all aircraft losses between 1995 and 2004, 33% of all accidents/incidents from 2004 to 2007 and 40–50% of delays at European airports.
- The severe weather-related accidents and incidents can be attributed to the following weather hazards: wind, wind shear and turbulence; in-flight icing; low visibility due to lithometers, fog or precipitation; lee waves; hail damage; thunderstorms; and volcanic ash.
- During the flight in the en-route area, aviation is only disrupted marginally by weather phenomena. However, during takeoff and landing, it is very sensitive to those effects, especially fog, snow and wind, which can disrupt the operations with even a low intensity.

Key Actions: besides collecting higher-quality and more comprehensive weather data with higher spatial and temporal resolution (see ACARE goals 5—Article Chapter 2—and 13—Article Chapter 4), its effects on aircraft dynamics must be modelled to identify effective prevention and corrective actions that must be simulated and further validated. Therefore, it is recommended that low-cost basic research on flights in adverse weather conditions (e.g. wind, rain, ash clouds,

lightning, icing, storms and weather fronts) is promoted and promising advances are selected for demonstration.

10.5.3 Integrating Drones in Manned Airspace

The record of aviation as one of the safest modes of transportation is based on the highest engineering standards and professional qualifications as regards aircraft and cannot be compromised for Unmanned Aerial Vehicles (UAVs) operating in the same airspace. These UAVs constitute a new threat to the European airspace, as demonstrated by the occurrence of several incidents involving conventional aircraft and UAVs. However, in 2050, the European air transport system must operate seamlessly through interoperable and networked systems, allowing manned and UAVs to safely operate simultaneously in the same airspace.

Key Findings

- Unmanned aerial system (UAS), of which the UAV is the airborne component, comprises two fundamental types: Remotely-Piloted Aircraft Systems (RPAS), a class of UAS that has a "pilot" operating the Remotely-Piloted Aircraft (RPA) from a Ground-Control Station (GCS); and UAS with no remote pilot, or autonomous air vehicles (AAVs). The term "drone", although possibly inaccurate or inappropriate, refers to all types of UAS.
- There are two types of UAS operations: (1) the professional use of drones for various security, safety, survey and other tasks and (2) the recreational use where the general public is using drones for fun and private activities. The Key Risk Areas identified in UAS operations are (a) airborne conflict/collision—the collision of UAS with aircraft in the air, (b) aircraft upset, (c) system failure and (d) third-party conflict—the collision of UAS with people or property.
- There is an increasing trend in the number of reported UAS occurrences (both incidents and accidents) per year from 2010 to May 2016 inclusive (which may be due to the increasing number of drones within the EU), of which 63% are related to Airborne Conflict, which is the main Key Risk Area. This means that airspace infringements and proximity of drones to other aircraft are causing a significant number of occurrences.
- Conventional ATM cannot be applied to unmanned aircraft, and therefore there is the need to develop a UAS Traffic Management (UTM) system that allows UAVs to fly jointly to manned aircraft and implement the regulatory framework regarding the integration of UAS into busy airspace. Also, the use of partially unused airspace could provide testing areas and additional capacity for UAVs to prove to be at least equal to manned aircraft in terms of safety.

Key Actions: it is recommended that:

1. The evolution of air traffic capacity compared with the growth of air transport is assessed to identify the spare capacity available to other users, like UAVs.

2. The qualifications required of operators of UAVs and other aircraft compared with airline pilots and air traffic controllers are established to ensure that aviation remains one of the safest means of transport.
3. The design, production, certification and maintenance procedures for UAVs and other aircraft are defined to preserve or improve on the safety levels of current airliners that operate in the same airspace.
4. The increased use of partially underused airspace is explored to enable the expansion of operations by new types of aircraft.

10.5.4 Comprehensive and Unobtrusive Security Measures

The recent societal threat of terrorist acts at airports or during flights implies the reinforcement of security measures to prevent those acts from happening, resulting in delays and queues, which are the most frequent sources of traveller dissatisfaction. While the patience and understanding of passengers are essential, there should be a minimum of delay, intrusion and disruption in the implementation of safety measures through the use of the most appropriate equipment and airport architectures. In 2050, efficient boarding and security measures must allow seamless and non-intrusive security for global travel, with minimum passenger and cargo impact.

Key Findings

- The main types of scanning and detection devices currently deployed by European airports are based on the following existing technologies: Advanced Imaging Technology (AIT), Advanced Technology (AT) X-Ray, Boarding Pass Scanners, Bottled Liquids Scanners (BLS), Chemical Analysis Devices (CAD), Enhanced Metal Detector (EMD), Explosives Detection Dogs (EDD) and Explosives Trace Detector (ETD);
- Passengers spend on average 20 minutes waiting in line to get to the security screening checkpoint. New technologies and procedures could significantly reduce these waiting times, allowing us to process about 360 passengers per hour instead of the approximately 150 passengers per hour that are processed nowadays;
- Examples of emerging trends in technology for threat detection that could improve airport security and efficiency while reducing the burden for passengers are biometrics screening, computed tomography (CT) scanning, facial scanning and behavioural analytics. Nevertheless, all of these new technologies are proceeding beyond the first level of maturity, i.e. in the concept phase;
- Until now, most risk-based decisions regarding the checkpoint have focused on assessing the risk of a particular item but considering all passengers as equals. A new risk-based differentiation concept is introduced, which focusses its attention on "the person" in the assessment of threats instead of focusing on the risk of the item. As a result, based on a reasoned process of selection, different people would be screened in different ways;

- The project Smart Security, a joint initiative of the International Air Transport Association (IATA) and Airports Council International (ACI), defines a future where passengers proceed through security checkpoints with minimal inconvenience, where security resources are allocated based on risk, and where airport facilities are optimized, through the implementation of new technologies and processes.

Key Actions: it is recommended that non-intrusive passenger screening methods and foolproof luggage checking that allow fast flow through registration, border and boarding procedures are developed.

10.5.5 Resilience to External and Internal Threats

Aviation is facing one of the greatest security challenges in its history. Threats are increasingly taking non-conventional forms, some physical, such as new forms of terrorism, and some using the digital space with complex cyberattack and others related to interference with navigation systems like GPS signals. Nevertheless, by 2050, air vehicles must be resilient by design to current and predicted on-board and on-the-ground security threat evolution, internally and externally to the aircraft.

Key Findings

- Current and emerging threats to aviation security have been clustered into the following eight threat categories: (1) Improvised Explosive Devices (IED), firearms and close range destructive threats; (2) Chemical, Biological, Radioactive, Nuclear and Explosive (CBRNE) threats; (3) Ground-to-air threats; (4) Ground-to-ground threats; (5) Cyber threats; (6) Electromagnetic threats; (7) Sabotage, seizure and hijacking and (8) Bluff threats and threats from social media.
- Currently, aviation security is primarily based on the preventive phase and is inflexible to new threats. This is also mirrored in the research landscape for aviation security, since most projects concentrate on preventive measures such as the detection of CBRNE-substances. However, the aviation security system should be resilient to the evolving threat situation and thus be based on the complete resilience cycle, which has the following phases: prepare (take into account), prevent (repel or thwart), protect against (absorb or mitigate), respond to (cope with) and recover from (and adapt to).

Key Actions: it is recommended that:

1. Aircraft are designed and procedures are established to (a) prevent unauthorised entry into the cockpit, (b) allow remote takeover up to safe landing in the case of an identified flight anomaly and (c) design the system to be immune to the most sophisticated hacking.
2. An independent observatory of external risks to aircraft overflights is set up to advise airlines, or failing that, warn passengers.

3. A worldwide airliner flight monitoring system and accident data recorders are designed to ensure that accident/incident data is available regardless of time and location of occurrence.

10.5.6 High-Bandwidth Data Resilient to Cyberattacks

The use of digital data and the level of interconnection of IT systems are strongly increasing in civil aviation. Consequently, stakeholders of the air transport system, like airlines, airport and air traffic control (ATC), are more and more interlinked and, thus, depend on secure means of data exchange. In the future, it is expected an increase in this interconnectivity, which means the air transport system will be even more vulnerable and exposed to multiple points of attack. Taking this into account, in 2050, the air transport system must have a fully secured global high-bandwidth data network, hardened and resilient by design to cyberattacks.

Key Findings

- More than a dozen wireless technologies are currently used by air traffic communication systems during different flight phases. From a conceptual perspective, all of them are insecure, as security was never part of their design. On the other hand, the L-band Digital Aeronautical Communications System (L-DACS) and Aeronautical Mobile Airport Communications System (Aero-MACS) that are supposed to replace the current Very High Frequency (VHS) system have begun to at least consider the issue of wireless security, and some corresponding designs are already included in the specifications or will be in the future. Incidents of interference with navigation signals like GPS are becoming more frequent, mainly is regions of geopolitical tensions.
- The assessment of cyber risk involves: (1) identification and inventory of key assets—data, systems and infrastructure—that are essential to operations; (2) revision of internal controls and digital profiles to identify internal vulnerabilities and external threats and (3) valuation of the cyber assets at risk using modelling and other data and technology tools.
- The vulnerabilities that need to be taken into account are: (a) in a large, complex interconnected system there have many entry points for cyber intrusion and many links to spread the cyber-attack; (b) the weakest node may be the preferred entry point, for example, small suppliers of equipment or codes well protected by large industries or government bodies.
- In order to face future cyberattacks, firstly, it would be necessary to identify the multiple threats that could compromise aviation security (e.g. phishing threats, jamming threats, remote hijacking, distributed-denial-of-service (DDoS) attacks and Wi-Fi-based attacks), as well as to identify the systems that could be vulnerable to attacks. Then, it would be required to develop strategies in order to mitigate the threats identified.

Key Actions: it is recommended that:

1. The evolution of bandwidth requirements required to cope with increasing telecommunication needs associated with improved navigation, on-board system monitoring, passenger connection and other services is assessed.
2. Evolving standards for protection against electronic signal and computer cyber-attacks are established, with different levels, the highest for flight systems and the lowest but non-trivial for ticketing, bearing in mind the risk of intrusion from lower levels.

For more information about these topics, please access the full chapter here.

(a) ***ACARE GOAL 14:*** *Overall, the air transport system has less than one accident per ten million commercial aircraft flights.* **Recommendation 14:** Consider accident causes by order of statistical occurrence, and for each class identify and implement appropriate safeguards. **Rationale:** The progress in aviation safety has been remarkable and steady on a timescale of decades by progressively identifying, investigating and virtually eliminating the major causes of accidents. **Stakeholders:** EU, MS, AA, AN, CA, AI, RC, UA, AR. **Relevance:** One of the safest modes of transport can only benefit from being made even safer, and this requires investigating accident classes, finding corrective actions and proving that they can be implemented. **Priority:** This is a long-term, sustained effort, not a peak-and-done activity. **Justification:** PARE report Section 5.1.

(b) ***ACARE GOAL 15:*** *Weather and other hazards from the environment are precisely evaluated and risks are properly mitigated.* **Recommendation 15*:** Promote low-cost basic research on flight in adverse weather conditions (wind, rain, ash clouds, lightning, icing, storms and weather fronts) and select promising advances for demonstration. **Rationale:** Besides collecting higher-quality and more comprehensive weather data with higher spatial and temporal resolution (ACARE goals 5 and 13), its effects on aircraft dynamics must be modelled to identify effective prevention/corrective actions that must be simulated/validated. **Stakeholders:** EU, MS, UA, RC, AI, AA, AN, CA, AR. **Relevance:** High-quality weather data and accurate knowledge of its effects on aircraft promote general timeliness of air transport (ACARE goal 5) and can prevent major incidents or accidents (ACARE goal 14). **Priority:** This must be a long-term, sustained effort of learning from incidents how to prevent them from becoming accidents. **Justification:** PARE report Section 5.2 and Topic T5.1.

(c) ***ACARE GOAL 16:*** *The European air transport system operates seamlessly through interoperable and networked systems, allowing manned and unmanned air vehicles to safely operate in the same airspace.* **Recommendation 16.1*:** Assess the evolution of air traffic capacity compared with the growth of air transport to identify the spare capacity available to other users like UAVs. **Recommendation 16.2:** Establish the qualifications required of operators of UAVs and other aircraft compared with airline pilots and air traffic controllers to ensure that aviation remains one of the safest means of transportation. **Recommendation 16.3:** Define the design, production, certification and maintenance procedures for UAVs and other aircraft to preserve or improve on the

safety levels of current airliners that operate in the same airspace. **Recommendation 16.4:** Explore the increased use of partially underused airspace to enable the expansion of operations by new types of aircraft. **Rationale:** The air traffic capacity needs to increase to keep up with the growth of air transport, and the operation of UAV's in manned airspace will require additional capacity within the total achievable. The record of aviation as one of the safest modes of transportation is based on the highest engineering standards and professional qualifications as regards aircraft, which cannot be compromised for UAVs operating in the same airspace. The use of partially unused airspace could provide testing area and additional capacity for UAV's to prove at least equal to manned aircraft in terms of safety, which is not the case at present. **Stakeholders:** EU, MS, AI, CA, RC, UA, AR, AA. **Relevance:** UAV's offer an immense potential to expand air operations that will be realized sooner if design and operational issues are addressed to prove they are at least as safe as manned aircraft. **Priority:** It is high for safety issues related to engineering standards of UAV's and qualifications of operators. It is medium for issues of sufficient airspace capacity to add UAV's as another class of users. The maturation of UAV technologies and operations to the safety standards of aircraft is the baseline fundamental effort on which other objectives depend. **Justification:** PARE report Section 5.3 and Topic T5.2.

(d) ***ACARE GOAL 17:*** *Efficient boarding and security measures allow seamless security for global travel, with minimum passenger and cargo impact. Passengers and cargo pass through security controls without intrusion.* **Recommendation 17:** Develop non-intrusive passenger screening methods and foolproof luggage checking that allow fast flow through registration, border and boarding procedures. **Rationale:** Unfailing detection of dangerous individuals and objects can be combined with fast and efficient passenger and luggage screening only by developing better technology at the level of detector sensitivity and data processing. **Stakeholders:** EU, MS, AP, AR, CA, PA. **Relevance:** Although passengers should understand that their safety is paramount, it is preferable not to test their tolerance and patience with intrusive and lengthy airport checking procedures. **Priority:** As with many other issues in aviation, it is again a matter of sustained, unrelenting effort to adopt the most recent effective technologies. **Justification:** PARE report Section 5.4 and Topic T5.3.

(e) ***ACARE GOAL 18:*** *Air vehicles are resilient by design to current and predicted onboard and on the ground security threat evolution, internally and externally to the aircraft.* **Recommendation 18.1:** Design aircraft and establish procedures to (a) prevent unauthorised entry into the cockpit, (b) allow remote takeover up to safe landing in the case of an identified flight anomaly and (c) design the system to be immune to the most sophisticated hacking. **Recommendation 18.2:** Set up an independent observatory of external risks to aircraft overflights to advise airlines, or failing that, warn passengers. **Recommendation 18.3:** Design a worldwide airliner flight monitoring system and accident data recorders to ensure that accident/incident data is available regardless of time and location of occurrence. **Rationale:** Threats to aircraft can come not only from

hijackers but also from aircrew disabilities, and in some cases, remote control may be the only solution that must be immune to hacking or interference. The ICAO mandated national reporting of risks to aviation safety in areas of conflict has proved inadequate in Ukraine and elsewhere, and an independent observatory is needed to advise airlines and warn passengers. The occurrence of accidents that cannot be explained or take years to be investigated should be avoided in the future by worldwide monitoring of aircraft flights and mandatory fitting of air data reorders that can survive nearly all conditions, e.g. separate from the aircraft, float in the ocean and emit location. **Stakeholders:** EU, MS, CA, AN, AR, AI, RES, AU. **Relevance:** Aircraft overflights of unsafe regions must be avoided. Interference with safe flight procedures must be prevented. If an accident occurs, its causes must be known to prevent recurrence. **Priority:** It is imperative to prevent interference with safe flight or overflight of conflict regions. Accident investigation should be feasible even in the most extreme cases. **Justification:** PARE report Section 5.5 and Topic T5.4.

(f) ***ACARE GOAL 19:*** *The air transport system has a fully secured global high-bandwidth data network, hardened and resilient by design to cyberattacks.* **Recommendation 19.1:** Assess the evolution of bandwidth requirements required to cope with increasing telecommunication needs associated with improved navigation, on-board system monitoring, passenger connection and other services. **Recommendation 19.2:** Establish evolving standards for protection against electromagnetic signal interference and digital computer cyber-attacks, with different levels, the highest for flight systems and the lowest but non-trivial for ticketing, bearing in mind the risk of intrusion from lower levels. **Rationale:** The increase in bandwidth is a prerequisite for more efficient and safer air transport in such aspects as navigation and air traffic management, systems and safety monitoring and general communications. More data on more flight-critical functions requires cyber protection to a higher level. The vulnerability to disruption by cyber-attacks goes to the lowest level of airline ticketing. Countering cyberattacks at all levels also requires blocking access up to the criticality chain. **Stakeholders:** EU, MS, AN, AR, AP, AI, RC, UA. **Relevance:** Aviation has been one of the preferred targets of malicious actions that could extend from hijacking to hacking. The cyberattacks should be prevented in the future better than hijacking has been in the past. **Priority:** It is important to ensure cyber resistance at all levels, and in particular flight-critical systems on board and on the ground. The challenges of larger bandwidth are shared by aviation with other sectors that may be competing for the electromagnetic spectrum. **Justification:** PARE report Section 5.6 and Topics T5.4, T5.5 and T5.6.

10.6 Research, Testing Capabilities and Education

In 2050, the aviation industry must be underpinned by world-class capabilities and facilities in research, testing and validation and education. Aviation should be supported by the world's leading research infrastructures covering the entire aviation system, from wind tunnels through simulation facilities to test aircraft. At the same time, students in aviation-related university courses, which should be academically challenging and support the evolving needs of industry and research, must perform highly.

To ensure that all of these goals are met, the Advisory Council for ACARE established the specific Flightpath 2050 goals 20 to 23, which are addressed in the 6th chapter of the PARE report, entitled "Prioritizing Research, Testing Capabilities and Education".

10.6.1 Research and Innovation Agenda

There is a large gap between the high-quality scientific research sponsored by the likes of the European Research Council (ERC) and the market-oriented near-term developments of the Joint Undertakings (JUs), e.g. "Clean Sky" and "SESAR", that needs to be filled by fundamental applied research with an aeronautical focus to ensure a source of new ideas that are the basis of innovation and long-term competitiveness. Taking this into account, by 2050, research and innovation strategies should be jointly defined by all stakeholders, public and private, and implemented in a coordinated way covering the entire innovation chain.

Key Findings

- Research is defined and funded in a coherent and agile way, to avoid duplication and inefficiencies, prioritising initiatives resulting from strategic roadmaps defined and agreed by all stakeholders, satisfying actual needs (industry pull) and potential future demands (technology push).
- The European Union (EU) aeronautics programme has started with a budget of €36 million in the 2nd Framework Programme (FP) for Research and Technological Development (FP2) and had a steady growth to 3.6 B€ in the 7th Framework Programme (FP7), maintained in follow on activities, which testifies its success and the growing importance of this initiative.
- The growth of the aeronautics programme has seen a shift from (i) basic research (less than 1 M€) to (ii) industrial cooperation (4–10 M€), to (iii) large-scale demonstration (20–120 M€) to (iv) integration activities or JUs (more than €1 billion). This growth should be considered an efficient element of integral European transport system growth that provides completely safe, secure and sustainable mobility for people and goods.
- Technological innovation can achieve a faster and cheaper transition to a more efficient and sustainable transport system by acting on three main factors: vehi-

cles' efficiency through new engines, materials and design; cleaner energy use through new fuels and propulsion systems; better use of network and safer and more secure operations through information and communication systems.
- The ERC has sponsored high-quality research in basic science, including mathematics and physics, with some underrepresentation of engineering.
- Research facilities, used for different disciplines and specialities, differ greatly in size and range of application but are often linked to one another through a complex immaterial network that transforms basic scientific knowledge into competitive products while integrating environmental, safety and security requirements.

Key Actions: it is recommended that the long-term competitiveness of aviation is safeguarded by supporting a Basic Research Programme (BRP) with a wide variety of low-cost applied basic research up to the 3rd Technology Readiness Level (TRL3), entitled "Experimental proof of concept", to bridge the gap between the fundamental research of the ERC and the near-market-driven focus of Joint Undertakings (JUs). This broad programme will ensure aviation does not miss out on the promising new ideas that could be the key to future progress.

10.6.2 Industry-Research-Academia Clusters

Aviation research programmes should be balanced over the full range from basic to applied research. This could be supported by a network of multidisciplinary technology clusters that should be created based on collaboration between Aerospace Industry (AI), Universities and Academia (UA), and Research Centres (RC), with Europe taken as an example.

Key Findings
- The creation of these technology clusters could be the result of three initiatives, two ongoing and one to be restored from the past: (iii) demonstration and (iv) integration activities existing in the JUs Clean Sky and SESAR; the fundamental research in mathematics, physics and engineering existing in the ERC; and restoring the (i) basic and (ii) industrial research that existed in the aeronautics programme since the beginning and lapsed with increasing scale.
- The following existing networks should be considered: Association of European Research Establishments in Aeronautics (EREA); Aerospace Engineering Universities (PEGASUS); Aviation Noise Research Network and Coordination (X-NOISE EV); and FORUM on Aviation and Emissions (FORUM-AE) project.
- From FP2 to FP7 and since then, the involvement of EU countries in FP projects suffered an evolution from a more uniform distribution towards a more concentrated one. Such change may be reasonably associated with the evolution of the relative importance of the different thematic categories (used to classify EU-funded projects related to the aeronautic sector) and the identification of a less fragmented and more specialized cooperation network.

- Considering all FPs, on average, an industrial actor participated in a mean number of 3.2 EU-funded projects, with a standard deviation of 14.6, a research organization in 3.0 (11.1) projects, and a university in 2.6 (6.1) projects.

Key Actions: it is recommended to create multidisciplinary technology clusters, which require a balanced and proportionate support of four levels of projects:

1. basic research (3–5%): having 50–100 UA up to 1 M€ each exploring up to TRL3 all sorts of novel, promising ideas.
2. collaborative industrial (15–17%): 20–40 industrial research projects (4−10 M€) joining AI, RC and UA develop further the more prospects.
3. large-scale demonstrators (20–30%): 5–10 large scale demonstrators (20−100 M€) to reach a practical scale on the best results at a lower level.
4. JUs (50–60%): 1–2 JUs lead by industrial shorter-term applications (1−2B€).

10.6.3 Test, Simulation and Development Facilities

The large simulation and test facilities are essential institutional support to the aeronautical industry, representing large investments that have been coordinated in some occasions (e.g. the joint Dutch-German aero-acoustic wind (DNW) tunnel and the joint British-French-German cryogenic pressurized wind tunnel or European Transonic Wind [ETW] tunnel). By 2050, strategic aerospace test, simulation and development facilities should be identified, maintained and continuously developed, and the ground and airborne validation and certification processes should be integrated where appropriate.

Key Findings

- Strategic aviation infrastructure is of the highest quality and efficiency, providing the basis for world-class research and competitive product development while supporting education. It ranges from wind tunnels via iron birds up to experimental aircraft and simulation capabilities for in-flight and airport operations.
- The data quality and operational efficiency of aviation infrastructure help the industry to minimise risks and development costs and help society to determine the impact of aviation in benefits such as fast transport as well as in penalties such as the impact on the atmosphere.
- The main topics of these facilities are:
  - Improved and validated fluid dynamics, aerodynamic control, combustion, noise and thermal modelling based on high-performance computation, covering all needs for the aircraft and its engines, external and internal.
  - Methods and tools facilitating the evaluation of aircraft and engine configurations.
  - Results from the demonstration allow us to assess not only improvements in vehicle development but also to verify and validate new modelling techniques.

Key Actions: it is recommended that (a) a list of simulation, testing and certification needs and (b) an inventory of existing facilities are compared in order to identify the needs (i) already met, (ii) those requiring upgrades to be met or (iii) those requiring new facilities.

10.6.4 Young Talent and Women in Aviation

Aeronautics requires mostly but not only hard skills in STEM (Science, Technology, Engineering and Mathematics) that are becoming less abundant and eagerly sought by other sectors. Moreover, the aviation sector is already facing and will face an increasing shortage of skilled aviation professionals in the future. Thus, the aviation industry must engage promising young talent of both genders as early as possible and sustain their interest. In 2050, students should be attracted to careers in aviation and courses offered by universities should match or overmatch the needs of the aeronautical industry, its research establishments and administrations and evolve continuously as those needs develop. Additionally, lifelong and continuous education in aviation should be the norm.

Key Findings

- The major demographic trend in developed countries is characterised by an ageing population and declining younger age cohorts. In 2010, the industry employment in Europe was already showing a concentration of age structures in the middle age range (35–50 years) and experiencing lower recruitment rates of youngsters—in part due to longer education and training periods—but also due to broad use of early retirement schemes. This demographic tendency, in addiction with lower proportions of qualified young people who were (and are) choosing for STEM-related careers, was (is) a concern for the aerospace industry.
- To attract more young talent and women to aviation, the following measures are highlighted: (i) promotion of diversity in types of education and training; (ii) implementation of awareness programmes regarding careers in aviation; (iii) organisation and promotion of scholarships, grants and prizes; (iv) improving knowledge transfer from experienced to young employees by, e.g. mentoring programmes and (v) improvement of recruiting processes as well as of the working environment.

Key Actions: it is recommended that a comprehensive programme of attraction of talent to aeronautics to all education levels be fostered, complemented by job satisfaction measures at the professional level, with special measures to promote gender equality and increase the participation of women.

For more information about these topics, please access the full chapter here.

(a) ***ACARE GOAL 20:*** *European research and innovation strategies are jointly defined by all stakeholders, public and private, and implemented in a coordinated way with individual responsibility.* **Recommendation 20:** Safeguard the

long-term competitiveness of European aviation by supporting a broad program with a wide variety of low-cost applied basic research up to TRL3 to bridge the gap between the fundamental research of the ERC and the near-market-driven focus of JUs, ensuring that the promising new ideas are duly exploited to foster progress in aviation. **Rationale:** The EU aeronautics programme has had a remarkable growth in the allocation of resources, from 36 M€ in FP2 to 3.6B€ in FP7. This has been accompanied by a shift in projects from basic research (less than 1 M€), to industrial cooperation (4–10 M€), to large-scale demonstration (20–120 M€) to Joint Undertakings (more than 1B€). This shift to large-scale near-market research and development has led to a decline of fundamental research essential to the long-term competitiveness of the aeronautical sector. **Stakeholders:** EU, MS, UA, RC, AI. **Relevance:** There is a large gap between the high-quality scientific research sponsored by the ERC and the market-oriented near-term developments of the JUs that needs to be filled by fundamental applied research with an aeronautical focus to provide a source of new ideas that are the basis of innovation and long-term competitiveness. **Priority:** The prosperous current state of the aeronautical sector could be undermined by a lack of long-term vision and new ideas that can be supported at a small fraction (1–3%) of what is invested in JUs. **Justification:** PARE report Section 6.1.

(b) ***ACARE GOAL 21:*** *Creation of a network of multidisciplinary technology clusters based on collaboration between industry, universities and research institutes.* **Recommendation 21:** The creation of multidisciplinary technology clusters requires a balanced and proportionate support of 4 levels of projects: (a) basic (3–5%); (b) collaborative industrial (15–17%); (c) large-scale demonstrators (20–30%); joint undertakings (50–60%). **Rationale:** A balanced aeronautical research programme should have 4 levels: (i) 50–100 basic research UA up to 1 M€ each exploring up to TRL3 all sorts of novel promising ideas; (ii) 20–40 industrial research projects (4–10€) joining AI, RC, UA develop further the more promosing prospects; (iii) 5–10 large scale demonstrators (20–100 M€) to reach practical scale on the best results at lower level; (iv) 1–2 joint undertakings (Clean Sky and SESAR) led by industrial shorter term applications (1–2 B€). Some aerospace research programmes have shifted from one end to the other and should be rebalanced. **Stakeholders:** EU, MS, AI, AN, RC, UA, AR, AP, CA. **Relevance:** The technology clusters could provide the filtering of results up the basic-industrial-demonstration-development chain. The basic projects as sources for new ideas should be based on peer review by UA as the ERC. The three higher levels would be based on selection by industry to ensure the focus of larger investments. **Priority:** Only a balanced allocation of resources at all 4 levels can promote the new ideas and links to practical application that can sustain competitiveness from the present to the future. **Justification:** Section 6.2 and Topic T6.1. **Remark:** The recommendation 21 is a means of implementation of recommendation 20.

(c) ***ACARE GOAL 22:*** *Identification, maintenance and ongoing development of strategic European aerospace test, simulation and development facilities. The ground and airborne validation and certification processes are integrated*

where appropriate. **Recommendation 22:** Compare (a) a list of simulation, testing and certification needs with (b) an inventory of existing facilities to identify the needs currently satisfied and those requiring upgraded or new facilities. **Rationale:** The large simulation and test facilities are an essential institutional support to the aeronautical industry. They represent large investments of the Member States that have been coordinated in some occasions (DNW, ETW). An overall approach should compare what is needed with what is available to prepare a program of upgraded and new facilities. **Stakeholders:** EU, MS, RC, AI, AU, AN. **Relevance:** Duplicate facilities may be justified by high demand at the European level or nationally funded objectives. Some facilities may have multiple uses besides aerospace. The cost, maintenance and benefits (or indispensability) of facilities are essential issues. **Justification:** PARE report Section 6.3.

(d) ***ACARE GOAL 23:*** *Students are attracted to careers in aviation. Courses offered by European universities closely match the needs of the aviation industry, its research establishments and administrations and evolve continuously as those needs develop.* **Recommendation 23:** Foster a comprehensive program of attraction of talent to aeronautics to all education levels, complemented by job satisfaction measures at the professional level, with special measures to promote gender equality and increase the participation of women. **Rationale:** The attraction of young talent to aeronautics should focus equally on both genders, start with the fascination of flying in youngsters, and continue with access to high-quality diversified university courses, followed by challenging and interesting careers in industry. **Stakeholders:** EU, MS, UA, AI, RC. **Relevance:** Aeronautics requires mostly but not only hard skills in STEM (Science, Technology, Engineering and Mathematics) that are becoming less abundant and eagerly sought by other sectors; thus, the aviation industry must engage promising young talent of both genders as early as possible and sustain interest through industry visits and contact with professionals. **Priority:** A good talent pool exists in students and educators that must be supported to continue to provide industry with the best engineers, as countries with large populations outnumber Europe in aerospace graduates. **Justification:** PARE report Section 6.4 and Topic T6.2.

10.7 Aircraft Markets

Objectives 24 to 28 are addressed in the 7th chapter of PARE's report, entitled "Aircraft Markets", which focuses on the following markets: large or long-range aircraft, regional jets, business jets, helicopters and convertibles and civil and military unmanned aerial vehicles (UAVs).

10.7.1 The Duopoly in Jet Airliners

The long-range airliner market is dominated by the Airbus–Boeing duopoly (since the 1990s) that arose at the end of a long competitive period in which Airbus steadily gained ground starting from a newcomer status. The airlines that want a competitive choice of aircraft are least interested in the Airbus–Boeing duopoly becoming a monopoly. However, this has not deterred the long-running dispute at the World Trade Organisation (WTO), which may become a permanent nuisance if not properly contained and may lose impetus due to the emergence of state-sponsored competitors from other parts of the world. Considering this, by 2050, a level playing field in the large aircraft market should be promoted.

Key Findings

- The market shares in the single-aisle narrow-body and twin-aisle wide-body markets are different, showing an appropriate balance: Airbus is leading the former and Boeing the latter. The third and fourth-largest airliner manufacturers in the world, Embraer (Brazilian) and Bombardier (Canadian), are about one-tenth of the size of the two world leaders and focus on regional and business aircraft.
- The reasons for the duopoly are multiple:

  1. Extremely high entry barriers: to ensure the entrance in the aerospace industry, there are four required ingredients: (i) strong financials; (ii) powerful science and engineering resources; and (iii) an efficient industrial production; (iv) worldwide spares and support organisation.
  2. Extreme concentration at the top of the market in terms of major revenue producers.

- Currently, both Airbus and Boeing are in the healthy situation as concerns having the largest order books in history and both face challenges in achieving higher production rates, In the case of Boeing persistent production quality issues and operational and certification issues have tarnished its reputation and damaged its financial situation. This market duopoly means that Boeing and Airbus are constantly fighting to gain an advantage over the other in terms of aircraft sales. And the fight extends to non-technical fields like legal actions.
- In 1992, a bilateral European Union (EU)-United States (US) agreement was made regarding the Trade in Large Civil Aircraft (TLCA), which banned support from governments in production financing and allowed up to 33% of the development programme costs to be met through government loans, which are to be fully repaid within 17 years with interest and royalties. In 2005, each party (Airbus and Boeing) filed complaints against the other at the WTO, claiming that each other's aircraft programs were unfairly subsidized (e.g. receiving unfair state aid from their respective governments). By doing this, the EU and USA withdrew from the TLCA agreement, starting a long-term dispute of no interest to airlines.
- It is possible that other competitors emerge in the long-range air transport market, for example from China, that has a non-transparent scheme of state subsidies.

Key Actions: to achieve PARE Objective 24, it is recommended that a strong legal, commercial, and technical basis is developed to (a) in any case, if necessary, deal with litigation at the WTO; (b) preferably, if possible, renew the large aircraft agreement between the EU and the US and (c) ensure it applies to new competitors or, in this or other ways, ensure a level playing field in the airliner market.

10.7.2 The Regional Jet and Turboprop Market

The regional jet and turboprop market, whose seating capacity ranges from 20 to 130, cannot be separated from long-distance air travel since it acts as its feeder to major hubs and as a direct link between smaller communities on shorter routes. It is also an important market, much more accessible to other entrants than the Airbus–Boeing duopoly of giants, and thus attracts increased competition. Taking this into account, by 2050, the position of the regional aircraft market should be strengthened as a contributor to air transport.

Key Findings

- The main market for regional aircraft is represented by regional carriers, which are carriers with an average stage range of around 500 kilometres (km) or fleet without narrow-body and wide-body aircraft (turboprops and/or regional jets only).
- In the regional market below 100 seats, the leading position of ATR (Franco-Italian aircraft manufacturer) faces fierce competition from traditional rivals from Canada and Brazil, as well as from newcomers from Japan, Russia and China, although some aircraft programs in this sector have had unstable periods.
- In the small turbofan aircraft with 50–120 seats, Embraer is the market leader with its E170/E190 family, its position being at this moment threatened seriously only by the A220 (ex-Bombardier C-Series) family. However, other potential competitors, like Comac ARJ21 (Chinese), Sukhoi Superjet 100 (Russian), and Mitsubishi MRJ 90 (Japanese), have different fortunes from selling mostly in the home market (first two) to cancellation (third).

Key Actions: to accomplish PARE Objective 25, it is recommended that the development of regional aircraft in a world with an increasing number of competitors be supported and that synergistic tie-ups between large and regional aircraft suppliers are also considered.

10.7.3 The Business Jet Market and Supersonic Prospects

There is strong competition not only on long-range and regional airliners but also in other categories like business aircraft, whose market extends from the largest airliners customized for heads of state to private aircraft flown by their owners. Dassault

Aviation (French), together with Gulfstream Aerospace (American) and Bombardier (Canadian), are world leaders in large business jets, and there are several other suppliers of smaller business jets. Moreover, the next civil supersonic transport following Concorde (the first and only effective supersonic commercial airliner co-developed by the British Aircraft Corporation (BAC) and Aerospatiale (French), now a part of Airbus) could be a supersonic business jet. Taking this into account, by 2050, the role of the business jet market could increase.

Key Findings

- According to the General Aviation Manufacturers Association (GAMA), three classes of business jets are identified by specific performances and price levels: (i) light jets (e.g. Cirrus SF50, Cessna Citation series lower end or Pilatus PC-24); (ii) midsized jets (e.g. Bombardier Challenger, Cessna Latitude or Embraer Legacy); and (iii) large business jets (e.g. Gulfstream Aerospace, Bombardier and Dassault Aviation's Falcon families).
- Concorde was designed in the 1960s, first flew in the 1970s, and ceased airline operations in 2003 due to high operating costs and difficulties in maintaining an old aircraft for lack of spares.
- The prospects for a commercial supersonic aircraft look dim because (i) the sonic boom could prevent flight overland, leading only overwater routes, with the transatlantic market small and the transpacific market requiring more range; and (ii) the overall number of aircraft, perhaps a few hundred, could hardly cover the high development cost of a supersonic airliner and a dedicated engine.
- Ongoing research, including flight testing of long-nose low-sonic boom designs, aims to enable low supersonic Mach number cruise overland with ground overpressures low enough to satisfy certification authorities.
- Of all aircraft manufacturers, Dassault Aviation would be best placed to design and produce a supersonic business jet since it has: (a) decades of experience with supersonic jet fighters; (b) a complete range of high-end, efficient business jets and (c) researched the critical aspects of a supersonic business jet. However it is in the US that some start-ups are attempting the development of supersonic and hypersonic vehicles with military application, besides supersonic airliner/business jet initiatives.
- The miniaturization of electronics allows business jets to be adapted to other missions like sensor platforms, patrol and surveillance that are high-value extensions of the baseline business jet market.

Key Actions: to achieve PARE Objective 26, it is recommended that the development of European business jets and their expanded use as sensor/surveillance/control platforms are supported.

10.7.4 Markets for Helicopters and Convertibles

Within the convertible aircraft (those capable of vertical take-off and landing [VTOL]), the helicopter market is one of Europe's major successes: Airbus Helicopters is the world leader, and Leonardo (an Italian manufacturer that currently owns the former British aircraft manufacturer renamed AgustaWestland) also holds a strong position in the market. The main competitors are Boeing–Vertol, Bell and Sikorsky from the USA and Mil and Kamov from Russia. The strong US investment in greater hot-and-high and high-speed capabilities must be matched if the EU wants to maintain long-term market share.

Key Findings

- The helicopter market has some stable elements like search-and-rescue (SAR), emergency medical services (EMS) and law and order protection. Other elements are more volatile and vulnerable to large fluctuations, such as offshore oil exploration and wars in inhospitable places lacking safe ground infrastructure or alternative means of transport (e.g. hot and dusty Iraq and high, hot and dusty Afghanistan), but are the main cause of helicopter's expansion.
- The decline in military operations in the countries named before and the reduction in oil exploration caused a reduction of both military and civil helicopter markets that are snow fully recovering. Consequently, faced with reducing order books, the American helicopter industry is pressing the US government to end decades of stagnation in helicopter technology.
- The USA has started a major programme, Future Vertical Lift (FVL), to design helicopters or tiltrotors with (i) twice the range, (ii) 50% higher speed and (iii) over twice the hover payload in demanding hot and high conditions, using engines with double power but similar fuel consumption, size and weight. Although it is a military programme, it could have civil spinoffs: (i) double-range for offshore oil exploration; (ii) higher speed for medical emergencies and executive transport; (iii) greater payload for rescue and transport missions. All this could challenge the position of Europe with over 50% of the world helicopter market.
- Russia is also funding the development of an advanced high-speed helicopters besides the production of current types to support the war in Ukraine. The Central Aerohydrodynamic Institute (TsAGI) confirmed in 2018 that the Kamov Design Bureau started working to create a flying laboratory on the basis of the Ka-52 helicopter. It is expected that new technologies will provide more speed (probably about 400 kilometres per hour (km/hr)) and range and better fuel efficiency.
- To compete with the USA and Russia, Airbus Helicopters has recently unveiled the aerodynamic configuration of the high-speed demonstrator it is developing as part of the Clean Sky 2 Joint Undertaking (CSJU) European research programme. The demonstrator codenamed Racer, for Rapid and Cost-Effective Rotorcraft, incorporates a host of innovative features and is optimised for a cruise speed of more than 400 km/hr, and has the world speed record for rotorcraft. The Airvus helicopter activities continue aiming at achieving the best trade-off between

speed, cost-efficiency, sustainability and mission performance. Agusta has also continued to develop tilt rotors beyond the long running civil AW609 program.

Key Actions: to accomplish PARE Objective 27, it is recommended that Europe keeps at least abreast of developments in high-power, high-speed helicopters and tiltrotor aircraft with enhanced hot-and-high lift capabilities.

10.7.5 Current UAV Market and Demand

Europe has the technology to develop all classes of UAVs that are increasingly relevant to a wide range of defence and civil missions. Nevertheless, Europe is far from being competitive in particular for large Unmanned Combat Aerial Vehicles (UCAVs), commonly known as combat drones, market due to a lack of coordination in the allocation of resources at EU level. There must be an end to the European dependence on foreign UAVs and a move to enter the international market since there is the technology to achieve both targets. By 2050, a European alternative to the drones used in Europe with the potential to also enter the world market must be provided.

Key Findings

- There were several prototype programmes across EU countries, such as (i) Taranis in the UK; (ii) Talarion in Germany; (iii) HammerHead in Italy; and (iv) multinational nEUROn implemented by Dassault Aviation in France. However, none of these programmes has reached production.
- In the meantime, Europe is buying the following drones from non-EU countries: Global Hawks, Reapers and Predators from the USA and Herons and Hermes from Israel.
- Other countries had progressed in this field, ahead of Europe. The reluctance of the USA to export armed drones has allowed China to take a leading position as the supplier of such systems in Asia and the Middle East, with attractive low costs and disappointing reliability.
- Europe is joining the drone effort with the multinational Eurodrone program.
- The development of "loyal wingman" as drones flying together with manned aircraft is part of nearly all new fighter programs.
- The number of countries supplying drones in significant quantities is increasing, for example, Turkey and Iran.

Key Actions: to achieve PARE Objective 28, it is recommended that the technological capabilities demonstrated in several prototype drones are leveraged into a coherent programme covering all levels to satisfy internal needs and compete in the world market.

(a) ***PARE OBJECTIVE 24:*** *Promote a level playing field in the large aircraft market*. **Recommendation 24:** Develop a strong legal, commercial and technical basis to (a) in any case, if necessary, deal with litigation at the WTO, and (b)

preferably, if possible, renew the large aircraft agreement between the EU and the USA or (c) by this or other means ensure a level playing field in the large civil aircraft market, also regarding possible new entrants. **Rationale:** The airlines that want a competitive choice of aircraft are least interested in the Airbus–Boeing duopoly becoming a monopoly. This has not deterred the long-running dispute at the WTO that may become a permanent nuisance. **Stakeholders**: EU, MS, AI, AR. **Relevance:** The distorted charges made at the WTO should not be allowed to question the fact that success in the airline market is and should be based on the merits of engineering quality and efficient operations. **Priority:** Bearing in mind mean that engineering excellence is the market winner, and legal challenges should not become an obstacle. **Justification:** sections 7.1 and 7.2 and topics T7.1, T7.2, T7.3 and T7.5.

(b) ***PARE OBJECTIVE 25:*** *Strengthen the position of the regional aircraft market.* **Recommendation 25*:** Support the development of regional aircraft in a world market with an increasing number of competitors, and additionally consider synergistic tie-ups between large and regional aircraft suppliers. **Rationale:** the majority stake of Airbus in the Bombardier C-series renamed A220 has extended the market reach to all jet airliners above 100 seats, with the breakdown of the Boeing-Embraer deal leaving Embraer alone as the only major competitor. In the propeller regional market below 100 seats, the leading position of ATR faces multiple challengers from Canada, Japan, Russia and China, all less successful so far. **Stakeholders:** EU, MS, AI, RC, UA, AR. **Relevance:** The under 100-seat regional aircraft market is relevant both as a feeder to major hubs and as a direct link between smaller communities, both in Europe and worldwide, amounting to a large and growing market that attracts increased competition. **Priority:** The leading position of ATR faces a number and variety of challengers. **Justification:** Sections 7.3 and 7.4 and topics 7.4 and 7.5.

(c) **PARE** ***OBJECTIVE 26:*** *Strengthen the position of the business jet market.* **Recommendation 26:** Support the development of European business jets and their expanded use as sensor/surveillance/control platforms. **Rationale:** Dassault, together with Gulfstream and Bombardier, is a world leader in large business jets, and other manufacturers of smaller business jets share of the rest of the market. **Stakeholders:** EU, MS, AI, RC, UA. **Relevance:** The miniaturization of electronics allows business jets to be adapted to other missions like sensor platforms, patrol and surveillance that are high-value extensions of the baseline business jet market. **Priority:** This is a successful area of aviation that needs to be sustained and could be expanded. **Justification:** Section 7.5 and topics 7.4 and 7.5.

(d) ***OBJECTIVE 27:*** *Maintain the leadership in the world helicopter market.* **Recommendation 27:** Ensure sustainment of developments in high-power, high-speed helicopters/convertibles with enhanced hot-and-high lift capabilities. **Rationale:** The USA has started a major program, FVL (Future Vertical Lift), to design helicopters/convertibles with (i) twice the range, (ii) 50% higher speed and (iii) over twice the hover payload in demanding hot and high conditions, using engines with double power but similar fuel consumption, size and

weight. Although it is a military program, it could have civil spinoffs: (i) double-range for offshore oil exploration; (ii) higher speed for medical emergencies and executive transport and (iii) greater payload for rescue and transport missions. All this could challenge the position of Europe with over 50% of the world helicopter market. **Stakeholders:** EU, MS, AI, RC and UA. **Relevance:** The FVL program in the USA is justified by the need to counter threats from near-peer adversaries in Europe and elsewhere; hence, it is relevant to the defence of Europe. The implications in the civil market could be to reverse the tables, passing dominance from Airbus Helicopters and AgustaWestland to Bell and Sikorsky. The FVL contenders are the Valor tiltrotor from Bell and the Defiant dual rotor plus pusher-propeller helicopter from Sikorsky; Europe has analogues in the Augusta-Bell AB609 and Airbus X3 that hold the world helicopter speed record, and competitive turboshaft engines from Turbomeca and Rolls-Royce. **Priority:** There is a need for a program with a minimum investment to ensure that Europe does not fall behind. It is not necessary to match the massive US funding of FVL. The result of FVL could be as expensive as the Bell V-22 Osprey with a small effect on the market, or it could, like the RAH-66 Comanche, lead to no significant production after years and billions of investments. The aim here is to safeguard against potential surprise breakthroughs that could change the leading market position without making large speculative investments and being ready to offer major improvements in helicopter/convertible performance and efficiency. **Justification:** PARE report Section 7.6 and topics 7.5 and 7.6.

(e) **PARE *OBJECTIVE 28:*** *Provide a European alternative to the drones used in Europe with potential to also enter the world market.* **Recommendation 28:** Leverage the technological capabilities demonstrated in several prototype drones into a coherent European Programme covering all levels to satisfy internal needs and complete in the world market. **Rationale:** The market for MALE (Medium Altitude Long Endurance) drones is a sad example of a lack of coordination in Europe. (i) of several prototype programs (Taranis in the UK, Mako in Germany, Hammerhead in Italy, Neuron Multinational led by France), none has yet reached operational status; (ii) in the meantime, several European nations have bought American and Israeli drones; (iii) in the international market, China has emerged as the major competitor of the USA through lower prices and fewer export restrictions, although it has a poor reliability record; (iv) other nations like Turkey and Iran have become drone suppliers; (v) Eurodrone is a large cooperative drone effort; (vi) most new fighter programs involve a drone "loyal wingman". **Stakeholders:** EU, MS, AI, RC, UA. **Relevance:** Europe has the technology to develop all classes of UAVs that are increasingly relevant to a wide range of defence and civil missions, so the issue is one of coordination in the allocation of resources. **Priority:** There must be an end to the European dependence on foreign UAVs and a move to enter the international market since there is the technology to achieve both targets. **Justification:** PARE report Section 7.6.

10.8 Emerging Aviation Technologies

Objectives 29 to 36 are addressed in the 8th chapter of PARE's report, entitled "Emerging Aviation Technologies", which concern new disrupting technologies that are reshaping the aviation sector.

10.8.1 Electric Propulsion and Electrified Aeroplane Components

Although the automobile sector may lead the electrification of transport vehicles, the specific needs of aeronautics and fast technological evolution, in particular advances in electric propulsion (EP) and systems, will have increasing importance from drones to aeroplanes, including convertibles (those capable of vertical take-off and landing [VTOL]). Moreover, many components/systems of conventional aircraft (e.g. actuation, de-icing, and air-conditioning), which use mechanical, hydraulic and pneumatic power sources, have been replaced throughout the years by electrical ones due to a lack of reliability and high maintenance costs. This provides a motivation to stay at the forefront of progress in the electrification of aircraft, including components.

Key Findings

- EP uses electrochemical power sources (ECPS) such as batteries or Hydrogen Fuel Cells (HFC). Currently, it is more effective for small vehicles, with short or medium ranges with a low weight, such as drones transporting measuring sensors. Nevertheless, the strongest developments of EP also take place at VTOL taxis, normally with a range between 50 and 100 kilometres (km) and partially larger. Under development are also regional and business aircraft with targets around 500 km.
- The flight range of all-electric aircraft depends on the specific energy (watt-hour per kilogramme—Wh/kg) of the ECPS, besides engineering parameters (such as lift over drag ratio, battery mass fraction—mass battery/total aeroplane mass—and efficiency of powertrain). Today all-electric aircraft are limited to the low total weight, i.e. small passenger numbers and/or low payload, as well as limited endurance and speed.
- Hybrid Electric Propulsion (HEP), for example, combining gas turbines and fuel cells, is suitable for regional aircraft.
- Long-range transportation of a large number of passengers or heavy cargo will rely for some time on conventional propulsion, which still has considerable development potential on its own and as part of hybrid systems with hydrogen or synthetic fuels.

Key Actions: to achieve PARE objective 29, it is recommended that a thorough assessment is made, followed by support measures on (a) emerging electric systems and propulsion technologies, (b) their potential to satisfy mission requirements and (c) the likely evolution of both.

10.8.2 Additive Manufacturing

Additive manufacturing allows the production of complex pieces with fewer parts and could replace spare part inventories with local production as limitations in series production, choice of materials and quality finish are overcome. Although it is an industry-wide activity, aviation may be the leader in some areas and should keep up with progress in other areas. Taking this into account, by 2050, advances in additive manufacturing must be promoted and exploited.

Key Findings

- Additive manufacturing has several attractive features: (i) since it consists of adding layers, it does not waste material unlike processes such as milling that carve a part out of a larger block; and (ii) it allows the manufacture of large complex specimens that would otherwise have to be builtup of smaller parts with junctions. Thus, additive manufacturing is one of the contributors to more efficient production.
- It also has limitations: (i) it is a slow process more suited to laboratory use or limited production of prototypes than to large-scale, high-rate serial production; (ii) the surface smoothness is dependent on the number of layers that cannot be excessive, thereby requiring subsequent operations to ensure a better finish and (iii) it is difficult to ensure the consistent quality standards required for certification.
- If the limitations of additive manufacturing can be overcome, there are promising prospects: (i) producing parts locally and as needed instead of transporting them from large stocks in a warehouse; and (ii) coupled with digital imaging and sensing, producing replicas and replacements for damaged parts already out-of-stock or unavailable. These features will become more useful as the range of materials suitable for additive manufacturing and their physical properties widen.

Key Actions: to accomplish PARE objective 30, it is recommended to consider the implications on prototyping, series production and spares supply of additive manufacturing regarding (a) usable materials, (b) quality standards and (c) life-cycle costs.

10.8.3 Efficient Production

The more complex world of aeronautics compared with other sectors, e.g. the automotive sector that has taken the lead on efficient production, provides opportunities for cross-fertilization in new technologies and their efficient utilization and management. The next stage in the evolution towards efficient production is known as "Industry 4.0", and aeronautics can benefit from 4.0 technologies, especially by integrating the full life-cycle design-production-operation-maintenance. Considering this, by 2050, the experience from other sectors should be incorporated to achieve more efficient and economical production in aviation.

Key Findings

- Some of the key concepts of efficient production read across production volume, value and rate and include: (i) just-in-time delivery, avoiding large and costly static inventories; (ii) work planning that maximizes human motivation and efficiency and (iii) the right mix of automation and human intervention.
- There are counter-factors that favour aeronautics regarding efficient production: (i) higher added value justifies bigger investments in quality and efficiency; and (ii) large order backlogs, if they are achieved, give a measure of long-term stability.
- The term Industry 4.0 refers to the combination of several major innovations in digital technology, all coming to maturity at present, all poised to transform the energy and manufacturing sectors. The following nine technology trends form the building blocks of Industry 4.0: (1) big data and analytics; (2) advanced robotics and artificial intelligence; (3) simulations; (4) horizontal and vertical system integration; (5) the industrial Internet of Things (IoT); (6) cybersecurity; (7) cloud computing; (8) additive manufacturing; and (9) augmented reality.
- Even though 4.0 factory requires significant investment, it will change traditional production relationships among suppliers, producers, and customers—as well as between humans and machines—and lead to a reduction on development time and costs of production and operation, thus leading to greater efficiency.

Key Actions: to achieve PARE objective 31, it is recommended to consider the best combination of 4.0 technologies, including automation and information, as applicable to various production rates and scales of equipment in aviation.

10.8.4 Telecommunications

Telecommunications is a major innovation and growth sector in modern societies that has become readily embedded in aeronautics technologies and the services they provide, including navigation and Air Traffic Management (ATM), on-board monitoring and preventive maintenance, flight safety and security, etc. The expected growth of air traffic and consequently the need for increased air traffic capacity and

safer and more efficient aircraft operations will require increased data sharing and larger telecommunications bandwidth protected from unintended or malicious interference. Taking this into account, by 2050, the telecommunications capacity needed for connected aircraft, navigation, monitoring and other services must be provided.

Key Findings

- The cluster of air traffic management (ATM), communication, navigation and surveillance (CNS), meteorological (MET) and aeronautical information (AIS) services are called air navigation services. Likewise, ATM services are divided into three different services: air traffic flow management (ATFM), airspace management (ASM) and air traffic services (ATS), which are composed in turn by air traffic control (ATC), flight information services (AIS) and alert services (AL).
- Telecommunications are the core of this network in such a way that they allow, for example, communications air-to-air, air to ground, ground to ground, as well as aircraft navigation or surveillance through the application of radio frequencies. Thereby, available communication and surveillance systems are essential for the correct functioning and effectiveness of ATC services, and equipment based on radio waves is essential for the correct aircraft navigation.
- As a consequence of the growth of air traffic and, likewise, of the growth of the number of Navigational Aids (NAVAIDs), the need for a higher number of radio channels was identified due to the congestion of the radio spectrum. This change in the frequency assignment has managed to increase the number of available channels in the very high frequency (VHF) band of aeronautical communications, which has allowed the creation of new control sectors and has contributed to increasing the ATM capacity.
- The future scenario of communications relies on the increasing use of the Aeronautical Mobile Satellite Service (AMSS) through the use of satellite communication (SATCOM) as well as the use of data links via very high frequency (VHF) and ultra high frequency (UHF).

Key Actions: to achieve PARE objective 32, it is recommended to assess the growth of capacity needs for navigation, systems monitoring and passenger services and how the required bandwidth can provided free from unintended or malicious interference.

10.8.5 Cybersecurity

The increasing reliance on computer systems tied by telecommunication links (more interconnected systems) multiplies the entry points for disruptive and malicious intrusions, which can compromise major systems and lead to unlawful physical acts like hijacking of aircraft, destruction of an aircraft in service or hostage-taking on board an aircraft or at airports. Considering this, the global aviation industry has many layers overseeing the safety of all the stakeholders involved, from aircraft

manufacturers to the passenger boarding a flight. Aeronautics needs no less and perhaps more than many other sectors ready to implement security measures to safeguard civil aviation against these acts of unlawful interference. Thus, by 2050, threats to cybersecurity must be monitored, and timely protection must be devised for all levels of aviation systems.

Key Findings

- Cybersecurity is about the prevention of and/or reaction to deliberate malicious acts undertaken via cyberspace (i.e. the domain of information flow and communication between computer systems and networks and includes physical as well as purely virtual elements) to either compromise the system directly or wherever systems play a key role. Airport and ATM cybersecurity is aimed at limiting the effects of such cyberthreats and the impact on airport organization and operations and the overall ATM network.
- One of the keys to fighting successfully against cyberattacks is to identify the potential cyberattackers, normally divided into four groups: state cyberforces (who have a large number of resources at their disposal, state backing and are very highly skilled); cybercriminals (who are interested in making money through fraud, blackmail or the sale of valuable information); terrorists (who are interested in obtaining and using sensitive information to launch a conventional attack); and hacktivists (who have a cause to fight for, such as political or ideological motives).
- Cybersecurity has become a real issue for aviation, driven by different factors such as:

  (i) The increasing interconnectivity of ATM, which means that the impact of an attack may extend across a growing number of interconnected systems
  (ii) The increased reliance on integrated data, which means a high potential for operational disruptions if connectivity is lost
  (iii) The migration towards common and Commercial Off The Shelf (COTS) components, underpinned by industry-standard protocols such as Internet Protocol (IP), with published vulnerabilities, which means that more people will have the technical background to launch attacks and more people will have access to core infrastructures through extended supply chains
  (iv) New methods of attack stemming from either criminal activities and/or state-sponsored actors of increasing levels of sophistication

- Cyber-vulnerabilities have also been exposed at industry and government level, even in the largest aerospace programmes, like the F-35 Lightning II-Joint Strike Fighter Programme. (a) industrial espionage has used small subcontractors with modest security resources to access up the supply chain to main contractors and (b) intergovernmental meetings among the representatives of participating nations have had uninvited, invisible "guests" from other nations.
- To ensure the identification of appropriate preventive security measures, the level of threat should be continually reviewed and risk assessments carried out, considering international, national and regional situations and environments. In this

manner, whenever a specific threat exists, selected and predetermined preventive security measures should be applied, commensurate with the associated risk assessment and the nature and severity of the threat.

Key Actions: to accomplish PARE objective 33, it is recommended to consider the levels of cyber protection needed for the various aeronautical activities and how to monitor and counter threats.

10.8.6 Big Data

The new generation of aircraft generates a large amount of data from multiple sources: flight tracking data, airport operations data, meteorological data, airline information, market information, passenger information, aircraft information and air safety. Some of these data can be obtained through thousands of sensors and sophisticated digitalised systems that are installed onboard an aircraft. Using big data analytics to process efficiently the large amount of data from different sources can have several applications in aviation: optimising fuel consumption, crew deployment and flight operations, reducing cost, improving maintenance, forecasting weather and air safety, and enhancing passenger experience; that will become standard to support real-time decision making. By 2050, the implications in aeronautics of advances in big data, including the use of what is already available, must be assessed.

Key Findings

- Currently, the aviation industry adopts condition/preventive maintenance procedures due to its operational efficiency, and it depends upon the failure mode calculations made after testing a part under specific circumstances. These conditions may fluctuate depending on the external factors/human errors, which may result in the variation in the lifetime of components, in turn reducing the operational efficiency of the aircraft.
- A study by the Federal Aviation Administration (FAA) states that during a year, a jet engine generates data equivalent to 20 terabytes (TB). Most of this data is not used for any of the analytics purposes since this data is unstructured. Big data analytics can be used to predict the fault in the component by analysing data obtained from various sensors and accounts of the specific component attached to an aircraft or fleet in which it is present.
- A typical aircraft has millions of parts. The amount of data output is immense and is increasing rapidly with a new generation of aircraft. Legacy aircraft used to capture more than 125 flight parameters. The Airbus A320, for example, produces 20,000 data parameters, but the latest A350 produces 400,000 and has a data output of approximately 250 gigabytes (GB) per flight. Boeing 787 captures more than 1000 flight parameters, with some reports claiming half a terabyte of data per flight.

- Nowadays, data mining, which is a broad field of data science used to make predictions on future data based on patterns found in collected data, can be used to improve aviation safety by analysing all the data collected from the reports and searching for patterns and anomalies that indicate potential incidents and hazardous situations before they happen. However, this method presents several limitations, e.g. the requirement of some knowledge of what should be searched for and found and, therefore, the necessity of a method that determines the relevant and important data. Besides, as the data included in reports is overly generalised, it is difficult to identify unknown patterns, and, in consequence, data mining usually discovers first already known trends.

Key Actions: to achieve PARE objective 34, it is recommended that the expected benefits versus the required investment in using big data techniques in aeronautical activities for which relevant data already exists or has to be newly sourced be considered.

10.8.7 Artificial Intelligence

Artificial intelligence (AI) has the advantage of systematic learning from large amounts of data ('Artificial Imitation'), and therefore it can lead to the more efficient and reliable implementation of known tasks. However, it lacks any form of imagination when facing unforeseen or untrained situations, and it cannot replace human imagination, resourcefulness or true and real intelligence. Considering this, by 2050, the potential benefits and risks of the use of AI in aeronautics must be assessed.

Key Findings

- Some existing AI use cases in aviation include:

    1. Flight management computer systems are used to assist the pilot with information and for handling purposes.
    2. Computer systems behind the autopilot feature steer the aircraft along a predefined trajectory without the need for pilot intervention.
    3. Computer systems supporting automated aircraft cabin pressurization to ensure the cabin environment is safe and comfortable for passengers.
    4. Analysis and prediction of passenger behaviour/demand.
    5. Seamless airport security processes, through facial recognition and biometrics, are currently being used to perform customer identity verification and to match passengers to their luggage through kiosks;
    6. Provision of support for the optimization of revenue management, route network, fleet management, and pricing strategies;
    7. Self-flying planes and autonomous in-flight services, e.g. meals, snacks, beverages, duty-free shopping and others, made available during a flight for the convenience of the passenger;

8. Autonomous airport processes, e.g. ground handling, loading, fuelling, cleaning, and aircraft safety checks;
9. AI assistants that answer customer enquiries and voice commands for domestic airline flight information and ticket availability through interactions using natural language; and
10. Aeroplane simulators, which process the data taken from simulated flights. One example is aircraft warfare, where computers can come up with the best success scenarios in these situations, as these machines can create strategies based on the placement, size, speed and strength of the forces and counterforces;

- Despite all the benefits and advantages of AI, there are still many problems to overcome before its widespread applications: (i) dependence on large datasets (that can still have limited training data and lead to the failure of the model) and lack of abstraction and generalisability; (ii) lack of interpretation of the obtained trained models and use of current AI approaches as a black box; and (iii) lack of robustness, since most current AI systems can be easily fooled, which is a problem that affects almost all machine learning techniques.

Key Actions: to achieve PARE objective 35, it is recommended that the situations in aeronautics in which the learning processes of AI can be efficient and safe are identified, and those where there is limited training material or potential dependence on imagination are distinguished.

10.8.8 New Materials and Structures

The development of new materials and structures employing them is driven both by the pull of aviation needs and by the progress of technologies. The contribution of materials and structures to progress in aeronautics has been steady and gradual, e.g. a 10–30% weight saving replacing a metal wing or fuselage by a full composite one, and the prospect of structures having a fraction of their current weight is brought by nanotube developments. Considering this, by 2050, the implications on aircraft design, certification, operation and maintenance of new developmental and breakthrough materials and structures, including nanotube structures and microelectromechanical devices, must be assessed.

Key Findings

- Progress in materials has been a key to advances in aeronautics since its birth, at least in two areas: (i) strong and light structures able to withstand flight and landing loads without demanding more than the essential lift and incurring excessive drag; (ii) high-temperature turbine blades of jet engines subject to large rotational and aerodynamic loads combined with high heat fluxes;
- Forty years ago, aluminium dominated the aerospace industry—as much as 70% of an aircraft structure, from the fuselage to main engine components. Materials

such as titanium, graphite, fibreglass or composites and alloys were also used, but only in very small quantities. Nowadays, a typical aeroplane is constructed as little as 20% pure aluminium, and most of the non-critical structural material consists of even lighter-weight carbon fibre-reinforced polymers (CFRPs) and honeycomb materials;

- While CFRPs represent the lion's share of composite material in both cabin and functional components, and honeycomb materials provide effective and light-weight internal structural components, next-generation materials include high temperature ceramic-matrix composites (CMCs), which are emerging in practical use after decades of testing;
- For engine parts and critical structural components related to safety, there is a simultaneous push for lower weight, increase strength and higher temperature resistance, intending to reduce the cost of aerospace manufacture, improve fuel economy through efficiency and light weighting. Considering that the melting point of current superalloys is around 1850 °C, to meet these temperature demands, heat-resistant superalloys (HRSAs), including titanium alloys, nickel alloys, and some non-metal composite materials such as ceramics, are now coming into use;
- The major drivers in current aerospace R&D are towards lighter construction materials and more efficient engines, with the overall goal being to reduce fuel consumption and carbon emissions associated with air travel and air freight. The significant interest in nanotechnology for the aerospace industry is justified by the potential of nanomaterials and nanoengineering to help the industry achieve this environmental goal.

Key Actions: to achieve PARE objective 36, it is recommended that the progress in advanced materials like composites, ceramics, and special alloys and also the prospects of less mature developments like nanotube structures and other nanotechnologies be considered:

(a) ***OBJECTIVE 29:*** *Keep at the forefront of progress in the electrification of aircraft.* **Recommendation 29:** Make a thorough assessment followed by support measures on (a) emerging electric systems and propulsion technologies, (b) their potential to satisfy mission requirements and (c) the likely evolution of both. **Rationale:** Although the automobile sector may lead the electrification of transport vehicles, the specific needs of aeronautics and fast technological evolution will have increasing importance from drones to aeroplanes. **Stakeholders:** EU, MS, AI, RC, UA, AR. **Relevance:** Small electric drones, emerging electric air taxis, more electric airliners with bleedless engines, and advances in electric propulsion and systems all point towards increasing electrification. **Priority:** Progress in electrification is rapid, leading to major market impact. **Justification:** PARE report Section 8.1.

(b) ***PARE OBJECTIVE 30:*** *Promote and exploit advances in additive manufacturing.* **Recommendation 30:** Consider the implications on prototyping, series production and spares supply of additive manufacturing regarding (a) usable

materials, (b) quality standards and (c) life-cycle costs. **Rationale:** Although additive manufacturing is an industry-wide activity, aeronautics may be the leader in some areas and should keep up with progress in other areas. **Stakeholders:** EU, MS, AI, RC, UA, AR, CA. **Relevance:** Additive manufacturing allows the production of complex pieces with less parts and could replace spare part inventories with local production, as limitations in series production, choice of materials and quality finish are overcome. **Priority:** Aeronautics needs to benefit from general progress and lead only where necessary. **Justification:** PARE report Section 8.2.

(c) ***PARE OBJECTIVE 31:*** *Incorporate the experience from other sectors to achieve more efficient and economical production in aeronautics.* **Recommendation 31**: Consider the best combination of 4.0 technologies, including automation and information, as applicable to various production rates and scales of equipment in aviation. **Rationale:** Aeronautics can benefit from 4.0 technologies, especially by integrating the full life-cycle design-production-operation-maintenance. **Stakeholders:** EU, MS, AI, AR, CA, RC, UA. **Relevance:** The 4.0 factory requires significant investment and reliance on big data and artificial intelligence and could lead to reductions in development time and cost of production and operation. **Priority:** Aeronautics is a user more than a leader except in specific areas, with the large-scale integration over a long-life time as the main issue. **Justification:** PARE report Section 8.3 and topic 8.1.

(d) ***PARE OBJECTIVE 32:*** *Provide the telecommunications capacity needed for connected aircraft, navigation, monitoring and other services.* **Recommendation 32*:** Assess the growth of capacity needs for navigation, systems monitoring and passenger services and how the required bandwidth can be provided free from unintended or malicious interference. **Rationale:** Increases in air traffic capacity and safer and more efficient aircraft operations will require increased data sharing and larger telecommunications bandwidth protected from degradation or interference. **Stakeholders:** EU, MS, AN, AR, AI, RC, UA, AP. **Relevance:** The growth of air traffic and progress in safety plus passenger services may require more than proportional increases in data sharing and telecommunications. **Priority:** This is an issue for society in general, with particular incidence in aviation. **Justification:** PARE report Section 8.4 and topic 8.2.

(e) ***PARE OBJECTIVE 33:*** *Monitor threats in cybersecurity and devise timely protection for all levels of aviation systems.* **Recommendation 33:** Consider the levels of cyber protection needed for the various aeronautical activities and how to monitor and counter threats. **Rationale:** The informatics incidents at specific air traffic control facilities and with airline reservation systems suggest possible vulnerabilities that could also be exploited in malicious attacks. **Stakeholders:** EU, MS, AN, AR, AP, AI, CA. **Relevance:** Security in aviation would be enhanced by a comprehensive cyber protection system with several levels, the highest for flight-critical tasks and the lowest for commercial ticket-

ing. **Priority:** The consequences of cyber-attacks have been seen elsewhere; aviation in the past has been a preferred target of malicious actions, and thus preventive measures are desirable. **Justification:** PARE report Section 8.5 and topics 8.3 and 8.4.

(f) ***PARE OBJECTIVE 34:*** *Assess the implications in aeronautics of advances in big data, including the use of what is already available.* **Recommendation 34:** Consider the expected benefits versus the required investment in using big data techniques in aeronautical activities for which relevant data already exists or can be newly sourced. **Rationale:** Large amounts of systems' data are already recorded by aircraft in flight, and its use for maintenance, safety or other purposes could be enhanced. **Stakeholders:** AR, AN, AP, EU, MS, AI, RC, UA. **Relevance:** More benefit could be taken out of currently collected data, and there is the prospect of gathering more data with a well-targeted strategy. **Priority:** This is part of a global trend shared by aviation. **Justification:** PARE report Section 8.6 and topics 8.5, 8.6 and 8.

(g) ***PARE OBJECTIVE 35:*** *Assess the potential benefits and risks of the use of artificial intelligence in aeronautics.* **Recommendation 35:** Identify the situations in aeronautics in which the learning processes of artificial intelligence can be efficient and safe and distinguish those where there is limited training material or potential dependence on imagination. **Rationale:** Artificial intelligence has the advantage of systematic learning from large amounts of data ('artificial imitation') but lacks any form of imagination when facing untrained situations. **Stakeholders:** EU, MS, AI, RC, UA, AR, AN, AP. **Relevance:** The use of artificial intelligence can lead to much more efficient and reliable implementation of known tasks, but care is needed concerning unforeseen or untrained situations when it cannot replace human imagination, resourcefulness or true and real intelligence. **Priority:** Part of a general trend that aviation may share without losing an eye on risks. **Justification:** PARE report Section 8.7 and topics 8.8 and 8.9.

(h) ***PARE OBJECTIVE 36****: Assess the implications on aircraft design, certification, operation and maintenance of new developmental and breakthrough materials and structures.* **Recommendation 36:** Consider the progress in advanced materials like composites, ceramics and special alloys and also the prospects of less mature developments like nanotube structures and other nanotechnologies. **Rationale:** Progress in materials, structures and production methods has a main steady gradual element (composites, ceramics, alloys) and there is also a potential for revolutionary breakthroughs (nanotube structures, microelectromechanical devices). **Stakeholders:** AI, RC, UA, EU, MS, AR, CA. **Relevance:** Research on materials, structures and production is a major topic, and a complement is some forward look on the maturation of long-term prospects. **Priority:** Maintaining the current high level of activity, perhaps with a little more emphasis on promising developments still lacking maturity. **Justification:** PARE report Sections 8.8 and 8.9, and topics 8.10, 8.11, 8.12 and 8.13.

10.9 International Cooperation

Objectives 37 to 42 are addressed in the 9th chapter of PARE's report, entitled "Cooperation Beyond Europe's Borders", which concerns areas of aeronautics that require cooperation between the worldwide communities. These include Air Traffic Management (ATM), harmonized certification rules, environmental effects, safety and security, and fair trade and open markets.

10.9.1 Air Traffic Management

Air traffic is expected to continue to grow at a rate of 2–7% per year, depending on the region of the world. The main potential limitation to this continuing growth is the capacity of the air traffic system, including airports and ATM. Airports are a local issue, although with far-reaching geographical implications in the case of major hubs. On the other side, ATM is a global issue in the sense that it should function seamlessly worldwide, over continents and across oceans, and in densely, sparsely populated or uninhabited regions.

Therefore, ATM is the major potential bottleneck to the growth of aviation on a global scale, with a particular incidence in developed regions with dense traffic like Europe and the north-eastern US but gradually spreading to other regions. Considering this, by 2050, the seamless compatibility of ATM systems must be promoted worldwide, across continents and oceans.

Key Findings

- When traffic approaches the available capacity, there is a combination of entirely undesirable consequences: (i) departure and arrival delays that cause passenger dissatisfaction and can hinder business activities; (ii) aircraft flying in holding patterns awaiting permission to land and takeoff queues of aeroplanes waiting to gain access to a runway; and (iii) increased fuel consumption, pollution and noise, precisely near the airport areas where these issues are more sensitive. The economic losses are not just increased fuel costs and lost revenues for airlines but also the loss of valuable time for passengers and business travel.
- Through advances in technology and procedures, ATM in Europe and the USA has mostly managed to stay ahead of the growth of air traffic, but not by a wide margin all the time, so there are still occasional delays, and the overall challenge remains. This challenge is recognized at a political level, as testified by the large programs SESAR (Single European Sky ATM Research) in Europe and NextGen (Next Generation Air Transportation System) in the USA that aim to keep air traffic capacity ahead of air transport growth and avoid the risk of large scale flight delays and cancellations.
- The market for ATM equipment and services includes radars, navigation and communication systems, satellite links, equipment for Air Traffic Control (ATC) centres and control towers, including operator consoles and other hardware and

operating systems incorporating sophisticated software. The market for ATM equipment and services is considerable and not to be underestimated compared with the market for aircraft and airline services, since they are all complementary and interdependent.

- The USA, by being a single nation, has a unique Air Navigation Services Provider (ANSP) and certification authority, which is the Federal Aviation Administration (FAA). On the contrary, in Europe, although the Single European Sky (SES) initiative was launched at the beginning of the present century, the national border division of the airspace leads to a diversity of ATM services, dominated by national ANSPs as well as national certification authorities. Despite all these factors, Europe betters the USA in most ATM performance indicators (e.g. timeliness of flights) and achieves the same or higher safety standards.
- Even though SESAR and NextGen have the same basic aim, the implementation frameworks for each are radically different, with the European approach based on a single, multi-stakeholder consortium, Eurocontrol and the US model requiring close internal coordination by the FAA between various government-led programmes to ensure interoperability of components delivered by a variety of consortia. To ensure compatibility, the EU and US systems don't need to be identical but have aligned requirements for equipment standards and technical interoperability, which should be extensible to other regions of the world with comparable or lower air traffic densities.

Key Actions: to achieve PARE objective 37, it is recommended that cooperation between SESAR in Europe and NextGen in the USA be supported to ensure compatibility across the North Atlantic and provide the basis for progress in the world ATM market as the growth of air travel increases capacity needs elsewhere.

10.9.2 Harmonized Certification

Aircraft certification is the process whereby an applicant requests approval from an aviation regulatory authority, such as the FAA and EASA, for manufacturing a new aircraft model or making changes to an existing aircraft. Some decades ago, different national certification authorities required different tests for the same purpose, duplicating effort and increasing cost with no benefit to safety or efficiency. The harmonization of certification standards avoids such costly duplications.

Since FAA and EASA are the leading certification authorities, the continuation of common or compatible certification standards and the mutual acceptance of certification results should continue as new technologies emerge and possibly new aircraft configurations as well. Taking this into account, by 2050, harmonized certification standards must be promoted worldwide as they already exist in other sectors to ensure the growth of aviation as one of the safest modes of transport.

Key Findings

- The certification of an airliner is the final stage of the development process and can also be the most complex, time-consuming and expensive, involving approved standards, guidance, tests, methods and procedures, as well as data submittals and plan documentation. It takes about 3000 flying hours for 3–5 years involving 3–6 prototype or pre-production aircraft, and it is difficult to compress without significantly increasing risks that could become delays and further costs.
- The overall framework for harmonized and coordinated certification between EASA and the FAA is currently established by the agreement between the USA and the EU on cooperation in the regulation of Civil Aviation Safety, which entered into force in 2011. The main purposes of the agreement are to automatically accept certain approvals issued within the other certification system and enable the reciprocal acceptance of findings of compliance during validation processes.

It is desirable that harmonized certification extends from the FAA and EASA as the references to other aviation authorities around the world if comparable safety standards can be assured; a good example was cooperation in the conditions for ungrounding the B737Max after two accidents (Chap. 4).

Key Actions: to accomplish PARE objective 38, it is recommended to strengthen the cooperation of EASA/FAA on common certification standards and their adoption worldwide to avoid duplication or degradation in specific regions.

10.9.3 Aviation Effects on the Environment

The effects of aviation on the environment can be considered at two levels: (i) locally as the emissions and noise near airports; (ii) globally as in-flight emissions worldwide. Aviation contributes a small percentage (about 3.5%) to global pollution of the environment caused by human activity, but its influence is concentrated and unfavourable locally, in the areas of airports. Nevertheless, contrary to airport noise (generated by takingoff and landing aircraft) that can be solved at a local level, emissions are a global issue and require considerable international negotiation. Considering this, by 2050, the effects of aviation on the environment on a global scale must be minimized, consistently with targets like Fit 55 in the EY and NetZero 2050 by ICAO.

Key Findings

- Emissions of aviation are the result of the combustion of fuel used to power the aircraft, and its level depends on the fuel quality and the process of combustion. The basic fuel used in modern civil aircraft is aviation kerosene. During the different phases of aircraft operation, several greenhouse gases (GHG) are emitted to the atmosphere, including (i) carbon and (ii) nitrogen oxides, (iii) water vapour, (iv) other substances, (v) particles and (vi) aerosols.

- In 2012, it was estimated that European aviation represented 22% of global aviation's CO_2 emissions. Similarly, aviation now comprises 14% of all EU transport NOX emissions and 7% of the total EU NOX emissions. From 2005 to 2014, CO_2 emissions increased by 5%. The increase in emissions is, however, less than the increase in passenger kilometres flown over the same period, due to an improvement in fuel efficiency driven by the introduction of new aircraft, the removal of older aircraft, and improvements in operational practice.
- Forecasts indicate that future technology improvements are unlikely to balance the effect of future traffic growth (and consequently emissions growth). Although alternative clean propulsion technologies are under development, such as electric-powered aircraft or cryogenic hydrogen fuel, these options are unlikely to be commercially ready before 2030. Alternatives include sustainable aviation fuels (SAFs), which are produced from various feedstocks that have lower GHG emissions. The main challenges with hydrogen and synthetic aviation fuels are (i) to produce the huge quantities required at (ii) costs not too much higher than fossil fuels, bearing in mind (iii) the available feedstocks and (iv) large investment in production facilities.
- Emission regulations can also play an important role in mitigating the environmental impact of aviation, but they need to be implemented on a global scale. The International Civil Aviation Organization (ICAO), created in 1944 as a specialized agency of the United Nations (UN), has approved a set of global market-based measures to address GHG emissions from international aviation, entitled CORSIA—Carbon Offsetting and Reduction Scheme for International Aviation.

Key Actions: to achieve PARE objective 39, it is recommended that the reduction of environmental effects of aviation on a global scale is a key point in worldwide cooperation.

10.9.4 Safety

Aviation authorities in some developed countries have banned some foreign airlines deemed not to meet adequate safety standards. This is necessary to protect the safety of those flying for business or leisure travel and also of the residents that could become the victims of eventual accidents. On one hand, the list of banned airlines could be of use to warn passengers that might be attracted to fly with cheap or unsafe airlines in other regions of the world. On the other hand, an effort to cooperate with aviation regulatory authorities worldwide, in particular in less developed regions, helping them to implement safety standards by providing technical assistance, would be a preventive measure, leaving bans as the necessary last resort in fewer cases. Taking this into account, by 2050, aviation safety must be promoted worldwide, including for all passengers flying with international, national or local airliners and flight providers.

Key Findings

- In principle, all airliners should be equally safe because they meet the same applicable EASA/FAA certification standards. Airbus/Boeing/Bombardier/Embraer and other manufacturers have comparable engineering skills and thoroughly develop operating and maintenance procedures. As a consequence, aviation remains one of the safest modes of transport, although with relatively large differences across the globe.
- The reasons for reduced relative safety in other regions of the world can be several: (i) persistence of extreme weather conditions in some regions, like arctic, tropical or deserts; (ii) operation of older aircraft requiring more careful maintenance; (iii) less adherence to maintenance and operating procedures that conditions (i) and (ii) require; (iv) weaker oversight by authorities. It must also be acknowledged that in all regions of the world, the safety standards also vary considerably depending on the type of operation: (i) airliners and business jets are much safer than private aircraft (ii) transport is safer than crop spraying or firefighting that involve low-altitude flying near obstacles and obscurants.
- Taking as an example Europe, the blacklist program began in 2005. First, an EU Member State (MS) identifies airliners subject to operating bans within their territory (after inspecting its aircraft that landed at the MS airport), and afterwards, the FAA and EASA, in coordination with the EC, evaluate the airliners on common criteria. The resulting EU Air Safety List, or "blacklist", which includes a list of all airlines banned from operating in Europe and another of airlines that are restricted from operating under certain conditions in Europe, is published at least every three months in the Official Journal of the European Union. These banned airliners are encouraged to improve their levels of safety and can request a compliance review from the EC to be removed from the list.
- ICAO is the primary forum for cooperation in all fields of civil aviation among its 193 MS. The UN agency promulgates Standards and Recommended Practices (SARPs) to facilitate harmonised regulations in aviation safety, security, efficiency and environmental protection on a global basis. As part of this effort, ICAO established in 2015 the Aviation Safety Implementation Assistance Partnership (ASIAP) to coordinate efforts for the provision of assistance to its MS and created the initiative "No Country Left Behind" (NCLB), which focuses on assisting all MS on a priority basis to provide support for the implementation of ICAO's SARPs.
- Despite the excellent safety performance of aviation in Europe, recent events remind the need to always remain vigilant and constantly search for weaknesses in the system before they manifest in an accident. EASA, MS and industry have been working closely together in safety risk management (SRM), namely hazards identification, risk assessment and decision making on the best course of action to mitigate those risks at the European level. EASA's 9th edition of the European Plan for Aviation Safety (EPAS), which covers the five years between 2020 and 2024, puts increased emphasis on the importance of the SRM process.

Key Actions: to achieve PARE objective 40, it is recommended to support activities raising the aviation safety standards to more uniform high levels across the globe, in particular helping the improvement of airlines banned from flying into Europe that may still carry European passengers elsewhere.

10.9.5 Security

Citizens often travel to countries with popular tourist destinations and important business hubs outside the jurisdiction of the authorities that apply stricter security standards (such as Europe, the USA or Japan) and are at greater risk of security threats like extremist groups that target aviation and foreigners. To boost security in these less developed regions and continue to attract business and tourism, which may be of interest to these regions, it is fundamental to support and cooperate with the local authorities. Thus, by 2050, aviation security must be promoted worldwide, including at airports and destinations frequently used for business and holiday travel.

Key Findings

- Integrating security systems and operations into the planning and design of airport construction and refurbishment projects can be a very complex task since (i) security systems involve equipment, technologies, procedures and operational approaches that need clear and concise guidelines; (ii) there is an environment of evolving threats, often accompanied by the implementation of new legal or regulatory requirements and operational updates to counter the changing threat conditions; (iii) security systems are inherently difficult to plan, design, and implement when applied to airports, which are designed to facilitate the fast and efficient movement of customers and goods.
- Airports tend to be in a constant state of change in terms of their physical layouts, operations, and tenants, and while the number of new airports being built is relatively small, many airports and terminals are being remodelled, expanded, and upgraded. The majority of changing security requirements will be accomplished in existing facilities that are often decades old, designed at a time when the threat profile and the security environment were dramatically less stringent than they are today.
- The airport operator has a responsibility to provide a safe and secure operating environment and infrastructure. The extent of necessary facility protection should be examined by the local Airport Security Committee based on the results of a comprehensive security assessment of the existing facility. High priority should be placed on the protection of the aircraft from the unlawful introduction of weapons, explosives, or dangerous substances.
- Secure air transport service enhances connectivity in trade, tourism and political and cultural links between states. ICAO has a Global Aviation Security Plan (GASeP) that seeks to unite the international aviation security community and help MS and stakeholders enhance the effectiveness of global aviation security.

Based on the main challenges that ICAO MS face, GASeP identifies key priority outcomes where ICAO, States and stakeholders should focus their urgent attention, resources and efforts: (1) enhance risk awareness and response; (2) develop security culture and human capability; (3) improve technological resources and foster innovation; (4) improve oversight and quality assurance; and (5) increase cooperation and support.

Key Actions: to accomplish PARE objective 41, it is recommended to support high-security levels at airports outside Europe by cooperating with authorities eager to keep European business/tourism travel and otherwise warn travellers of risk.

10.9.6 Fair Trade and Open Markets

Aircraft manufacturing is arguably one of the most contested industries in international trade governance, mainly because the traditional tools of trade governance are particularly ill-suited to aircraft manufacturing. Considering this, promoting a level playing field is not easy, and desirability and quality may function more effectively and subtly than harsh political pressures. Regarding open markets, major exceptions in civil (all non-military) aviation do exist, generally uncontested, being the most visible and long-standing the Japanese airliners' tendency to buy Boeing aircraft more often than from Airbus, though the difference has reduced over time. To help undermine protectionist and biased local choices, safe and efficient aircraft should be made available. Taking into account this, by 2050, an open and fair market for aircraft must be promoted as far as possible, at least in the civil sector.

Key Findings

- Trade disputes are not unknown in aviation at the highest level of the World Trade Organization (WTO). Since Airbus emerged some 40 years ago to challenge Boeing's position as the world's dominant aircraft manufacturer, the US and EU governments on behalf of Boeing and Airbus, respectively, have been accusing one another of illegitimately propping up their respective national champions while simultaneously professing their own innocence in providing support. The Canadian government on behalf of Bombardier has also accused the Brazilian government of subsidizing Embraer.
- The basic principle of a "level playing field" is that no competitor should be at an advantage from government intervention, such as subsidies, loans at below market rates and tax breaks with regard to production and export. The issue becomes more serious in countries with less transparent financing methods.
- The interest of airlines and their passengers is open competition on performance, comfort, safety, efficiency and environmental protection regardless of manufacturer or country of origin, both for the complete aircraft and its components and systems.

Key Actions: to achieve PARE objective 42, it is recommended a competition based on quality rather than other interests, which is supported by the advances in efficiency and compliance with the highest environmental, safety and security standards.

(a) ***PARE OBJECTIVE 37:*** *Promote the seamless compatibility of Air Traffic Management (ATM) systems worldwide, across continents and oceans.* **Recommendation 37:** Support cooperation between SESAR in Europe and NextGen in the USA to ensure compatibility across the North Atlantic and provide the basis for progress in the world ATM market as the growth of air travel increases capacity needs elsewhere. **Rationale:** Air Traffic Management is a major potential bottleneck to the growth of aviation on a global scale, with particular incidence in developed regions with dense traffic like Europe and the Eastern US but gradually spreading to other regions. **Stakeholders:** EU, MS, AN, AR, AP, AI, RC, UA. **Relevance:** Air Traffic Management should be seamless across continents and oceans, and the needed increases in capacity combined with safety and security are a major market for equipment suppliers worldwide. **Priority:** It is essential to avoid the congestion and delays of the past and cope with traffic growth and new users like UAVs. **Justification:** PARE report Section 9.1 and topics 9.1 and 9.2.

(b) ***PARE OBJECTIVE 38:*** *Promote harmonized certification standards worldwide, as already exist in other sectors, to ensure the growth of aviation as one of the safest mode of transport.* **Recommendation 38:** Strengthen the cooperation of EASA/FAA on common certification standards and their adoption worldwide to avoid duplication or degradation in specific regions. **Rationale:** The coordinated and mutually accepted certification by either the FAA or EASA is a breakthrough in avoiding costly duplication and preventing misuse of certification as a trade barrier. The Russian example of local certification is being followed by China, whose aircraft have faced long delays and major difficulties in obtaining EASA or FAA certification. Resorting to 'local certification' leads to lower safety standards that can affect not only locals but also others travelling in those countries. The export of such EASA/FAA-uncertificated aircraft could damage the unique overall safety record of aviation. **Stakeholders:** The EU and MS, possibly with US coordination, since there is a common interest in supporting EASA/FAA standards. **Relevance:** The EU and MS could insist on cooperation being conditional on progress towards worldwide certification standards. Although aircraft not certificated by EASA or FAA cannot operate in Europe, USA or other developed regions, their use as cheap, unsafe transport elsewhere cannot be encouraged and puts foreign visitors at risk. **Priority:** It is prudent to prevent the emergence of a parallel market of local or Third-World aviation with degraded safety standards that are already lower elsewhere than in Europe/USA. This may require diplomatic and negotiation skills. **Justification:** PARE report section 9.2 and topics 9.6 and 9.7.

(c) ***PARE OBJECTIVE 39:*** *Minimize the effects of aviation on the environment on a global scale.* **Recommendation 39:** The reduction of environmental effects

of aviation on a global scale should be a key point in international cooperation. **Rationale:** The EU has played a leading and exemplary role as champion of environmental protection in all areas, including aviation and this is a worthy effort to pursue. **Stakeholders:** EU, MS, AN, CA, AI, RC, UA, AP. **Relevance:** Although aviation is not the biggest polluter in general or among means of transport, the objective should be to reduce its percentage by doing better than average progress. **Priority:** There is already a significant effort, although airport noise and on-route emissions are increasing concerns. **Justification:** PARE report Section 9.3 and topic 9.5.

(d) ***PARE OBJECTIVE 40****: Promote aviation safety worldwide, including for national and other passengers flying with foreign airlines.* **Recommendation 40:** Support activities raising the aviation safety standards to more uniform high levels across the globe, in particular helping the improvement of airlines banned from flying into Europe that may still carry European passengers elsewhere. **Rationale:** The differences in aviation safety standards among airlines from different countries justify banning some from flying into Europe or the USA, and the provision of some technical assistance could help lifting the bans and improving safety in less developed regions. **Stakeholders:** EU, MS, CA, AN. **Relevance:** Besides ensuring safety (and security) of air travel in Europe and the USA, it is desirable to promote similar standards in less developed regions of the world, both for the locals and for European and US business or leisure travel. **Priority:** This is an extension of current preventive bans to cooperative improvement for non-European and non-US airlines and aviation authorities. **Justification:** PARE report Section 9.4 and topics 9.6 and 9.7.

(e) ***PARE OBJECTIVE 41:*** *Promote aviation security worldwide, including at airports and destinations frequently used for business and holiday travel.* **Recommendation 41:** Support high security levels at airports worldwide by cooperating with authorities eager to keep business/tourism travel and otherwise warn travellers of risk. **Rationale:** Travellers may be at greater risk in less developed regions where extremist groups target aviation and foreigners, and thus support for cooperative local authorities is important to boost security. **Stakeholders:** EU, MS, CA, AP, AI. **Relevance:** Local authorities in Third World countries may have an interest in attracting business and tourism and welcome assistance in making foreigners and their own citizens safer. **Priority:** This is a continuous, gradual and moderate effort. **Justification:** PARE report Section 9.5 and topic 9.8.

(f) ***PARE OBJECTIVE 42:*** *Promote as far as possible an open and fair market for aircraft, at least in the civil sector.* **Recommendation 42:** The advances in efficiency and compliance with the highest environmental, safety and security standards can contribute to a competition based on quality rather than other interests. **Rationale:** Making available safe and efficient aircraft may help undermine protectionist and biased local choices. **Stakeholders:** EU, MS, AI, RC, UA. **Relevance:** Promoting a level playing field is not easy, and desire and quality may function in a more effective and subtle way than harsh political pres-

sures. **Priority:** It is a matter of good business practice. **Justification 43:** PARE report Section 9.6 and 9.7 and topics 9.9 and 9.10.

10.10 Attracting Young Talent to Aeronautics

PARE objectives 43 to 48 are addressed in the 10th chapter of PARE's report, entitled "Attracting Young Talent to Aeronautics", which analyses the reasons and factors that influence the career inclinations of young people and the measures to consider attracting them to aeronautics.

According to the Social Cognitive Career Theory (SCCT), the academic and career choices of young people are influenced by personal, cognitive and contextual factors (namely prior experience, social support, self-efficacy and outcome expectation) and form gradually more consciously from childhood and primary school through teenage and secondary school to adulthood and university. The career choices also depend on the image and reputation of the employer and its ability to provide good living and working conditions and to foster the commitment and fidelity of the workforce as a second family.

To attract young people to aeronautics, it is also important to understand that aeronautics involves diverse job types (e.g. engineers, senior technicians and operators), with several requiring the so-called STEM (Science, Technology, Engineering and Mathematics) skills.

10.10.1 Childhood and Primary School

In the history of aviation, many of the pioneers and major contributors started their interest at an early age. It is therefore essential to stimulate the interest of children in aeronautics (objective 43), which can be done in a more informal context by parents or more formally in primary schools, through specific programmes or educational materials developed for children.

Key Findings

- At the beginning of childhood, children's attention span depends on their age and the interest that objects and tasks awaken in them, in particular their ability to manipulate these objects nearby. It is very difficult for kids to focus on a monotonous and less attractive activity, and consequently, they tend to easily move on from one activity to another frequently.
- Fortunately, flying-related stories are usually entertaining for children. In this manner, they usually enjoy doing activities such as playing with kites, listening to tales that take place in space, watching flying-related cartoons, and making paper gliders. These kinds of funny activities related to aviation can act as an incentive to awaken in kids the feeling of loving this thrilling sector, and, in this way, they could grow up willing to work in a job related to aviation.

Key Actions: to stimulate children's interest in aeronautics, it is necessary to make STEM fields funny and enjoyable for kids. To achieve this objective, PARE recommends making available online and accessible to primary schools and parents, children's stories and cartoons involving flying that are both entertaining and educational.

10.10.2 Teenage and Secondary School

The basic teaching of physics, natural science and mathematics, even at the secondary school level, can lead to some understanding of flying in the atmosphere. On a more practical side, 'toy' drones are now so common and inexpensive that they can be readily used to give real flying experience, train piloting skills and also teach responsible use. Moreover, these more relaxed activities, which can occur in leisure times at school or outside the school, can be complemented by visits to university laboratories and presentations from aeronautics professionals. This combination of advanced technologies and bright prospects should be employed to motivate teenage and secondary school students to choose university degrees in aerospace engineering (objective 44).

Key Findings

- A variety of informal and complementary learning activities can be promoted among students and STEM educators in a way to connect them with the STEM community and workforce, such as teaching youth at science summer camps or after-school programs; getting students to join STEM-related clubs, namely Space and Aeronautics Clubs; promoting and supporting students' participation in science fairs and competitions; creating job shadow opportunities; promoting visits to aerospace companies and giving them books and magazines on STEM topics.
- The student finishing secondary school has a wide and sometimes bewildering choice of university degrees to apply for. Even restricted to aerospace engineering, there are several universities with promising and high-quality degrees that will provide the necessary knowledge, skills and abilities that allow them to work efficiently within the aerospace field.

Key Actions: to achieve this objective, PARE recommends to:

1. Make available from the early teens, on-line and to secondary schools a set of easy-to-implement flight experiments and challenges, such as drones now so commonplace and cheap.
2. Give secondary school students at later stages the opportunity to come to presentations and laboratories at a university together with a parent/mentor or trusted friend.

10.10.3 Adulthood and University

The strong analytical and problem-solving skills of the graduates from the best aerospace engineering degrees are in much demand from other sectors in and out of engineering (e.g. civil construction, automotive domain, consultant, and financial services, etc.). Although an aerospace engineer would have as first-choice aeronautics or space, some consultant companies may be quicker to offer professionally enticing and well-paid job opportunities, often advertising before the university degree is complete. Considering this and the increasing competitiveness from other sectors, the aeronautical industry should do its best to attract the brightest graduates in aeronautical engineering to industry before they are lured away by attractive, well-paid offers elsewhere (objective 45).

Key Findings

Today, aeronautics and air transport are among the main drivers' cohesion and competitiveness, representing in Europe 220 billion euros and providing 4.5 million jobs in Europe (2019 data), a figure that is expected to double in 2050. These data reveal that aeronautics plays a key role in facilitating European economic growth and social inclusion, providing income to regions that are otherwise isolated and helping people to broaden their horizons.

Employment opportunities for which graduates in Aerospace Engineering are specifically trained lie mainly in the aerospace field: (i) major European aerospace industries; (ii) small- and medium-size industries that supply the former; (iii) agencies and contractors responsible for aircraft maintenance; (iv) airline companies; (v) ATM authorities; (vi) the air force and other military aviation sectors; and (vii) public and private bodies for testing in the aerospace field.

European-level data show that about 50% of aerospace engineers are employed in other industrial sectors, even in regions where aerospace activities are most strongly established and offer the greatest employment opportunities.

The famous "pipeline" problem in STEM fields, which is the need to attract more students and workers too and to retain them (especially women and minorities), can be related to self-efficacy expectations and has a special impact for aeronautics. Self-efficacy itself is linked directly to persistence and can be improved through increased self-awareness of the sector and formalized mentoring programs and initiatives that involve students in research or activities.

To ensure the persistence of students in the aeronautical sector after graduation, industry and other employers should engage them at an early stage through professional stays and follow-up with attractive job offers without delay after graduation. For example, the Partnership of a European Group of Aeronautics and Space Universities (PEGASUS) was established with the aim of attracting the best students and to offer highly relevant educational and research programmes. As of 2014, it included 25 members of academic institutions, and most PEGASUS engineering programmes also include one or several periods of practical training in laboratories or industrial structures.

Key Actions: to achieve this objective, PARE recommends providing industrial stays for students of aeronautical engineering with mentoring that values their skills and keeps track of the most promising for employment after graduation.

10.10.4 Careers in Aeronautics and Space

The aeronautical sector can be divided into two major areas: industry and air transport, both of which offer a diverse range of interesting career opportunities. However, young people may not be fully aware of all of them and/or need more insights and guidance to decide to pursue a career in aviation. Regarding the first area, to ensure promising students' choice (over other sectors, as stated before) or to make careers in the aeronautical industry interesting relative to other alternatives, the aeronautical industry must focus on fascinating technology with adequate reward (objective 46) and invite them during their university course, thereby establishing early links that ensure their choice before other attractions arise.

Key Findings

- The aeronautics industry develops and manufactures civil and military aircraft, helicopters, drones, aero-engines and other systems and equipment, which involves designing components and systems, generating CAD models and drawings and undertaking fluids and thermal analysis, etc. The industry work also involves testing the systems and equipment manufactured and supporting the products in service afterwards. For this, it is necessary a range of job categories:

  1. Engineers (system engineer, mechanical/electrical/thermal design and others)
  2. Senior technicians (production, logistics, quality assurance and others)
  3. Operators (assembly, testing, check-out and others)

- Air transport includes the transportation of people, goods and mail on regular lines and non-scheduled activities (charter, taxi, plane rental with the pilot, flight training and others) and relies on aeronautical maintenance activities and airport assistance activities (track and field operations, onboard trades and civil aviation professions). Considering this, job categories in this area include:

  1. Airlines or handling agents: flight crew (pilots and cabin crew), ground handling
  2. Airport operators in airport management, maintenance, security and operators: maintenance, repair and overhaul market (MRO) technicians, airport security and passenger screening operators, etc.
  3. On-site employees at retail outlets, restaurants, hotels, etc.
  4. Air navigation service providers: air traffic controllers

Key Actions: to achieve this objective, PARE recommends bringing an aeronautical engineering student together with a mentor/relative/trusted friend, who can

play a major role in advising a mature choice, to a one-day visit to industry, with briefings and access to facilities, to be remembered for life as a career choice.

10.10.5 Motivating and Rewarding the Workforce

Besides attracting people to the aerospace sector, it is necessary to have favourable conditions and create measures to motivate and reward the existing workforce in the sector to promote dedication, ingenuity, efficiency and loyalty (objective 47). The best motivation and reward for the workforce is the openness to new ideas and giving an opportunity for sensible innovation that also benefits the company's interests. Moreover, some companies (i) share a part of the profits with the employees; and (ii) allow employees to buy a limited number of shares at favourable prices with the condition that they cannot be sold for a number of years. Measures like these motivate employees to work harder for the aims of the company they share, knowing that the success due in part to their efforts will be recognized.

Key Findings

- According to the HiPAir project, a strategic partnership co-funded by the ERASMUS+ Programme of the European Commission, one of the new forms of work organisation that have been developed is the concept of High-Performance Work Practices (HPWP). Though many of the practices referred to as high performance are commonly used by most organisations to motivate workers, such as financial incentives, flexible job descriptions and continuous skill development programs, the concept of HPWP is fairly unknown. HPWP can be defined as modern management practices designed to stimulate employee and organisation performance and include, for example, recruitment and integration, employee involvement, internal communication, training and reward and commitment.

Key Actions: to achieve this objective, PARE recommends giving employees opportunities for innovation, letting them share important and interesting work, and recognising their contribution to the success of the company.

10.10.6 Retaining the Fidelity of Employees

The traditional model of a society of lifetime work in a company still exists in countries like Japan but is presented by some as a 'lack of mobility'. As an alternative to the modern high-mobility, short-term employment, and not excluding intermediate cases, there is nothing wrong with the old model of a company-family, which supports a dedicated workforce with long experience and attracts the next generation. The mobility of employment has more to do with cultural traditions than with company efficiency. Therefore, aerospace companies should retain a capable and

faithful workforce by providing stable employment, interesting work, reliable benefits, and a friendly environment (objective 48).

Key Findings

- Employee's satisfaction is a measure of how happy workers are with their job and working environment and can be seen as an emotional state that results from the evaluation of one's job or experience. To enrich the levels of employee satisfaction, some measures include (a) foster and reward the training of employees; (b) facilitate schedule flexibility; (c) give opportunities for professional improvement; (d) value the opinions expressed by the employees; (e) create a good work environment and conditions for workers.
- Employee fidelity can be defined as employees who are devoted to the success of their organization and believe that being an employee of this organization is in their best interest. Not only do they plan to remain with the organization, but they do not actively seek alternative employment opportunities. The maintenance of a faithful, long-experienced workforce can also attract new ideas and support further recruitment by proving a stable and progressive environment for progress.
- A significant percentage (approximately 30%) of the current aerospace workforce in developed countries is at or near retirement age. Considering this, companies have to ensure they maintain current expertise while forecasting future hiring needs, such as what skill sets will be needed and when to hire for them, and transferring the talent from older to younger employees, an issue that is becoming more acute as the older employees retire.
- This lopsided demographic mix, coupled with high attrition rates and increased labour mobility, poses serious risks to the industry. Companies can mitigate the risks with employee retention and succession planning. Besides wage, other factors determine the loyalty of the workforce in an aerospace organisation: perceived high job security (reduced risk of being made redundant), a good and reliable pension system, an effective health insurance scheme and an optional extension of the retirement age could only increase the attractiveness and excitement of employment in aerospace and retain talent, especially people with STEM skills.

Key Actions: to achieve this objective, PARE recommends having competent staff, able to adapt to change and to contribute to progress, sufficiently well integrated not to wish other careers and serve as an example to relatives and friends.

(a) ***PARE OBJECTIVE 43:*** *Stimulate the interest of children with entertaining stories about flying.* **Recommendation 43:** Make available online and accessible to primary schools and parents, children's stories and cartoons involving flying that are both entertaining and educational. **Rationale:** Flying and space can be fascinating subjects for children and can be used to make interesting stories. **Stakeholders:** EU, MS, AI, families, primary schools. **Relevance:** In the history of aviation, many of the pioneers and major contributors started their interest at an early age. **Priority:** Normal activity to be sustained. **Justification:** PARE report Section 10.1 and 9.7 and topics 10.2 and 10.3.

(b) ***PARE OBJECTIVE 44:*** *Motivate secondary school students to choose university degrees in aerospace engineering.* **Recommendation 44.1:** Make available to the early teens, online and in secondary schools, a set of easy-to-implement flight experiments and challenges, such as drones now so commonplace and cheap. **Recommendation 44.2:** Give secondary school students at later stages the opportunity to come to presentations and laboratories at a university, together with a parent/mentor or trusted friend. **Rationale:** There are readily available inexpensive 'toy' drones that give real flying experience and can also teach responsible use. Complemented by visits to university laboratories and presentations, this would continue to attract good candidates of both genders. **Stakeholders:** EU, MS, UA, secondary schools, families, friends. **Relevance:** Given the variety of university degrees that university candidates can choose from and the demands of 'hard skills' in STEM (Science, Technology, Engineering and Mathematics) aeronautics can attract the best as a combination of advanced technologies with bright future prospects. **Priority:** In addition to suitable dissemination, the visits to universities with aeronautical education can make all the difference in attracting the best candidates. **Justification:** PARE report Section 10.2 and topic 10.3.

(c) ***PARE OBJECTIVE 45:*** *Attract the brightest graduates in aeronautical engineering to industry before they are lured away by attractive, well-paid offers elsewhere,* e.g. *by consultants and financial services.* **Recommendation 45:** Provide industrial stays for students of aeronautical engineering with mentoring that values their skills and keeps track of the most promising for employment after graduation. **Rationale:** In order to attract bright graduates in aeronautical engineering with many alternative enticing job offers, industry and other employers should engage them at an early stage through professional stays and follow-up with attractive job offers without delay after graduation. **Stakeholders:** EU, MS, AI, RC, UA, AN, AR, AP. **Relevance:** The strong analytical and problem-solving skills of the graduates from the best aerospace engineering degrees are in much demand from other sectors in and out of engineering, so the aeronautical industry cannot afford to be slow in attraction measures. **Priority:** Attracting the best engineers is the key to progress in aviation and an absolute must to remain a leader as other developed, more populous or lower-cost regions and sections strengthen their competitive challenge. **Justification:** PARE report Section 10.3 and topics 10.4 and 10.5.

(d) ***PARE OBJECTIVE 46:*** *Make careers in the aeronautical industry interesting relative to other alternatives by focussing on fascinating technology with adequate reward.* **Recommendation 46:** Bring an aeronautical engineering student together with a mentor/relative/trusted friend to a one-day visit to industry to be remembered for life as a career choice. **Rationale:** An in-depth one-day visit to industry with briefings and access to facilities can be a day to be remembered for life and make a career decision. The company of a mentor/relative/trusted friend can play a major role in advising a mature choice. **Stakeholders:** EU, MS, AI, RC, UA, AN, AR, AP, PA, other employers. **Relevance:** Perhaps the most important question when choosing a career is "What will I do or be my

work?" and convincing answers with supporting evidence is the best reply. **Priority:** This kind of initiative already exists, and the idea of a memorable one-day visit could be expanded. **Justification:** PARE report Section 10.4 and topics 10.6 and 10.7.

(e) ***PARE OBJECTIVE 47:*** *Motivate and reward the workforce to promote dedication, ingenuity, efficiency and loyalty.* **Recommendation 47:** Give employees opportunities for innovation, let them share important and interesting work, and recognise their contribution to the success of the company. **Rationale:** Giving opportunities for the workforce to be creative and proactive rewards the commitments of employees, creates links with receptive management, and promotes the prosperity of the enterprise. **Stakeholders:** EU, MS, AI, RC, UA, AN, AR, AP, CA. **Relevance:** The best way to keep a skilled and up-to-date workforce is to follow-up on their initiatives for progress with suitable, mature support. **Priority:** Should be normal company practice and is well worth reminding and keeping in mind. **Justification:** PARE report Section 10.5 and topics 10.8 and 10.9.

(f) ***PARE OBJECTIVE 48:*** *Retain a capable and faithful workforce by providing stable employment, interesting work, reliable benefits and a friendly environment.* **Recommendation 48:** A worthwhile alternative to job mobility is to have competent staff, able to adapt to change and to contribute to progress, sufficiently well integrated not to wish other careers and serve as an example to relatives and friends. **Rationale:** As an alternative to the modern high-mobility, short-term employment, and not excluding intermediate cases, there is nothing wrong with the old model of a company-family, which supports a dedicated workforce with long experience and attracts the next generation. **Stakeholders:** EU, MS, AI, UA, AR, AN, AP, CA, other employers. **Relevance:** The maintenance of a faithful, long-experienced workforce can also attract new ideas and support further recruitment by proving a stable and progressive environment for progress. **Priority:** Should be normal company practice worth keeping. **Justification:** PARE report Section 10.6 and topics 10.10 and 10.11.

10.11 Increasing the Participation of Women in Aerospace

PARE objectives 49 to 58 are addressed in chapter 11 of PARE's report, entitled "Increasing the Participation of Women in Aerospace", which analyses the factors that influence the participation of women in this sector and proposes measures to attract and retain more women.

Some of these factors are: women's interest in the aerospace sector and the women's self-confidence in STEM (Science, Technology, Engineering and Mathematics) subjects; the gender stereotypes that women are exposed to in the educational context and the gender imbalance in STEM education; the inequality of employment opportunities, the gender imbalance in STEM jobs; and the unequal treatment and labour conditions.

10.11.1 Generating Interest in Aerospace and Building Confidence

Evidence suggests that women are underrepresented in some areas of work, notably those where some knowledge of STEM subjects is required. Main reasons for this pattern include a lack of encouragement from friends, family and teachers; a lack of awareness—that is, reduced prior knowledge of STEM as a career option; and a lack of young women's self-confidence in STEM roles, which in turn is partially caused by the societal bias that girls face since childhood. Considering this, it is fundamental to counter family and societal bias discouraging girls from the interest in vehicles, be it cars or planes (objective 49).

Key Findings

- Girls and boys also start understanding gender stereotypes or societal bias regarding gender as early as age two, and they learn to adjust their behaviour according to internalised gender stereotypes by age four. The assumption that "girls play with dolls and boys play with cars/planes" and other predominant stereotypes with relation to gender and STEM (e.g. "boys are better at math and science than girls" or "science is for men, not for women") makes it particularly challenging for girls to see STEM as a potential career choice and, on the other hand, may equip boys with easily available and pre-established roles in science and technology.
- A study commissioned by the European Commission's (EC) Directorate-General for Mobility and Transport (DG MOVE) identified ten specific communication good practices and strategies able to be transferable across the full spectrum of transport sectors in the Member States and believed to promote transport jobs effectively to young women and men. These are: (1) Using research to confirm the approach; (2) Taking a strategic approach (a long term plan with specific targets); (3) Going into schools, colleges and universities; (4) Providing opportunities to experience the job; (5) Showcasing real people as role models; (6) Working with men to engage women; (7) Communicating with young people on their terms; (8) Building in careers advice provision to promotional strategies; (9) Using networks & mentoring to support female retention; (10) Using existing resources & networks to increase cost-effectiveness.
- Policymakers can help fight gender stereotypes and bring behavioural change regarding gender equality through effective gender mainstreaming, which is a strategy that involves the integration of a gender perspective into the preparation, design, implementation, monitoring and evaluation of policies, regulatory measures and spending programmes.

Key Actions: to achieve this objective, PARE recommends making available online and accessible to primary schools and parents, children's stories and cartoons involving flying that are both entertaining and educational.

10.11.2 Implementing Changes in the Educational Context

In most countries, despite an average of half of the pupils that enter the primary schools being female and continue studying further into secondary education, in tertiary education/university men are more likely to obtain a degree in a STEM field of studies. Research suggests that there is little to no difference in boys' and girls' average ability at STEM subjects, which means that to attract more girls to STEM subjects at schools and universities, the solution is to tackle the stereotypes that they are exposed to from primary and secondary education until university graduation. In practice, it is necessary to give girls and boys in primary schools the same opportunities to choose their games and entertainment (objective 50) and to encourage more girls to take aeronautical engineering degrees (objective 51).

Key Findings

- As said before, teachers play an important role in influencing girls and young women to study and pursue STEM fields. Some measures that have been adopted so far include:

  1. Making available on the internet and to primary school's children's stories and cartoons where girls drive cars and fly aeroplanes as much as boys do, and let them play with vehicle models or ask for them as presents
  2. Developing a toolkit for primary and secondary school teachers to fight gender stereotypes and raise awareness about transport professions among young people
  3. Have a gender-neutral environment in classes that help young women to participate and feel engaged.

- The article "Girl Power", published in 2018 in the *Social Sciences* open access journal, states that women in engineering majors enter college with the same levels of interest and intent to persist in the major as male peers, but the majority face negative experiences such as male favouritism, differential treatment and sexual harassment from classmates and faculty that make it difficult to persist and succeed in their majors. Those who have positive experiences via supportive faculty members and peers, research experiences or participation in engineering organizations are more likely to continue taking STEM classes, complete degrees and continue to post-baccalaureate STEM careers in comparison to those who do not.

  Key Actions: to achieve these two objectives, PARE recommends (respectively)

1. Including flight experiments equally accessible to boys and girls in primary and secondary school programmes and activities.
2. Reinforcing and accelerating visits to universities and industry, role models of success stories, and the same fascinating technologies.

10.11.3 Improvement of the Employment Context

As the traditional masculine beliefs and values have been rooted in the aviation industry for a long period, despite several efforts made, the percentage of women pursuing a career in the field remains low, particularly in technical positions that require STEM skills. To oppose this, it is necessary to:

- Provide women with attractive careers in aeronautics in industry and academia (objective 52)
- Discourage and prevent the continuation of abuse based on gender (objective 53)
- Ensure that the protection of the family, maternity and parenthood is effectively implemented with its legal basis as a minimum (objective 54)
- Give equal recognition of achievements regardless of gender, taking into account the circumstances (objective 55)
- See the differences between the genders as an opportunity for a symbiosis of distinct talents that furthers smooth progress (objective 56)
- Increase the participation of women in aeronautics in the most effective way (objective 57)
- Recognise the historic achievements of women, including in aeronautics, in biased or unfavourable circumstances (objective 58)

Key Findings:

- Women feel more supported in environments that recognize the range of skills they have, provide opportunities for progression, and take a firm line on sexist behaviour. Concerning on-the-job treatment and preventing abuse, there are three main factors to take into consideration:

    1. *Women mentoring programs:* women mentors can support in efforts to make the industry and job functions more transparent, giving entrants a realistic depiction of how the industry works and what it is like in the work environment.
    2. *Women job satisfaction determinants:* on-the-job treatment should also consider women's determinants for job satisfaction and design their jobs ensuring these factors.
    3. *Women networking:* establishing networks of women to share experiences and promote opportunities is perceived to be an important element of improving working conditions.

- The protection of the family and children is a fundamental value of society that the law tries to ensure in all situations, including employment. Instead of dismissing someone using collateral arguments, a change of type of work or a different assignment can make more compatible company priorities and family needs by adjusting schedules or timetables.
- The recognition of professional achievements must be objective and fair, using the same criteria applied in the same way, regardless of gender, age or belief. However, fairness also means equal opportunities, and while applying the same final criteria, all should have the same opportunities to attain those objectives.

The adjustment of working conditions, schedules or responsibilities to account for special individual circumstances of women or other groups is not a favour, but rather having an even playing field inside the company as the company would like to have in the market versus its competitors, promoting a balanced division of tasks within the organization.

- Women were present in aviation activities since the first flights. However, until the 1970s, their presence was, with rare exceptions, mainly as passengers or cabin crew onboard of male-piloted vehicles. Biased opinion persisted over decades, determining a large delay in pioneer women penetrating in male-dominated aviation professions. Examples of women that had remarkable successes in aviation despite modest recognition: Amelia Earhart, Hanna Reich, Jaqueline Cochran and Jaqueline Auriol. These notable exceptions were the effect of a combination of factors: outstanding qualities, the encouragement of a political system using women images for propaganda objectives, powerful support from families and luck.
- Women and men can have different sensibilities, distinct approaches to the same problem, and complementary abilities that can be of benefit to the balanced and efficient performance of many tasks. Greater participation of women in aeronautics is not only an enlargement of the workforce in numbers, but it is also an enrichment in quality and talent, which are the foundations of inventiveness and competitiveness, on which depends the continuing European leadership in an ever more competitive world with new challenges.

Key Actions: to achieve these two objectives, PARE recommends (respectively)

1. Making all aspects of job recruitment, from the announcements to the interview to the benefits, gender-equal, and try to compensate for eventual gender differences
2. Taking gender abuse as seriously as gross incompetence or major financial misconduct as concerns the consequences and leaving no doubts in anyone's mind about this policy
3. Taking family, maternity and parenthood into consideration in the assignment of tasks and giving a suitable working environment
4. Avoiding direct and reverse discrimination or bias by judging and rewarding achievements in an even, transparent and fair way that is not seen as gender bias
5. Assigning positions and tasks using the best talents and skills available in both genders to promote creativity and efficiency
6. Considering the greater numbers of women in aeronautics not just as a numerical enlargement of the workforce but also as a broadening of the talent available
7. Considering the lives of outstanding women, including aviators and astronauts, not only from a biographical point of view but also recognising the challenges they had to overcome to realise their achievements

(a) ***PARE OBJECTIVE 49:*** *Counter family and societal bias discouraging girls from interest in vehicles, be it cars or planes.* **Recommendation 49:** Make available on the internet and to primary school's children stories and cartoons

where girls drive cars and fly aeroplanes as much as boys do, and let them play with vehicle models or ask for them as presents. **Rationale:** The bias since childhood that "girls play with dolls and boys play with cars/planes" needs to be countered by gender-neutral stories about boys and girls flying or collaborating in making origami's planes. **Stakeholders:** EU, MS, primary education, families, friends. **Relevance:** Flying has always been an inspiration for mankind from childhood through adulthood to old age, regardless of gender. **Priority:** This should be a normal, steady activity. **Justification:** PARE report Section 11.1 and topics 11.1 and 11.3.

(b) ***PARE OBJECTIVE 50:*** *Give girls and boys in primary schools the same opportunities to choose their games and entertainment.* **Recommendation 50:** The primary and secondary school programmes and activities could include flight experiments equally accessible to boys and girls. **Rationale:** Gender bias may be countered in a natural, non-dramatic, non-political fashion by simple neutral education since primary school by providing educators with suitable material that is attractive to children of both genders. **Stakeholders:** EU, MS, primary schools, families. **Relevance:** The best way to avoid problems is not to create them by adopting a neutral attitude, and this also applies to gender in the education of children. **Priority:** Should be normal primary school practice and deserves further encouragement. **Justification:** PARE report Section 11.2 and topic 11.5.

(c) ***PARE OBJECTIVE 51:*** *Encourage more girls to take aeronautical engineering degrees.* **Recommendation 51:** Reinforce and accelerate this slowly growing trend by visits to universities and industry, role models of success stories, and the same fascinating technologies. **Rationale:** The measures to attract young talent to aeronautical careers are similar, and some extra effort needs to be put in compensating gender bias. **Stakeholders:** EU, MS, UA, AI, RC, PA, other educators and employers. **Relevance:** Women perform at least as well as men in aeronautical engineering degrees and enlarge the talent pool both in quantity and diversity. **Priority:** There is a vast potential of talent that is only modestly tapped. **Justification:** PARE report Section 11.2 and topic 11.5.

(d) ***PARE OBJECTIVE 52:*** *Provide women with attractive careers in aeronautics in industry and academia.* **Recommendation 52:** Make all aspects of job recruitment, from the announcements to the interview to the benefits, gender equal, and try to compensate for eventual gender differences. **Rationale:** The outdated lingering view that aviation is a job for men except for flight attendants should be countered by recruiting campaigns that are carefully gender-neutral and cleverly undermine any gender bias. **Stakeholders:** EU, MS, AI, RC, UA, PA, AN, AR, AP. **Relevance:** Women bring similar competence and different skills, and having both genders is a human resources asset. **Priority:** Should be normal practice and offer scope for improvement. **Justification:** PARE report Section 11.3 and topic 11.8.

(e) ***PARE OBJECTIVE 53:*** *Discourage and prevent continuation of abuse based on gender.* **53b. Recommendation 53:** Take gender abuse as seriously as gross incompetence or major financial misconduct as concerns the consequences, and

leave no doubts in anyone's mind about this policy. **Rationale:** The activity of a company and its hierarchical structure should not tolerate gender abuse as much as it should not tolerate gross incompetence or financial fraud. **Stakeholders:** EU and MS legislation, AI, RC, UA, PA, AR, AN, AP and all other employers. **Relevance:** The tendency to ignore gender abuse or look the other way if it comes from up the chain is among the examples of non-acceptable attitudes. **Priority:** This is a human rights issue that is fundamental principle. **Justification:** PARE report Section 11.3.

(f) ***PARE OBJECTIVE 54:*** *Ensure that the protection of family, maternity and parenthood is effectively implemented with its legal basis as a minimum.* **Recommendation 54:** Take family, maternity and parenthood into consideration in the assignment of tasks and giving a suitable working environment. **Rationale:** In addition to faithfully implementing the legislation on the protection of family, more can be done by goodwill towards dedicated employees. **Stakeholders:** EU, MS, employers in general. **Relevance:** Respect for family values and culture should be part of company ethics. **Priority:** Normal practice to be encouraged. **Justification:** PARE report Section 11.3.

(g) ***PARE OBJECTIVE 55:*** *Give equal recognition of achievements regardless of gender, taking into account the circumstances.* **Recommendation 55:** Avoid direct and reverse discrimination or bias by judging and rewarding achievements in an even, transparent and fair way that is not seen as gender bias. **Rationale:** Perhaps the best way to achieve gender balance is to avoid direct or reverse discrimination and to recognise and compensate for legitimate differences. **Stakeholders:** EU, MS, employers in general. **Relevance:** Women need no favours, only fair and equal treatment so that their merits are recognised and cannot be dismissed as reverse discrimination. **Priority:** Should be normal practice, and it is still not always the case. **Justification:** PARE report Section 11.3.

(h) ***PARE OBJECTIVE 56:*** *See the differences between the genders as an opportunity for a symbiosis of distinct talents that furthers smooth progress.* **Recommendation 56:** Assign positions and tasks using the best talents and skills available in both genders to promote creativity and efficiency. **Rationale:** The equality of men and women does not exclude different talents that bring most benefit in a symbiotic collaboration. **Stakeholders:** EU, MS, all employers. **Relevance:** Choosing the right person on the team for the task should be based not only on experience but also the balance of skills, in which gender can be one of many factors. **Priority:** Should be normal practice. **Justification:** PARE report Section 11.3.

(i) ***PARE OBJECTIVE 57:*** *Increase the participation of women in aeronautics in the most effective way.* **Recommendation 57:** The greater numbers of women in aeronautics should be regarded as not just as a numerical enlargement of the workforce but also as a broadening of the range of the talent available. **Rationale:** The rationale for the increased participation of women in aeronautics is not just a potential increase in numbers but also an opportunity for broader individual and team skills. **Stakeholders:** EU, MS, AI, RC, UA, other

employers. **Relevance:** A broad view of the added talent brought by more women in aviation is needed to obtain the greatest benefit. **Priority:** A general policy issue. **Justification:** PARE report Section 11.3 and topic 11.2.

(j) ***PARE OBJECTIVE 58:*** *Recognise the historic achievements of women, including in aeronautics, in biased or unfavourable circumstances.* **Recommendation 58:** Consider the lives of outstanding women, including aviators and astronauts, not only from a biographical point of view but also recognising the challenges they had to overcome to realise their achievements. **Rationale:** The achievements of great women in aeronautics and the gender unequal challenges they had to overcome are an indicator of the obstacles that should be removed. **Stakeholders:** EU, MS, employers. **Relevance:** The role models serve as examples for women who should be assured to have more of a gender equal treatment. **Priority:** It is an activity to counter unfounded and untrue preconceptions that could reduce the value of women's contributions to aviation. **Justification:** PARE report Section 11.3 and topics 11.11 to 11.14.

10.12 Conclusions

This overview of the current state of aeronautics and future prospects is open to discussion in all aspects:

- If the key findings focus on the most relevant issues.
- If the key actions are adequate to address those findings.
- If the recommendations point towards an appropriate course of action.

The overall assessment is based on the present and a preview of the future, which will need to be updated over time as new developments emerge.

References

The main documents relating to the contents of this chapter are the ACARE and PARE reports.

1. PARE, Report Year I, https://www.pareproject.eu/publications.
2. PARE, Report Year II, https://www.pareproject.eu/publications
3. PARE, Report Year III, https://www.pareproject.eu/publications
4. ACARE, https://www.acare4europe.org/
5. ACARE, SRIA, https://www.acare4europe.org/sria
6. European Commission, 2011, "Flightpath 2050 Europe's Vision for Aviation: Report of the High Level Group on Aviation Research", Publications Office of the European Union, EUR 098 EN, Luxembourg, https://doi.org/10.2777/50266, https://www.acare4europe.org/documents/latest-acaredocuments/acare-flightpath-2050
7. "PARE - Perspectives for the Aeronautical Research in Europe", Coordination and Support Action, 2017-2020, EC Project Reference: 769220 [Online] Available: https://www.pare-project.eu

The manufacturer's authorised representative in the EU is Springer Nature Customer Service Centre GmbH, Europaplatz 3, 69115 Heidelberg, Germany. If you have any concerns regarding our products, please contact ProductSafety@springernature.com

Printed and bound by CPI Group (UK) Ltd, Croydon, CR0 4YY
01/10/2026
02225932-0016